T0262017

Gel Electrophoresis: Novel Approaches

Gel Electrophoresis: Novel Approaches

Edited by **Jill Clark**

New York

Published by Callisto Reference,
106 Park Avenue, Suite 200,
New York, NY 10016, USA
www.callistoreference.com

Gel Electrophoresis: Novel Approaches
Edited by Jill Clark

© 2015 Callisto Reference

International Standard Book Number: 978-1-63239-349-4 (Hardback)

Printed in the United States of America.

Contents

Preface   VII

Part 1   **Electrophoresis Application in Enzymology**   1

Chapter 1   **Applications of Zymography (Substrate-SDS-PAGE)
for Peptidase Screening in a Post-Genomic Era**   3
Claudia M. d'Avila-Levy, André L. S. Santos, Patrícia Cuervo,
José Batista de Jesus and Marta H. Branquinha

Chapter 2   **Polyacrylamide Gel Electrophoresis an
Important Tool for the Detection and Analysis
of Enzymatic Activities by Electrophoretic Zymograms**   27
Reyna Lucero Camacho Morales, Vanesa Zazueta-Novoa,
Carlos A. Leal-Morales, Alberto Flores Martínez,
Patricia Ponce Noyola and Roberto Zazueta-Sandoval

Part 2   **Temporal Temperature Gel Electrophoresis**   45

Chapter 3   **Temporal Temperature Gel Eleetrophoresis
to Survey Pathogenic Bacterial Communities:
The Case of Surgical Site Infections**   47
Romano-Bertrand Sara, Parer Sylvie, Lotthé Anne,
Colson Pascal, Albat Bernard and Jumas-Bilak Estelle

Part 3   **Two-Dimensional Gel Electrophoresis (2-DE)**   69

Chapter 4   **Two-Dimensional Gel Electrophoresis Reveals Differential
Protein Expression Between Individual *Daphnia***   71
Darren J. Bauer, Gary B. Smejkal and W. Kelley Thomas

Chapter 5   **Two Dimensional Gel
Electrophoresis in Cancer Proteomics**   83
Soundarapandian Kannan, Mohanan V. Sujitha,
Shenbagamoorthy Sundarraj and
Ramasamy Thirumurugan

Chapter 6   **Two-Dimensional Gel Electrophoresis**
             and Mass Spectrometry in Studies of
             Nanoparticle-Protein Interactions   **115**
             Helen Karlsson, Stefan Ljunggren, Maria Ahrén,
             Bijar Ghafouri, Kajsa Uvdal, Mats Lindahl
             and Anders Ljungman

Part 4   **Other Applications of Gel Electrophoresis Technique**   **147**

Chapter 7   **Enzymatic Staining for Detection of**
             Phenol-Oxidizing Isozymes Involved in Lignin-
             Degradation by *Lentinula edodes* **on Native-PAGE**   **149**
             Eiji Tanesaka, Naomi Saeki, Akinori Kochi and
             Motonobu Yoshida

Chapter 8   **Protection Studies by Antioxidants Using Single Cell Gel**
             Electrophoresis (Comet Assay)   **169**
             Pınar Erkekoglu

Chapter 9   **Gel Electrophoresis as a Tool to Study**
             Polymorphism and Nutritive Value of the
             Seed Storage Proteins in the Grain Sorghum   **203**
             Lev Elkonin, Julia Italianskaya and Irina Fadeeva

Chapter 10   **Gel Electrophoresis as Quality Control Method**
              of the Radiolabeled Monoclonal Antibodies   **219**
              Veronika Kocurová

Chapter 11   **Extraction and Electrophoresis of DNA from**
              the Remains of Mexican Ancient Populations   **235**
              Maria de Lourdes Muñoz, Mauro Lopez-Armenta,
              Miguel Moreno-Galeana, Alvaro Díaz-Badillo,
              Gerardo Pérez-Ramirez, Alma Herrera-Salazar,
              Elizabeth Mejia-Pérez-Campos, Sergio Gómez-Chávez
              and Adrián Martínez-Meza

              Permissions

              List of Contributors

Preface

As a fundamental concept, gel electrophoresis is a biotechnology method in which macromolecules such as DNA, RNA or protein are fragmented according to their physical characteristics such as molecular mass or charge. These molecules are enforced throughout a porous gel matrix in the influence of electric field enabling countless functions and utilizations. This book is not all-inclusive but still covers a majority of applications of this technique in the varied fields of medical and life sciences. This book contains three sections: Electrophoresis Application in Ecological and Biotechnological Aspects, Electrophoresis Application in Bacteriology, Parasitology, Mycology and Public Health, and Analysis of Protein-Nucleic Acid Interaction and Chromosomal Replication. We have made an attempt to keep the data of the book detailed and wide-ranging, and we hope that it will benefit the readers.

Significant researches are present in this book. Intensive efforts have been employed by authors to make this book an outstanding discourse. This book contains the enlightening chapters which have been written on the basis of significant researches done by the experts.

Finally, I would also like to thank all the members involved in this book for being a team and meeting all the deadlines for the submission of their respective works. I would also like to thank my friends and family for being supportive in my efforts.

Editor

Part 1

Electrophoresis Application in Enzymology

Applications of Zymography (Substrate-SDS-PAGE) for Peptidase Screening in a Post-Genomic Era

Claudia M. d'Avila-Levy[1], André L. S. Santos[2], Patrícia Cuervo[1],
José Batista de Jesus[1,3] and Marta H. Branquinha[2]
[1]Instituto Oswaldo Cruz, Fundação Oswaldo Cruz, Rio de Janeiro
[2]Departamento de Microbiologia Geral, Instituto de Microbiologia Paulo de Góes,
Universidade Federal do Rio de Janeiro, Rio de Janeiro
[3]Universidade Federal de São João Del Rei, São João Del Rei
Brazil

1. Introduction

Peptidases are enzymes that catalyze the hydrolysis of peptide bonds in proteins or peptides. The hydrolysis can be specific or unspecific, leading to highly regulated cleavage of specific peptide bonds, or to complete degradation of proteins to oligopeptides and/or amino acids. Peptidases can be classified as endo- or exopeptidases, the latter only act near the ends of the polypeptide chain. Endopeptidases are divided into six major families by virtue of the specific chemistry of their active site: aspartic, serine, metallo-, cysteine, glutamic and threonine peptidases (Rawlings et al. 2010).

Zymography is an electrophoretic technique, based on sodium dodecyl sulfate polyacrylamide gel electrophoresis (SDS-PAGE) and a substrate (e.g. gelatin, casein, albumin, hemoglobin, etc.) co-polymerized with the polyacrylamide matrix. Proteins are prepared by the standard SDS-PAGE buffer under non-reducing conditions (no boiling and no reducing agent), and are separated by molecular mass in the standard denaturing SDS-PAGE co-polymerized with a protein substrate. After electrophoresis, peptidases are re-natured by the removal of the denaturing SDS by a non-ionic detergent, such as Triton X-100, followed by incubation in conditions specific for each peptidase activity (time, temperature, ions, ionic strength), when the enzymes hydrolyze the embedded substrate, then proteolytic activity can be visualized as cleared bands on a Coomassie stained background (Heussen and Dowdle, 1980). Therefore, only endopeptidases can be detected by substrate-SDS-PAGE, which requires a considerable degradation of the substrate for visualization of the degradation haloes. Alternatively, an overlay with specific chromogenic or fluorogenic peptide substrate can be done after SDS-PAGE separation of the proteins and renaturation with Triton X-100, which allows the detection of specific peptidases in complex biological samples.

This technique has many benefits: (1) it is relatively inexpensive, requires short assaying times, and peptidases with distinct molecular masses can be detected on a single gel; (2)

separation of proteins by molecular mass through non-reducing electrophoretic migration allows a presumptive correlation with known peptidases; (3) incubation with proteolytic inhibitors provides powerful information about enzyme classification; (4) pH and temperature changes help to assess peptidase characteristics; (5) several substrates can be co-polymerized to assess peptidase degradation capacity; (6) densitometry can be used for quantitative analysis. Ultimately, in organisms with complete genome sequences, bioinformatic analysis provides rich information on putative peptidases, such as: peptidase classification, approximated molecular mass, possible cellular localization through classical motifs, evolutionary and functional relationships, and so on. However, it cannot be ascertained if the described ORFs are indeed expressed and active. Therefore, a zymographic assay coupled with bioinformatic analysis may allow the detection of functionally active enzymes.

The advantages of this technique are exemplified by its application nowadays to unveil peptidases in biological systems, which possesses genome information, but still zymography is the method of choice for peptidase screening, identification and characterization. Wilder and colleagues, for instance, report that zymography can selectively distinguish cathepsins K, L, S and V in cells and tissues by its electrophoretic mobility and by simply manipulating substrate and pH. The sequence homology among these cathepsins leads to a substrate promiscuity, which precludes desired specificity for in solution assays with specific chromogenic or fluorogenic peptide substrate (Wilder et al. 2011). Zymography allows the detection of a 37 kDa (cathepsin K), 35 kDa (cathepsin V), 25 kDa (cathepsin S) and 20 kDa (cathepsin L). Cathepsin K activity disappeared and V remained when incubated at pH 4.0 instead of 6.0, allowing the visualization of each enzyme (Wilder et al. 2011). Kupai and colleagues also highlighted that substrate zymography is the method of choice, among several analyzed, to detect the activity of the different matrix metallopeptidase (MMP) isoenzymes from a wide range of biological samples (Kupai et al. 2010). Also, it allows high throughput screening of specific MMP inhibitors, especially because the nature of the residues in the enzyme's active site is highly conserved among the different MMPs, therefore, once again, in solution enzymatic assays are not applicable (Devel et al. 2006, Kupai et al. 2010). Also, for the screening of tissue inhibitors of metallopeptidases (TIMPs), reverse zymography is a powerful approach. This technique is based on the ability of the inhibitors to block gelatinase activity of a MMP, usually MMP-2. A calibrated solution of gelatinase-A (MMP-2) is co-polymerized with gelatin in the polyacrylamide gel. The samples possibly containing TIMPs are then separated by electrophoresis, SDS is removed and the gel is incubated in a buffer that allows the gelatinase to digest the gelatin, except where it is inhibited by TIMP proteins. After staining with Coomassie blue, the result is a gel with a pale blue background (where gelatin was degraded by the gelatinase) with blue bands showing the positions and relative amounts of TIMPs (Snoek-van Beurden and Von den Hoff, 2005).

In view of this, below we will present comments on peptidase screening through zymography discussing possible protocol variations and its implications, and then we present and discuss practical examples of the application of zymography to generate critical data in organisms that still do not possess genome information. Finally, we will discuss the possibility of direct peptidase identification through two-dimensional zymography coupled to mass spectrometry.

2. Comments on peptidase screening through substrate-SDS-PAGE

Several research groups perform substrate-SDS-PAGE to assess, screen and characterize peptidases in complex or purified biological preparations. After the original publication from Heussen and Dowdle (1980), several adaptations have been implemented to improve the detection of a specific peptidase class. Below we will present a generalized protocol indicating possible variations.

2.1 Sample homogenization

The preparation of the biological sample is critical for the success of the zymography, all the procedure must be performed at 4°C, the addition of detergents such as Triton X-100, SDS or CHAPS (3-[(3-cholamidopropyl)dimethylammonio]-1-propanesulfonate) for the solubilization and recovery of hydrophobic enzymes is necessary, if one is interested in such enzymes, also addition of proteolytic inhibitors to undesired peptidase classes is also an interesting strategy. Alternatively, the separation of hydrophobic from hydrophilic proteins can be achieved during phase partition in solutions of Triton X-114, which occurs at 37°C preserving enzyme integrity (Figure 1) (Bouvier et al. 1987). After sample preparation, SDS-PAGE sample buffer is added to the biological sample (62.5 mM Tris HCl, pH 6.8, 2% SDS, 10% (v/v) glycerol, and 0.002% bromophenol blue). A concentrated sample buffer can be used to avoid sample dilution, which is critical for the detection of low abundant enzymes. The proteins are not denatured since sample is kept at 4°C, there is no sample boiling, nor the addition of reducing agents such as dithiothreitol (DTT) or 2-mercaptoethanol, as usual in sample preparation for standard SDS-PAGE analysis. The sample must maintain its native form due to the the the further step of substrate degradation.

2.2 Polyacrylamide gels containing sodium dodecyl sulfate and co-polymerized substrates

Here, to the standard Laemmli protocol (Laemmli, 1970), a substrate can be co-polymerized to the gel (Heussen and Dowdle, 1980). Alternatively, an overlay with fluorogenic or chromogenic peptide substrates can be done (Cadavid-Restrepo et al. 2011). The acrylamide concentration in gels varies more commonly from 7 to 15%, which impact on protein separation; low molecular mass proteins usually require higher acrylamide concentration for better protein resolution. The co-polymerized substrate can be virtually any protein. Gelatin is commonly used as a protein substrate because it is easily hydrolyzed by several peptidases and does not tend to migrate out of the resolving gel in electrophoretic tests performed at 4°C, and is inexpensive (Michaud et al. 1996). In addition to gelatin, several other proteins have been used, such as: casein, bovine serum albumin, human serum albumin, hemoglobin, mucin, immunoglobulin, and collagen (d'Avila-Levy et al. 2005; Pereira et al. 2010a). Also, complex mixtures of proteins can be used, which may reflect a functional role of the enzyme. For instance, our research group employed gut proteins from an insect to co-polymerize in acrylamide gels. Then, extracts from a protozoan believed to interact with the insect gut were assayed, revealing the peptidases capable of degrading the insect gut proteins (Pereira et al. 2010a). An example of zymographies performed with a set of eight distinct proteinaceous substrates, as well as, a densitometric measure of the degradation halos can be seen in Figure 2.

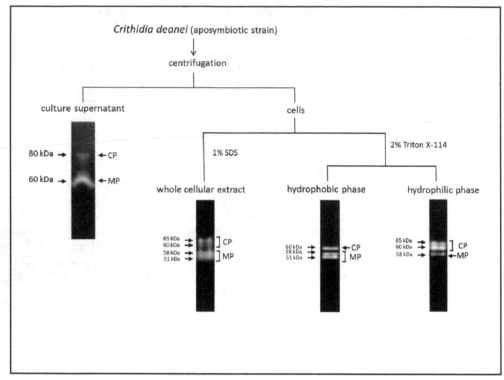

Fig. 1. Extracellular and cell-associated proteolytic enzymes of *Crithidia deanei* cells, an insect trypanosomatid. Parasites were cultured in a complex medium (brain heart infusion) for 48 h at 28°C. Then, cells were harvested by centrifugation, the culture supernatant was filtered in Millipore membrane 0.22 μm and concentrated 50-fold by dialysis (cut-off 9000 Da) against polyethylene glycol 4000 overnight at 4°C. The cells were lysed by: the addition of SDS, generating whole cellular extract, or by Triton X-114 to obtain the hydrophilic (cytoplasmatic and intravesicular fraction) and hydrophobic (membrane fraction) phases. The extracellular and cellular extracts were applied on gelatin-SDS-PAGE to evidence the proteolytic enzymes. The gels were incubated in 50 mM sodium phosphate buffer pH 5.5 supplemented with DTT 2 mM at 37°C for 24 h. MP, metallopeptidase and CP, cysteine peptidase. For experimental details see d'Avila-Levy et al. 2001, 2003.

2.3 Enzyme renaturation and proteolysis

After electrophoresis, the enzymes are renatured by replacement of the anionic detergent SDS by the non-ionic detergent Triton X-100, through gel washing. Then, gels are incubated under conditions ideal for detection of the desired peptidase. For instance, metallopeptidases are known to require neutral to basic pH for activity, while cysteine peptidases require an acidic pH and a reducing agent, usually DTT (Branquinha et al. 1996). However, it is common to screen biological samples, where there is no previous clue on what peptidase class shall be present, nor the best conditions for proteolysis. Therefore, it is necessary to assess several parameters, such as incubation time, pH, temperature, influence of ions or reducing agents and finally assess the inhibition profile. A general flowchart for establishing such conditions is shown in Figure 3, and a general view of Gelatin-SDS-PAGE screening in an uncharacterized organism can be seen in Figure 4.

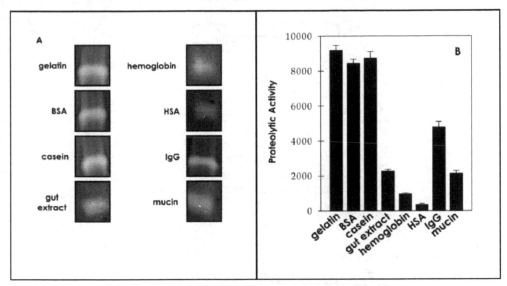

Fig. 2. Degradation of different proteinaceous substrates co-polymerized to SDS-PAGE by a surface metallopeptidase from *Herpetomonas samuelpessoai*, an insect trypanosomatid. The following substrates were individually incorporated into SDS-PAGE to evidence the proteolytic activity: gelatin, bovine serum albumin, human serum albumin, casein, immunoglobulin G (IgG), hemoglobin, mucin and gut extract from *Aedes aegypti*. The gels were incubated for 20 h at 37ºC in 50 mM sodium phosphate buffer pH 6.0 supplemented with DTT 2 mM (A). The degradation halos, which correlate with degradation capability, were densitometric measured and expressed as arbitrary units of proteolytic activity (B). For experimental details see Pereira et al. 2010b. Reprinted with permission of *Protist*.

3. Practical examples of peptidases screening through SDS-PAGE-substrate

3.1 A first glance on *Bodo* sp. peptidases

Bodo sp. is a free-living flagellate that belongs to the family Bodonidae, order Kinetoplastida. This bodonid isolate still has its taxonomic position unsolved, but it is phylogenetically related to *Bodo caudatus* and *Bodo curvifilus*, which are considered ancestral to the trypanosomatids. The Trypanosomatidae family comprises parasites that are of particular interest due to their medical importance, such as the etiologic agent of Chagas' disease (*Trypanosoma cruzi*), African trypanosomiasis (*Trypanosoma brucei* complex) and the various forms of leishmaniasis caused by *Leishmania* spp.. Due to their medical relevance, this family has been the focus of extensive research (Wallace, 1966; Vickerman, 1994). Peptidase characterization in *Bodo* sp. and comparison to peptidases from closely related pathogenic protozoa may help to understand peptidase function and evolution in general. The gold standard approach for such comparison would be a bioinformatic analysis of the *Bodo*

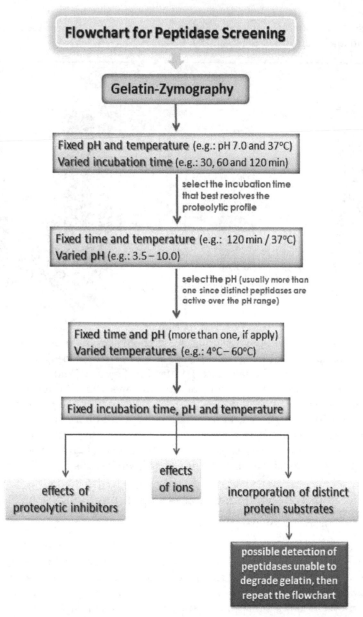

Fig. 3. Flowchart for peptidase screening. Several parameters must be assessed to resolve the proteolytic profile in an unknown biological sample. This scheme represents a suggestion of a step-by-step analysis of these parameters, which are: amount of sample, incubation time, pH, temperature, effect of ions, effect of reducing agents, effect of peptidase inhibitors, and ability to degrade distinct proteinaceous substrates. Usually, gelatin is the proteinaceous substrate of choice for initial screening because it is easily hydrolyzed by several peptidases,

does not tend to migrate out of the resolving gel and is inexpensive, then with an arbitrary pH and temperature (usually neutral pH at 37°C), the incubation time is varied from minutes to even 72 h, depending on the sample. After selecting the incubation time that allows the detection of the higher number of enzymes without band overlapping, variations on the pH allows the determination of this biochemical characteristic of each band. After this assay, the peptidase(s) of interest can be tested over a range of temperatures, ions, reducing agents or proteolytic inhibitors. Finally, distinct proteinaceous substrates can be co-polymerized to the gels, revealing either the ability of the detected peptidases to degrade other substrates, which ultimately gives a glance on peptidase function, or even revealing enzymes not capable of degrading gelatin (see figure 2). The proper combination of these parameters may reveal interesting enzymes, such as peptidases strictly dependable on metal ions, stimulated by reducing agents, active only at acidic or alkaline conditions and so on.

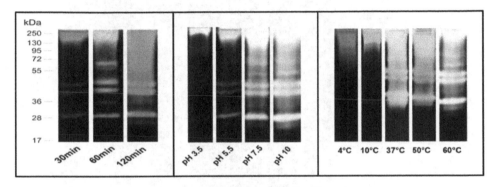

Fig. 4. Gelatin-SDS-PAGE screening of peptidases in homogenates from *Aedes albopictus* pupa. The following parameters were assessed: incubation time, pH and temperature. Peptidase activities were detected after incubation of the gels for 30, 60 or 120 min at 37°C in 100 mM Tris-HCl buffer pH 7.5. The numbers on the left indicate apparent molecular masses of the active bands expressed on kiloDaltons (kDa). Afterwards, 60 min incubation was selected and the gels were incubated in reaction buffer containing 100 mM sodium acetate at pH 3.5 or 5.5 or 100 mM Tris-HCl at pH 7.5 or 10.0. Finally, the effect of temperature on the proteolytic activities was assayed by incubating of the gels for 60 min at 4, 10, 37, 50 or 60°C in reaction buffer containing 100 mM Tris-HCl at pH 7.5. Saboia-Vahia et al. unpublished data.

genome coupled to more defined biochemical characterization of individual peptidases. However, in the absence of a *Bodo* genome, we have employed substrate-SDS-PAGE to assess peptidases in this bodonid, which presents serine peptidases ranging from 250 to 75 kDa, with a slight preference for acidic pH. This finding is dissimilar to what has been described in related pathogenic protozoa (Figure 5) (d'Avila-Levy et al., 2009). Curiously, all the analyzed closely related parasitic trypanosomatids, as well, as *Cryptobia salmonistica*, reveal through gelatin-SDS-PAGE only metallo- and cysteine peptidases, which are prototypal peptidases and virulence factors. In trypanosomatids, for instance, serine peptidases can only be detected by in-solution assays or after enrichment processes (Grellier et al. 2001). It is somewhat intriguing that cysteine and metallopeptidases are either not resistant to the denaturation/refold process and/or are not abundantly expressed by this free living bodonid, because they were not detected by zymography. This may reflect substantial differences among the peptidases from these organisms. The raw data revealed by susbtrate-zymography provided the first observation on possible differences in peptidase profile among the families Bodonidae, Trypanosomatidae, and Cryptobiidae, forming the basis for future research.

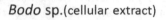

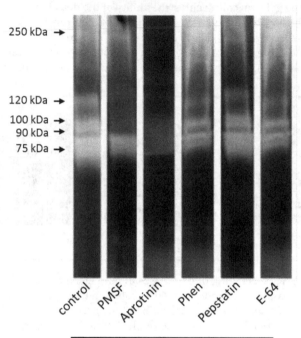

Proteolytic inhibitors

Fig. 5. Inhibition profile of cellular peptidases of *Bodo* sp. in gelatin–SDS-PAGE. In order to determine the enzymatic class, after electrophoresis, the gels were incubated for 48 h at 28°C in 50mM phosphate buffer pH 5.5 in the absence (control) or in the presence of the following proteolytic inhibitors: 1mM phenylmethylsulfonyl fluoride (PMSF), 1 mg/ml aprotinin, 10mM 1,10-phenanthroline (Phen), 1 mM pepstatin A, or 10 µM *trans*-epoxysuccinyl L-leucylamido-(4-guanidino) butane (E-64). Numbers on the left indicate relative molecular mass of the peptidases. For experimental details see d'Avila-Levy et al. 2009. Reprinted with permission of *The Journal of Eukaryotic Microbiology*.

3.2 Identification of peptidases in *Herpetomonas* spp. and possible biological functions proposed by overlay gel approaches

In addition to the heteroxenic parasites that are of particular interest in the Trypanosomatidae family due to their medical importance, several genera are composed of monoxenic parasites of the gut of a wide range of insects. The *Herpetomonas* genus is composed of insect trypanosomatids that display promastigote, paramastigote and opisthomastigote developmental stages during its life cycle (McGhee and Cosgrove, 1980), being used as a model to study the complex events of cell differentiation process. Also, these traditionally "non-mammalian and non-pathogenic" microorganisms have been used as experimental models of the Trypanosomatidae family for exploring their basic mechanisms at the genetic, physiological, ultrastructural and biochemical levels. In the same way, several research groups have described common structures/molecules produced by monoxenous and heteroxenous parasites belonging to the Trypanosomatidae family (Lopes et al. 1981; Breganó et al. 2003; Santos et al. 2006, 2007; Elias et al. 2008). Interestingly, *Herpetomonas* species have been detected not only in insects, but repeatedly in plants and mammals, including immunosuppressed humans, mainly in HIV-infected individuals, in whom the parasites caused either visceral or cutaneous lesions (reviewed by Chicharro and Alvar, 2003; Morio et al. 2008), showing the ability to develop digenetic life style under certain conditions. Collectively, these studies emphasize the need for further investigation in the biochemical machinery of these intriguing insect trypanosomatids.

Whole cellular extracts of *H. samuelpessoai* promastigotes when analyzed by gelatin-SDS-PAGE revealed two major peptidase classes: a prominent metallopeptidase of 66 kDa (actually a broad hydrolytic activity ranging from 60 to 80 kDa), inhibited by 10 mM 1,10-phenanthroline, and a minor cysteine peptidase activity of 45 kDa, restrained by 1 µM E-64 and leupeptin (Santos et al. 2003).

The 66 kDa metallopeptidase activity was detected in the parasite membrane fraction after Triton X-114 partition (Etges, 1992; Schneider and Glaser, 1993; Santos et al. 2003) or after treatment of living cells with phospholipase C (Santos et al. 2002; Santos et al. 2006) as well as in the extracellular environment as the major secreted peptidase component (Santos et al. 2001, 2003, 2006; Elias et al. 2006). This metallopeptidase produced by *H. samuelpessoai* cells shares common biochemical and immunological properties (Elias et al. 2006; Santos et al. 2006) with the major metallopeptidase expressed by *Leishmania* species, called leishmanolysin or gp63, a virulence factor that participates in different stages of the parasite life cycle such as adhesion and escape from host immune response (Yao, 2010). The incorporation of different proteinaceous substrates into SDS-PAGE demonstrated that leishmanolysin-like molecule from *H. samuelpessoai* was able to degrade hemoglobin, casein, immunoglobulin G, mucin, human and bovine albumins as well as the gut protein extract from *Aedes aegypti* (Figure 2) (Pereira et al. 2010a), an experimental model to study the trypanosomatids-insect interplay (reviewed by Santos et al. 2006), culminating in the generation of peptides and amino acids required for parasite growth and development, as well as it might cleave structural barriers in order to improve its dissemination. Also, the pH dependence of the 66 kDa metallopeptidase of *H. samuelpessoai* was also determined by overlay gels, presenting a broad spectrum of pH (ranging from 5 to 10) and temperature (26 to 50°C), showing maximum hydrolytic activity at pH 6.0 at 37°C (Pereira et al. 2010a). These large spectra of pH and temperature retain maximum flexibility for the trypanosomatid to survive under different environmental conditions. In this sense, the surface leishmanolysin-like molecules of *H. samuelpessoai* cells participate in adhesive properties during the interaction with invertebrate gut (Pereira et al. 2010a) and mammalian macrophages (Pereira et al. 2010b). Other *Herpetomonas* species, including *H. megaseliae* and *H. anglusteri*, produce at least one metallopeptidase similar to the leishmanolysin, which is a conserved molecule with ancestral functions during the insect colonization.

The 45 kDa cysteine peptidase synthesized by *H. samuelpessoai* cells had its activity reduced during the parasite growth at 37°C in comparison to 26°C, and when cultured up to 72 h in the presence of the differentiation-eliciting agent, dimethylsulfoxide. The modulation in the 45 kDa cysteine peptidase expression is connected to the differentiation process, since both temperature and dimethylsulfoxide are able to trigger the promastigote into paramastigote transformation in *H. samuelpessoai* (Santos et al. 2003; Pereira et al. 2009). In contrast, the expression of leishmanolysin-like molecules was not modulated during the differentiation in *H. samuelpessoai* (Pereira et al. 2009, 2010b).

The cultivation of *H. megaseliae* and *H. samuelpessoai* in different growth media induced the production of distinct profiles of both cellular and extracellular peptidases as revealed by a simple inspection using substrate-SDS-PAGE (Branquinha et al. 1996; Santos et al. 2002, 2003; Nogueira de Melo et al. 2006). In addition, the incorporation of different proteinaceous substrates into SDS-PAGE allowed the identification of substrate specific proteolytic activity in a complex cellular extract. For example, cellular cysteine peptidase (115–100, 40 and 35 kDa) and metallopeptidase (70 and 60 kDa) activities of *H. megaseliae* were detected in both casein and gelatin zymograms (Nogueira de Melo et al. 2006). Additionally, the use of casein in the gel revealed a distinct acidic metallopeptidase of 50 kDa when the parasite was cultured in the modified Roitman's complex medium. However, no proteolytic activity was detected when hemoglobin was used as co-polymerized substrate (Nogueira de Melo et al. 2006).

3.3 Proteases produced by *Herpetomonas* species: Taxonomic marker

Insect trypanosomatids have been traditionally allocated to a number of genera that were described based on morphological features, host and geographical origin (Wallace et al. 1983; Momen, 2001). However, for identification purposes, these criteria proved to be impractical and insufficient, because the same trypanosomatid species may be recovered from diverse species of insects and the same insect species may harbor various species of trypanosomatids. In addition, the morphology of trypanosomatid cells can be modified by environmental factors (Podlipaev, 2001; Momen, 2001, 2002). Therefore, there is a need to develop more effective means of trypanosomatid identification. With this task in mind, the expression of proteolytic activities in the Trypanosomatidae family was explored as a potential marker to discriminate between the morphologically indistinguishable flagellates isolated from insects and plants (Branquinha et al. 1996; Santos et al. 1999, 2005, 2008). For instance, many trypanosomatids have been erroneously placed in the genus *Herpetomonas* or, conversely, many *Herpetomonas* spp. may remain hidden in other genera. Santos and co-workers (2005) proposed an additional tool for trypanosomatid identification, including species belonging to the *Herpetomonas* genus by using *in situ* detection of proteolytic activities on gelatin-SDS-PAGE, in association with specific peptidase inhibitors. The results showed that nine distinct *Herpetomonas* species (*H. anglusteri*, *H. samuelpessoai*, *H. mariadeanei*, *H. roitmani*, *H. muscarum ingenoplastis*, *H. muscarum muscarum*, *H. megaseliae*, *H. dendoderi* and *Herpetomonas* sp. isolated from the salivary gland of a phytophagous insect) produced species specific cellular peptidase profiles (Table 1), which can be useful in the correct identification of these parasites. The exception for this observation was seen in *H. samuelpessoai* and *H. anglusteri*, which presented a similar cell-associated proteolytic pattern. However, these two *Herpetomonas* species excreted distinct proteolytic activities, which may be a reflection of changes in the nutritional requirements during the life-cycle of the flagellates. Therefore, the authors infer that profiles of both cellular and extracellular peptidases represent an additional criterion to be used in the identification of trypanosomatids (Santos et al. 2005).

Herpetomonas species	Host		Predominant evolutive stage in culture	Number of cell-associated peptidases	Molecular masses of peptidases in kDa	
	Family	Species			Metallo-peptidases[a]	Cysteine peptidases[b]
Herpetomonas sp.	Hemiptera: Coreidae	*Phthia picta*	Promastigote	4	72, 60	45, 40
H. anglusteri	Diptera: Sarcophagidae	*Liopygia ruficornis*	Promastigote	2	60	45
H. dendoderi	Diptera: Culicidae	*Haemagogus janthinomys*	Promastigote	5	130, 110, 95	60, 45
H. mariadeanei	Diptera: Muscidae	*Muscina stabulans*	Promastigote	2	nd[c]	42, 38
H. megaseliae	Diptera: Phoridae	*Megaselia scalaris*	Promastigote	8	100, 80, 67, 60	95, 45, 40, 35
H. muscarum ingenoplastis	Diptera: Calliphoridae	*Phormia regina*	Promastigote	2	80, 67	nd
H. muscarum muscarum	Diptera: Muscidae	*Musca domestica*	Promastigote	6	100, 80	95, 50, 45, 40
H. roitmani	Diptera: Syrphidae	*Ornidia obesa*	Opisthomastigote	1	50	nd
H. samuelpessoai	Hemiptera: Reduviidae	*Zelus leucogrammus*	Promastigote	2	60	45

[a] The metallopeptidase activities were completely blocked by 10 mM 1,10-phenanthroline.
[b] The cysteine peptidases were inhibited by 10 µM E-64.
[c] Non detected (nd).

Table 1. Peptidase profiles in different *Herpetomonas* species detected in gelatin-SDS-PAGE.

3.4 Peptidase screening in *Crithidia*

Among the insect trypanosomatids, the genus *Crithidia* comprises monoxenic trypanosomatids of insects that were originally characterized by the presence of choanomastigote forms in their life cycles (Hoare and Wallace, 1966). The first studies employing zymograms in order to detect proteolytic activity in *Crithidia* spp. were performed by Frank and Ashall in 1990. In the two studies published in that year, the activity in *Crithidia fasciculata* extracts was compared to *T. cruzi*. In this sense, it is worth mentioning that *C. fasciculata*, among all non-pathogenic trypanosomatid species, has been considered an excellent model organism for many studies concerning trypanosomatids, because it can be cultivated in high yields and do not require specific bio-safety precautions (Vickerman, 1994).

In a first approach (Ashall, 1990), parasite extracts were made by the use of 0.5% Nonidet P-40 and mixed with SDS-PAGE sample buffer in non-denaturing conditions. After electrophoresis in SDS-PAGE, gels were overlaid with 0.75% agarose containing the chromogenic substrate Bz-Arg-pNA at 0.5 mM and at pH 8.0 and incubated for 4-6 h at 37°C, and then photographed using a blue filter to reveal yellow bands containing *p*-nitroaniline. A single component with molecular mass >200 kDa that hydrolyzed this substrate was detected in *C. fasciculata* as well as in *T. cruzi* crude extracts. A modified procedure was also employed (Ashall et al. 1990), in which electrophoresis was followed by shaking the gels with 2% Triton X-100 and then the incubation of gels for 10 min in the presence of a range of amidomethylcoumarin substrates containing arginine adjacent to the amidomethylcoumarin moiety, at pH 8.0. Fluorescent bands were visualized in gels by ultraviolet light by the hydrolysis of each substrate. A single band of substrate hydrolysis occurred with all six substrates tested, in both *T. cruzi* and *C. fasciculata*, with the same electrophoretic mobility (150-200 kDa). Incubation of gel strips with various peptidase inhibitors showed that the enzyme was strongly inhibited by diisopropyl phosphorofluoridate (DFP), N-α-tosyl-L-lysinyl-chloromethylketone (TLCK), leupeptin and a peptidyldiazomethane containing lysine at P_1, but not by E-64, PMSF, pepstatin A, 1,10-phenanthroline and a peptidyldiazomethane containing methionine at P_1. This enzyme was characterized as an alkaline peptidase, probably from the serine-type, that cleaves peptide bonds on the carboxyl side of arginine residues at pH 8.0 (Ashall, 1990).

Following this set of experiments, Etges (1992) employed surface radioiodination of living cells, fractionation by Triton X-114 extraction and phase separation, and zymogram analysis by fibrinogen-SDS-PAGE in order to describe the presence of a surface metallopeptidase in *C. fasciculata* with biochemical similarities to the gp63 from *Leishmania* spp. This peptidase is one of major surface molecules in all *Leishmania* species and play vital roles in the different stages of *Leishmania* life cycle, being suggested its participation in many aspects of the infection inside the mammalian host (Yao, 2010). The presence of a similar neutral-to-alkaline metallopeptidase at the surface of *C. fasciculata* led to the suggestion that gp63 should not be involved in the infection of the mammalian host by *Leishmania*, but rather contributes to the survival of the trypanosomatid inside the digestive tract of the insect (Santos et al. 2006).

The work of Etges (1992) opened the possibility to use the same technique in order to analyze the proteolytic profiles in different members of distinct trypanosomatid genera. With this task in mind, our group has analyzed the proteolytic profiles of a great number of species from 8 different genera of trypanosomatids by the use of SDS-PAGE containing 0.1% co-polymerized gelatin as substrate (Branquinha et al. 1996; Santos et al. 2005; 2008). In those studies, it became clear that two distinct proteolytic activities can be detected in total cell lysates: cysteine- and metallopeptidases. For detection of cysteine peptidase activity, the optimal conditions were established to be an acidic pH value (5.0-6.0) and the presence of a reducing agent, such as DTT, which was essential for detection of this activity. The use of specific inhibitors, such as E-64, prevented the development of cysteine peptidase activity bands. Metallopeptidases were consistently observed in a broad pH range (5.0-10.0), and the zinc-chelator 1,10-phenanthroline completely inhibited their activity.

In our first work (Branquinha et al. 1996), three *Crithidia* species were studied: *C. fasciculata*, *C. guilhermei* and *C. luciliae*. Cells were lysed by the addition of SDS-PAGE sample buffer in non-denaturing conditions, and peptidases were characterized by electrophoresis on 7-15% gradient SDS-PAGE with 0.1% gelatin co-polymerized as substrate. Cell lysates of the three species produced similar patterns of proteolysis at 28°C: two cysteine peptidase bands in the 80-110 kDa range and a minor cysteine peptidase activity detected at 45 kDa; and a metallopeptidase band detected in the 55-66 kDa range. When cells were lysed with the non-ionic detergent Triton X-114, cysteine peptidases were detected in the aqueous phase, whereas the metallopeptidase partitioned into the detergent-rich phase, which suggested that the latter is membrane-associated. Interestingly, cell lysates of these species were also employed by our group in another comparative study (Santos et al. 2005), in which a single cysteine peptidase was found at 50 kDa and two metallopeptidases were detected at 70 kDa and at 90 kDa. In the latter study, peptidases were analyzed in 10% linear polyacrylamide gels containing gelatin, and the temperature of incubation after electrophoresis was 37°C. As explained by Martinez and Cazzulo (1992), the apparent molecular mass of each band varies depending on the experimental conditions, including acrylamide concentration and temperature of incubation, which may explain this discrepancy. Despite this fact, both studies highlighted the common proteolytic profile between these species, possibly reflecting their phylogenetic proximity.

Unlike the similarities detected in the three *Crithidia* spp. described above, heterogeneous proteolytic profiles were observed in different members of this genus. For instance, *Crithidia acantocephali* produced 4 cysteine peptidases of 80, 75, 70 and 50 kDa, while *Crithidia harmosa* presented 3 metallopeptidases at 63, 50 and 45 kDa (Santos et al. 2005). These data suggested the value of proteolytic enzymes in distinguishing between trypanosomatid species that cannot be differentiated on structural grounds (Santos et al. 2005). In order to study the distribution of metallopeptidases in trypanosomatids, our group also investigated cell-associated proteolytic activities in distinct species by gelatin-SDS-PAGE in conditions that favor the detection of this subgroup, specifically alkaline conditions (pH 9.0) and proteolytic inhibitors that putatively identified these enzymes, such as 1,10-phenanthroline (Santos et al. 2008). The analysis confirmed the previous results in all the species cited above, showing a great heterogeneity of expression of metallopeptidases not only in *Crithidia* spp. but in a wide range of trypanosomatids as well.

In a similar approach, our group described the differential expression of peptidases in endosymbiont-harboring *Crithidia* species in comparison to members of this genus that naturally lacks a bacterium in the cytoplasm (d'Avila-Levy et al. 2001). In this genus, the trypanosomatids *Crithidia deanei*, *Crithidia desouzai* and *Crithidia oncopelti* have been described to contain a bacterium symbiont in the cytoplasm, known as endosymbiont, which can be eliminated by the use of antibiotics, leading to the generation of cured strains (reviewed by De Souza and Motta, 1999). Gelatin-SDS-PAGE analysis was used to characterize the cell-associated and extracellular peptidases in these organisms, and our survey showed that a similar proteolytic profile was observed in cells of *C. desouzai* and in wild and cured strains of *C. deanei*: two cysteine peptidases migrating at 60-65 kDa and two metallopeptidases at 51-58 kDa. An additional cysteine peptidase was detected in wild strains at 100 kDa. A subsequent study from our group showed that, after Triton X-114 extraction preformed in *C. deanei* cells, a 65-kDa cysteine peptidase partitioned exclusively in the aqueous phase, possibly present in intracellular compartments, and a 51-kDa metallopeptidase was only detected in the detergent-rich phase (d'Avila-Levy et al. 2003). The remaining enzymes, at 60 kDa and at 58 kDa, which corresponds to a cysteine-type and to a metallo-type peptidase, respectively, were found in both aqueous and detergent-rich phases. In cells of *C. oncopelti*, two metallopeptidases were detected in 59-63 kDa range (d'Avila-Levy et al. 2001) (Figure 1).

The analysis of the spent culture medium showed a similar profile among the above-mentioned species: *C. desouzai* and both strains of *C. deanei* displayed an 80-kDa cysteine peptidase and a 60-kDa metallopeptidase, and *C. oncopelti* showed four bands of protein degradation migrating at 101 kDa, 92 kDa, 76 kDa and 59 kDa, all belonging to the metallopeptidase class. For comparison, *C. fasciculata* displayed a more complex extracellular profile, comprising five metallopeptidases migrating at 101 kDa, 92 kDa, 76 kDa, 60 kDa and 43 kDa (d'Avila-Levy et al. 2001). In summary, the proteolytic profiles of *C. deanei* and *C. desouzai* are identical, and distinct from *C. oncopelti*, which is in accordance to a revision in *Crithidia* taxonomy proposed previously by Brandão et al. (2000) and d'Avila-Levy et al. (2004) and recently confirmed by molecular phylogenetic analyses (Teixeira et al. 2011). In this sense, this genus must be subdivided into three groups: the first one (*Angomonas*) must include *C. deanei* and *C. desouzai*, the second one (designated as *Strigomonas*) must include *C. oncopelti* and the remaining *Crithidia* spp. would remain in the originally described genus.

In the same work (d'Avila-Levy et al. 2001), the availability of *C. deanei* wild and cured strains allowed us to study whether the presence of the endosymbiont induces any alteration in the proteolytic profile. The absence of the cell-associated 100-kDa cysteine peptidase in the cured strain was the only qualitative difference found, and may possibly be related to the absence of the endosymbiont. In addition, the activity of extracellular peptidases was enhanced in the cured strain, which provides evidence that the presence of the endosymbiont diminishes the secretion of proteolytic enzymes, mainly the metallopeptidase (d'Avila-Levy et al. 2001).

Extracellular peptidases were also the focus of studies in some species belonging to the genus *Crithidia*. Unlike cell-associated enzymes, qualitative differences were observed when extracellular proteolytic enzymes were analyzed. In all the species tested, only metallopeptidases were detected, and 3 bands in the 60-80 kDa range were common to *C. fasciculata*, *C. guilhermei* and *C. luciliae*. Nevertheless, bands with lower molecular mass (30-40 kDa) were found exclusively in *C. fasciculata*, while higher molecular mass bands (90-100 kDa) were only detected in *C. fasciculata* and *C. guilhermei* (d'Avila-Levy et al. 2001; Santos et al. 2005). Interestingly, the extracellular proteolytic profile of *C. luciliae* was also analyzed by Jaffe and Dwyer (2003), but only two metallopeptidases were detected at 97 kDa and at 50 kDa, which could be explained by the smallest amount of spent culture medium employed as well as by the reduced incubation period for proteolysis development.

Melo et al. (2002) characterized the extracellular peptidases from *C. guilhermei* through the incorporation of different protein substrates into SDS-PAGE. When cells were grown in yeast extract-peptone-sucrose medium, the extracellular proteolytic zymogram comprised four bands with gelatinolytic activity migrating at 80 kDa, 67 kDa, 60 kDa and 55 kDa. All bands were inhibited by 1,10-phenanthroline, which classified these enzymes as metallopeptidases, and these gelatinases remained active over a broad pH range, being the maximum activity reached at pH 5.0, which is in accordance to their proper activity in the insect gut. Interestingly, these enzymes were mainly detected at 37°C; when gels were incubated at 28°C, which corresponds to the room temperature and to the expected value in the insect gut, the proteolytic activity was reduced and the 55-kDa band was not detected, possibly reflecting the adaptations of the parasite to the different environments it might confront during its life cycle. Besides the gelatinolytic activity, the 60-, 67- and 80-kDa bands were also able to degrade casein incorporated into SDS-PAGE, but with minor activity, and no proteolytic activity was detected when bovine serum albumin incorporated into the gel. A distinct pattern of degradation was observed when hemoglobin was used as substrate: a 43-kDa metallopeptidase was exclusively detected in these conditions. These hemoglobinases are possibly involved in supplying exogenous iron and heme for the parasite.

Besides the characterization of these peptidases when *C. guilhermei* cells were grown in yeast extract-peptone-sucrose medium, log-phase cells grown in different culture medium composition were obtained and analyzed. The proteolytic zymograms displayed no qualitative difference, only quantitative variations. In this sense, the replacement of sucrose by glucose enhanced the proteolytic activity of the four bands, while either the replacement of sucrose by glycerol or the cultivation of cells in BHI decreased the proteolytic detection. These results pointed out to the influence of the culture medium composition in the production of extracellular peptidases in this microorganism (Melo et al. 2002).

4. Two-dimensional zymography coupled to peptidase identification through mass spectrometry: Possibilities and technical difficulties

For decades, one-dimensional (1D) zymographic gel systems have been broadly used for the analysis and characterization of proteolytic activities in several organisms. Especially in protozoa parasites, this technique has been extensively useful to detect and identify peptidases involved in virulence of pathogenic protozoa (North and Coombs 1981; Coombs and North 1983; Lockwood et al. 1987; Williams and Coombs 1995; Cuervo et al. 2006; De Jesus et al, 2009). Also, through this technique, crucial roles of these enzymes during the cell cycle of parasites have been revealed (Brooks et al. 2001; De Jesus et al. 2007). In the post-genomic era, this methodology is shedding light on the biochemical traits of organisms of unknown genomes (Santos et al. 2005; Pereira et al. 2009; d'Avila-Levy et al. 2001), and has the potential of increasing the functional annotation of the genome for those organisms yet sequenced. However, information regarding on isoforms of proteolytic enzymes, isoelectric point of peptidases, and even a higher resolution of complex proteolytic profiles cannot be obtained by 1D zymographic systems. In superior eukaryotes, a broader analysis of functional peptidases has been achieved by combining zymographic techniques with proteomic technologies, specifically two-dimensional electrophoresis (2D) and mass spectrometry that enable a better resolution of peptidase arranges and the direct identification of peptidase species (Ong and Chang 1997; Park et al. 2002; Zhao and Russell 2003; Wilkesman and Schröder 2007; Lee et al. 2011). Nevertheless, this combined approach has been little used in the study of protozoan parasites (De Jesus et al. 2009).

Proteomic approaches intend to produce the widest possible resolution of individual proteins from a protein mixture, followed by protein identification by mass spectrometry (MS). The fractionation of complex cellular extracts by 2D is attained by combining two independent electrophoretic separations, the isoelectric focusing (IEF) in the first dimension and SDS-PAGE in the second dimension (MacGillivray and Rickwood 1974; O'Farrell 1975). After, protein spots are excised from the gel, submitted to enzymatic digestion and the resulting peptides are analyzed by MS. The developments of soft ionization sources for protein MS analysis, such as matrix-assisted laser desorption/ionization (MALDI) and electrospray ionization (ESI) enabled the reliable identification of proteins (Karas and Hillenkamp, 1988; Tanaka et al. 1988; Fenn et al. 1989). In this way, the combination of MALDI or ESI with several different mass analyzers and increasingly powerful bioinformatics tools allows the identification of thousands protein components from a complex biological sample. Although protein identification relies on genome sequences data, several algorithms based on homology analyses yet permit to identify proteins of organisms with unknown genomes (Shevchenko et al. 2001; Waridel et al. 2007). The expressive contribution of 2D and MS approaches to the understanding of several aspects of the biology of protozoan parasites such as pathogenic trypanosomatids has been recently reviewed (Cuervo et al. 2010). In these parasites, proteomics studies have contributed to catalogue global protein profiles, provide experimental evidence for gene expression, reveal changes in protein expression during development, assign potential functions to the hypothetical proteins, elucidate the subcellular localization, and determine potential drug and vaccine targets (Cuervo et al. 2010, 2011). Despite all the advantages of 2D, the determination of enzymatic activity in this technique is hampered due to the use of chaotropic agents and additional denaturant components present in the sample buffer used for IEF.

The potentialities of both approaches, i. e., the capability to resolve complex protein mixtures by 2D and the capability to reveal functional (active) peptidases by zymography are merged in the two-dimensional zymography (2DZ) methodology (Figure 6). This technique, coupled with mass spectrometry for protein identification make possible the broader mapping of active proteolytic enzymes present in a protein extract (Zhao et al. 2004; De Jesus et al. 2009; Saitoh et al. 2007; Paes-Leme et al. 2009; Larocca et al. 2010; Lee et al. 2011). Two main strategies are used for 2DZ analysis: the first one consist on the separation of protein sample by 2DZ or 2D reverse zymography in parallel with separation by denaturing or non-denaturing 2D followed by staining with MS compatible stain. After migration, the comparison and overlapping of both gel images, using appropriated gel image analysis software, allow the assigning of proteolytic spots to protein spots which are carefully excised from the gel and further identified by MS (Métayer et al. 2002; Park et al. 2002; Choi et al. 2004; Taiyoji et al. 2009; Lee et al. 2011). Using this strategy our group identified active cysteine peptidases in whole extracts of the two *Trichomonas vaginalis* isolates exhibiting high and low virulence phenotypes (De Jesus et al. 2007) (Figure 6). Whole extracts analyzed by 2DZ gels showed both qualitative and quantitative differences in the cysteine peptidase spots between the isolates. According to the pH distribution across the gel strip, proteolytic spots displayed p*I* values between 4.2 and 6.5, a biochemical characteristic that cannot be obtained from 1DZ. It was also observed that the qualitative and quantitative differences in the cysteine peptidases (CP) expression revealed by 2DZ may be related to the virulence pattern of the *T. vaginalis* isolate (Figure 6). After identification of the active "cysteine peptidase fingerprint" expressed by each *T. vaginalis* isolate by tandem MS analysis (MS/MS) it was corroborated that distinct isoforms of CP4 are expressed between the isolates, specifically differentiated by a change in one amino acid of a main peptide. Whereas low-virulence parasites expressed NSWGTAWGEK-containing CP4 isoforms, the virulent isolate expressed a NSWGTTWGEK-containing CP4 isoform (De Jesus et al. 2007). The NSWGTTWGEK-containing CP4 isoform is present in several virulent isolates, is secreted and can induce apoptosis in the epithelial cells (Sommer et al. 2005; De

Jesus et al. 2007). Another important contribution of 2DZ analysis in this work was to reveal that only a limited number of active gelatinase-CPs are expressed *in vitro*, which contrast to a high number of CP genes present in the parasite genome. Alternatively, the employment of different substrates may reveal other peptidase activities. Additionally, in this work, 2DZ allowed preliminary mapping of active forms of low-abundance CPs, which are not easily visualized in 2D Coomassie-stained gels (De Jesus et al. 2007).

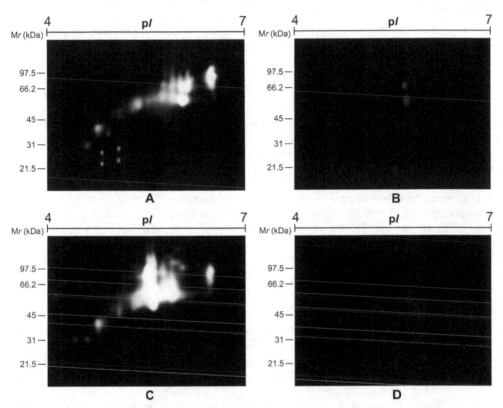

Fig. 6. Two-dimensional-substrate gel electrophoresis showing the profiles of active cysteine peptidase detected in whole extracts of *Trichomonas vaginalis* isolates displaying low (A, B) and high (C, D) virulence phenotypes. Assays were performed in the absence (A, C) or presence (B, D) of cysteine peptidase inhibitor E-64. For experimental details see De Jesus et al., 2009. Reprinted with permission of *Journal of Proteome Research*.

The second strategy consists on the electrophoretic separation in SDS-substrate gels and direct MS analysis from 2DZ gels. However, the major challenge of this approach consists on having "to fish" a specific protein in a "protein sea". To overcome this drawback, fluorescent substrates are used (Zhao et al 2004; Thimon et al. 2008). Proteins are separated by 2D, gels are further incubated with fluorescent peptide substrate and the emitted fluorescence is observed under an UV transilluminator. As the substrate is not embedded in the gel, it can be easily washed, and the protein spot can be excised from the gel for MS/MS analysis. It is clear that the 2DZ-MS techniques should be preceded by broad biochemical characterization of the proteolytic profile of the organism as suggested in the flowchart (Figure 1). The use of 2DZ approaches combined with MS/MS analysis might be a shortcut in the identification of the active degradome and, associated to conventional 2D mapping, might allow the identification of active and inactive peptidases without the use of specific antibodies or laborious purification methods.

5. Conclusion

Substrate-SDS-PAGE has been described in 1980 by Heussen and Dowdle, 30 years after description of this technique, several and important advances in methodological approaches to unveil biological systems have been achieved, such as automated DNA sequencing and protein identification through mass spectrometry just to cite a few. In spite of this, this simple and inexpensive methodology still provides powerful and unique information about peptidases. In organisms with complete genome sequences, bioinformatic analysis provides rich information on putative peptidases, however, it cannot be ascertained if the ORFs are indeed expressed and active. It is common to observe an elevated number of putative peptidases in the organisms' genome, with a limited number of active peptidases. While in organisms without genome information, this technique allows the detection and assessment of several biochemical characteristics of the enzymes, such as preferable pH, temperature, catalytic type and substrate preference. Finally, two-dimensional zymography coupled with mass spectrometry for protein identification make possible the broader mapping of active proteolytic enzymes present in a protein extract, allowing the detection of distinct isoforms of peptidases differentiated by a single change in one amino acid of a main peptide. Therefore, thirty years after its first description, zymography still is a powerful approach to unveil peptidases.

6. References

Ashall, F. (1990). Characterisation of an alkaline peptidase of *Trypanoma cruzi* and other trypanosomatids. *Molecular and Biochemical Parasitology* Vol.38, No.1, pp.77-87.

Ashall, F., Harris, D., Roberts, H., Healy, N. & Shaw, E. (1990). Substrate specificity and inhibitor sensitivity of a trypanosomatid alkaline peptidase. *Biochimica et Biophysica Acta* Vol.1035, pp. 293-299.

Bouvier, J., Etges, R., & Bordier, C. (1987). Identification of the promastigote surface protease in seven species of *Leishmania*. *Molecular and Biochemical Parasitology* Vol. 24, pp. 73-79.

Brandão, A.A., Miranda, A., Degrave, W.M. & Sousa, M.A. (2000). The heterogeneity of choanomastigote-shaped trypanosomatids as analyzed by their kDNA minicircle size: taxonomic implications. *Parasitology Research* Vol.86, pp. 809-812.

Branquinha, M.H., Vermelho, A.B., Goldenberg, S. & Bonaldo, M.C. (1996). Ubiquity of cysteine- and metalloproteinase in a wide range of trypanosomatids. *Journal of Eukaryotic Microbiology* Vol.43, pp. 131-135.

Breganó, J.W., Picão, R.C., Graça, V.K., Menolli, R.A., Jankevicius, S.I., Filho, P.P. & Jankevicius, J.V. (2003). *Phytomonas serpens*, a tomato parasite, shares antigens with *Trypanosoma cruzi* that are recognized by human sera and induce protective immunity in mice. *FEMS Immunology and Medical Microbiology* Vol.39, pp. 257-264.

Brooks, D.R., Denise, H., Westrop, G.D., Coombs, G.H. & Mottram, J.C. (2001). The stage-regulated expression of *Leishmania mexicana* CPB cysteine proteases is mediated by an intercistronic sequence element. *The Journal of Biological Chemistry* Vol. 276, pp. 47061-47069.

Cadavid-Restrepo, G., Gastardelo, T.S., Faudry, E., de Almeida, H., Bastos, I.M., Negreiros, R.S., Lima, M.M., Assumpção, T.C., Almeida, K.C., Ragno, M., Ebel, C., Ribeiro, B.M., Felix, C.R. & Santana, J.M. (2011). The major leucyl aminopeptidase of *Trypanosoma cruzi* (LAPTc) assembles into a homohexamer and belongs to the M17 family of metallopeptidases. *BMC Biochemistry* Vol. 12, pp. 46.

Chicharro, C. & Alvar, J. (2003). Lower trypanosomatids in HIV/AIDS patients. *Annual Tropical Medical Parasitology* Vol.97, pp. 75–80.

Choi, N.S., Yoo, K.H., Yoon, K.S., Maeng, P.J. & Kim, S.H. (2004). Nano-scale proteomics approach using two-dimensional fibrin zymography combined with fluorescent SYPRO ruby dye. *Journal of Biochemistry and Molecular Biology* Vol.37, pp. 298-303.

Coombs, G.H. & North, M.J. (1983). An analysis of the proteinases of *Trichomonas vaginalis* by polyacrylamide gel electrophoresis. *Parasitology* Vol.86, pp. 1-6.

Cuervo, P., Sabóia-Vahia, L., Costa Silva-Filho, F., Fernandes, O., Cupolillo, E., D & Jesus, J.B. (2006). A zymographic study of metalloprotease activities in extracts and extracellular secretions of *Leishmania (Viannia) braziliensis* strains. *Parasitology* Vol. 132, pp. 177-185.

Cuervo, P., Domont, G.B. & De Jesus, J.B. (2010). Proteomics of trypanosomatids of human medical importance. *Journal of Proteomics* Vol.10, pp. 845-867.

Cuervo, P., Fernandes, N. & de Jesus, J.B. (2011). A proteomics view of programmed cell death mechanisms during host-parasite interactions. *Journal of Proteomics* Vol. 75, pp. 246-256.

d'Avila-Levy, C.M., Melo, A.C.N., Vermelho, A.B. & Branquinha, M.H. (2001). Differential expresson of proteolytic enzymes in endosymbiont-harboring *Crithidia* species. *FEMS Microbiology Letters*, Vol.202, pp. 73-77.

d'Avila-Levy, C.M., Souza, R.F., Gomes, R.C., Vermelho, A.B. & Branquinha, M.H. (2003). A metalloproteinase extracellularly released by *Crithidia deanei*. *Canadian Journal of Microbiology* Vol.49, pp. 625-632.

d'Avila-Levy, C.M., Araújo, F.M., Vermelho, A.B., Branquinha, M.H., Alviano, C.S., Soares, R.M. & Santos, A.L.S. (2004). Differential lectin recognition of glycoproteins in choanomastigote-shaped trypanosomatids: taxonomic implications. *FEMS Microbiol Lett.* Vol. 231, pp. 171-176.

d'Avila-Levy, C.M., Araújo, F.M., Vermelho, A.B., Soares, R.M., Santos, A.L.S. & Branquinha, M.H. (2005). Proteolytic expression in *Blastocrithidia culicis*: influence of the endosymbiont and similarities with virulence factors of pathogenic trypanosomatids. *Parasitology* Vol. 130, pp. 413-420.

d'Avila-Levy, C.M., Volotão, A.C.C., Araújo, F.M., De Jesus, J.B., Motta, M.C.M., Vermelho, A.B., Santos, A.L.S. & Branquinha, M.H. (2009). *Bodo* sp., a free-living flagellate, expresses divergent proteolytic activities from the closely related parasitic trypanosomatids. *The Journal of Eukaryotic Microbiology* Vol. 56, pp. 454–458.

De Jesus, J.B., Cuervo, P., Junqueira, M., Britto, C., Silva-Filho, F.C., Soares, M.J., Cupolillo, E., Fernandes, O. & Domont, G.B. (2007). A further proteomic study on the effect of iron in the human pathogen *Trichomonas vaginalis*. *Proteomics* Vol. 7, pp. 1961-1972.

De Jesus, J.B., Cuervo, P., Britto, C., Sabóia-Vahia, L., Costa E Silva-Filho, F., Borges-Veloso, A., Barreiros Petrópolis, D., Cupolillo, E. & Barbosa Domont, G. (2009). Cysteine peptidase expression in *Trichomonas vaginalis* isolates displaying high- and low-virulence phenotypes. *Journal of Proteome Research* Vol. 8, pp. 1555-1164.

De Souza, W. & Motta, M.C.M. (1999). Endosymbiosis in protozoa of the Trypanosomatidae family. *FEMS Microbiology Letters* Vol.173, No.1, pp. 1-8.

Devel, L., Rogakos, V., David, A., Makaritis, A., Beau, F., Cuniasse, P., Yiotakis, A.& Dive, V. (2006). Development of selective inhibitors and substrate of matrix metallo proteinase-12. *Journal of Biological Chemistry* Vol. 281, pp. 11152-11160.

Elias, C.G.R., Pereira, F.M., Silva, B.A., Alviano, C.S., Soares, R.M.A. & Santos, A.L.S. (2006). Leishmanolysin (gp63 metallopeptidase)-like activity extracellularly released by *Herpetomonas samuelpessoai*. *Parasitology* Vol.132, pp. 37–47.

Elias, C.G.R., Pereira, F.M., Dias, F.A., Silva, T.L., Lopes, A.H., d'Avila-Levy, C.M., Branquinha, M.H. & Santos, A.L.S. (2008). Cysteine peptidases in the tomato trypanosomatid *Phytomonas serpens*: influence of growth conditions, similarities with cruzipain and secretion to the extracellular environment. *Experimental Parasitology* Vol. 120, pp. 343–352.

Etges, R. (1992). Identification of a surface metalloproteinase on 13 species of *Leishmania* isolated from humans, *Crithidia fasciculata* and *Herpetomonas samuelpessoai*. *Acta Tropica* Vol.50, pp. 205–217.

Fenn, J., Mann, M., Meng, C.K., Wong, S.F. & Whitehouse, C.M. (1989). Electrospray ionization for mass spectrometry of large biomolecules. *Science* Vol. 246, pp. 64-71.

Grellier, P., Vendeville, S., Joyeau, R., Bastos, I.M., Drobecq, H., Frappier, F., Teixeira, A.R., Schrével J, Davioud-Charvet E, Sergheraert C & Santana JM. (2001). *Trypanosoma cruzi* prolyl oligopeptidase Tc80 is involved in nonphagocytic mammalian cell invasion by trypomastigotes. *Journal of Biological Chemistry* Vol. 276, pp. 47078-47086.

Heussen, C. & Dowdle, E.B. (1980). Electrophoretic analysis of plasminogen activators in polyacrylamide gels containing sodium dodecyl sulphate and copolymerized substrates. *Analytical Biochemistry* Vol. 102, pp. 196–202.

Hoare, C.A. & Wallace, F.G. (1966). Developmental stages of trypanosomatid flagellates: a new terminology. *Nature*, Vol.212, pp. 1385-1386.

Jaffe, C.L. & Dwyer, D.M. (2003). Extracellular release of the surface metalloprotease, gp63, from Leishmania and insect trypanosomatids. *Parasitology Research* Vol.91, pp. 229-37.

Karas, M. & Hillemkamp, F. (1988). Laser desorption ionization of proteins with molecular masses exceeding 10000 kDa. *Analytical Chemistry* Vol. 60, pp. 2299-2301.

Kupai, K., Szucs, G., Cseh, S., Hajdu, I., Csonka, C., Csont, T. & Ferdinandy, P. (2010). Matrix metalloproteinase activity assays: Importance of zymography. *Journal of Pharmacological and Toxicological Methods* Vol. 61, pp. 205-209.

Laemmli, U.K. (1970). Cleavage of structural proteins during the assembly of the head of bacteriophage T4. *Nature* Vol. 227, pp. 680-685.

Larocca, M., Rossano, R. & Riccio, P. (2010). Analysis of green kiwi fruit (Actinidia deliciosa cv. Hayward) proteinases by two-dimensional zymography and direct identification of zymographic spots by mass spectrometry. *Journal of the Science of Food and Agriculture* Vol. 90, pp. 2411-248.

Lee, K.J., Kim, J.B., Ha, B.K., Kim, S.H., Kang, S.Y., Lee, B.M. & Kim, D.S. (2011). Proteomic characterization of Kunitz trypsin inhibitor variants, Tia and Tib, in soybean [Glycine max (L.) Merrill]. *Amino Acids*, in press.

Lockwood, B.C., North, M.J., Scott, K.I., Bremner, A.F. & Coombs, G.H. (1987). The use of a highly sensitive electrophoretic method to compare the proteinases of trichomonads. *Molecular and Biochemical Parasitology* Vol. 24, pp. 89-95.

Lopes, J.D., Caulada, Z., Barbieri, C.L. & Camargo, E.P. (1981). Cross-reactivity between *Trypanosoma cruzi* and insect trypanosomatids as a basis for the diagnosos of Chagas' disease. *American Journal of Tropical Medicine and Hygine* Vol.30, pp. 1183-1188.

MacGillivray, A.J. & Rickwood, D. (1974). The heterogeneity of mouse-chromatin nonhistone proteins as evidenced by two-dimensional polyacrylamide-gel electrophoresis and ion-exchange chromatography. *European Journal of Biochemistry* Vol. 41, pp. 181-190.

Martinez, J. & Cazzulo, J.J. (1992). Anomalous electrophoretic behaviour of the major cysteine proteinase (cruzipain) from *Trypanosoma cruzi* in relation to its apparent molecular mass. *FEMS Microbiology Letters* Vol.74, pp. 225-229.

McGhee, R.B. & Cosgrove, W.B. (1980). Biology and physiology of the lower Trypanosomatidae. *Microbiology Reviews* Vol.44, pp. 140–173.

Melo, A.C.N., d'Avila-Levy, C.M., Branquinha, M.H. & Vermelho, A.B. (2002). *Crithidia guilhermei*: gelatin- and haemoglobin-degrading extracellular metalloproteinases. *Experimental Parasitology* Vol.102, pp. 150-156.

Métayer, S., Dacheux, F., Dacheux, J.L. & Gatti, J.L. (2002). Comparison, characterization, and identification of proteases and protease inhibitors in epididymal fluids of domestic mammals. Matrix metalloproteinases are major fluid gelatinases. *Biology of Reproduction* Vol. 66, pp. 1219-1229.

Michaud, D., Cantin, L., Raworth, D.A. & Vrain, T.C. (1996). Assessing the stability of cystatin/cysteine proteinase complexes using mildly-denaturing gelatin-polyacrylamide gel electrophoresis. *Electrophoresis* Vol.17, pp. 74-79.

Momen, H. (2001). Some current problems in the systematics of trypanosomatids. *International Journal for Parasitology* Vol.31, pp. 640-642.

Momen, H. (2002). Molecular taxonomy of trypanosomatids: some problems and pitfalls. *Archives of Medical Research* Vol.33, pp. 413–415.

Morio, F., Reynes, J., Dollet, M., Pratlong, F., Dedet, J.P. & Ravel, C. (2008). Isolation of a protozoan parasite genetically related to the insect trypanosomatid *Herpetomonas samuelpessoai* from a human immunodeficiency virus positive patient. *Jounal of Clinical Microbiology* Vol.46, pp. 3845-3847.

Nogueira de Melo, A.C., d'Avila-Levy, C.M., Dias, F.A., Armada, J.L.A., Silva, H.D., Lopes, A.H.C.S., Santos, A.L.S., Branquinha, M.H. & Vermelho, A.B. (2006). Peptidases and gp63-like proteins in *Herpetomonas megaseliae*: possible involvement in the adhesion to the invertebrate host. *International Journal for Parasitology*, Vol.36, pp. 415–422.

North, M.J. & Coombs, G.H. (1975). Proteinases of *Leishmania mexicana* amastigotes and promastigotes: analysis by gel electrophoresis. *Molecular and Biochemical Parasitology* Vol. 3, pp. 293-300.

O'Farrell, P.H. (1975). High resolution two-dimensional electrophoresis of proteins. *The Journal of Biological Chemistry* Vol. 250, pp. 4007-4021.

Ong, K.L. & Chang, F.N. (1997). Analysis of proteins from different phase variants of the entomopathogenic bacteria *Photorhabdus luminescens* by two-dimensional zymography. *Electrophoresis* Vol. 18, pp. 834-839.

Paes Leme, A.F., Kitano, E.S., Furtado, M.F., Valente, R.H., Camargo, A.C., Ho, P.L., Fox, J.W. & Serrano, S.M. (2009). Analysis of the subproteomes of proteinases and heparin-binding toxins of eight Bothrops venoms. *Proteomics* Vol. 9, pp. 733-745.

Park, S.G., Kho, C.W., Cho, S., Lee, D.H., Kim, S.H. & Park, B.C. (2002). A functional proteomic analysis of secreted fibrinolytic enzymes from *Bacillus subtilis* 168 using a combined method of two-dimensional gel electrophoresis and zymography. *Proteomics*. Vol. 2, pp. 206-211.

Pereira, F.M., Elias, C.G., d'Avila-Levy, C.M., Branquinha, M.H. & Santos, A.L.S. (2009). Cysteine peptidases in *Herpetomonas samuelpessoai* are modulated by temperature and dimethylsulfoxide-triggered differentiation. *Parasitology*, Vol.136, pp. 45-54.

Pereira, F.M., Dias, F.A., Elias, C.G., d'Avila-Levy, C.M., Silva, C.S., Santos-Mallet, J.R., Branquinha, M.H. & Santos, A.L.S. (2010a). Leishmanolysin-like molecules in *Herpetomonas samuelpessoai* mediate hydrolysis of protein substrates and interaction with insect. *Protist* Vol.161, pp. 589-602.

Pereira, F.M., Santos-Mallet, J.R., Branquinha, M.H., d'Avila-Levy, C.M. & Santos, A.L.S. (2010b). Influence of leishmanolysin-like molecules of *Herpetomonas samuelpessoai* on the interaction with macrophages. *Microbes and Infection* Vol.12, pp. 1061-170.

Podlipaev, S. (2001). The more insect trypanosomatids under study-the more diverse Trypanosomatidae appears. *International Journal for Parasitology* Vol.31, pp. 648–652.

Rawlings, N.D., Barrett, A.J. & Bateman, A. (2010). MEROPS: the peptidase database. *Nucleic Acids Research*. Vol. 38, pp. D227-D233.

Saitoh, E., Yamamoto, S., Okamoto, E., Hayakawa, Y., Hoshino, T., Sato, R., Isemura, S., Ohtsubo, S. & Taniguchi, M. (2007). Identification of cysteine proteases and screening of cysteine protease inhibitors in biological samples by a two-dimensional gel system of zymography and reverse zymography. *Analytical Chemistry Insights* Vol. 18, pp. 51-59.

Santos, A.L.S., Ferreira, A., Franco, V.A., Alviano, C.S. & Soares, R.M.A. (1999). Characterization of proteinases in *Herpetomonas anglusteri* and *Herpetomonas roitmani*. *Current Microbiology* Vol.39, pp. 61–64.

Santos, A.L.S., Abreu, C.M., Alviano, C.S. & Soares, R.M.A. (2002). Activation of the glycosylphosphatidylinositol anchored membrane proteinase upon release from *Herpetomonas samuelpessoai* by phospholipase C. *Current Microbiology* Vol.45, pp. 293–298.

Santos, A.L.S., Batista, L.M., Abreu, C.M., Alviano, C.S., Angluster, J. & Soares, R.M.A. (2001). Developmentally regulated protein expression mediated by dimethylsulfoxide in *Herpetomonas samuelpessoai*. *Current Microbiology* Vol.42, pp. 111–116.

Santos, A.L.S., Rodrigues, M.L., Alviano, C.S., Angluster, J. & Soares, R.M.A. (2003). *Herpetomonas samuelpessoai*: dimethylsulfoxide-induced differentiation is influenced by proteinase expression. *Current Microbiology* Vol.46, pp. 11–17.

Santos, A.L.S., Abreu, C.M., Alviano, C.S. & Soares, R.M.A. (2005). Use of proteolytic enzymes as an additional tool for trypanosomatid identification. *Parasitology* Vol.130, pp. 79–88.

Santos, A.L.S., Branquinha, M.H. & d'Avila-Levy, C.M. (2006). The ubiquitous gp63-like metalloprotease from lower trypanosomatids: in the search for a function. *Anais da Academia Brasileira de Ciências* Vol.78, pp. 687–714.

Santos, A.L.S., d'Avila-Levy, C.M., Elias, C.G.R., Vermelho, A.B. & Branquinha, M.H. (2007). *Phytomonas serpens*: immunological similarities with the human trypanosomatid pathogens. *Microbes and Infection* Vol.9, pp. 915–921.

Santos, A.L.S., Soares, R.M.A., Alviano, C.S. & Kneipp, L.F. (2008). Heterogeneous production of metallo-type peptidases in parasites belonging to the family Trypanosomatidae. *European Journal of Protistology* Vol.44, pp.103–113.

Schneider, P. & Glaser, T. A. (1993). Characterization of a surface metalloprotease from *Herpetomonas samuelpessoai* and comparison with *Leishmania major* promastigote surface protease. *Molecular and Biochemical Parasitology* Vol.58, pp. 277–282.

Shevchenko, A., Sunyaev, S., Loboda, A., Shevchenko, A., Bork, P., Ens, W. & Standing, K.G. (2001). Charting the proteomes of organisms with unsequenced genomes by MALDI-quadrupole time-of-flight mass spectrometry and BLAST homology searching. *Analytical Chemistry* Vol.73, pp. 1917-1926.

Sommer, U., Costello, C.E., Hayes, G.R., Beach, D.H., Gilbert, R.O., Lucas, J.J. & Singh, B.N. (2005). Identification of *Trichomonas vaginalis* cysteine proteases that induce apoptosis in human vaginal epithelial cells. *Journal of Biological Chemistry* Vol. 280, pp. 23853-23860.

Snoek-van Beurden, P.A. & Von den Hoff, J.W. (2005). Zymographic techniques for the analysis of matrix metalloproteinases and their inhibitors. *Biotechniques*. Vol. 38, pp. 73-83.

Taiyoji, M., Shitomi, Y., Taniguchi, M., Saitoh, E. & Ohtsubo, S. (2009). Identification of proteinaceous inhibitors of a cysteine proteinase (an Arg-specific gingipain) from *Porphyromonas gingivalis* in rice grain, using targeted-proteomics approaches. *Journal of Proteome Research* Vol. 8, pp. 5165-5174.

Tanaka, K., Waki, H., Ido, Y., Akita, S., Yoshida, Y. & Yoshida, T. (1988). Protein and polymer analyses up to m/z 100000 by laser ionization time-of-flight mass spectrometry. *Rapid Communications in Mass Spectrometry* Vol. 2, pp.151-153.

Teixeira, M.M., Borghesan, T.C., Ferreira, R.C., Santos, M.A., Takata, C.S., Campaner, M., Nunes, V.L., Milder, R.V., de Souza, W. & Camargo, E.P. (2011). Phylogenetic validation of the genera *Angomonas* and *Strigomonas* of trypanosomatids harboring bacterial endosymbionts with the description of new species of trypanosomatids and of proteobacterial symbionts. *Protist* Vol.162, pp. 503-524.

Thimon, V., Belghazi, M., Labas, V., Dacheux, J.L. & Gatti, J.L.(2008). One- and two-dimensional SDS-PAGE zymography with quenched fluorogenic substrates provides identification of biological fluid proteases by direct mass spectrometry. *Analytical Biochemistry* Vol. 375, pp. 382-384.

Vickerman, K. (1994). The evolutionary expansion of the trypanosomatid flagellates. *The International Journal for Parasitology*, Vol.24, pp. 1317-1331.

Wallace, F.G. (1966). The trypanosomatid parasites of insects and arachnids. *Experimental Parasitology* Vol. 18, pp. 124-193.

Wallace, F.G., Camargo, E.P., McGhee, R.B. & Roitman, I. (1983). Guidelines for the description of new species of lower trypanosomatids. *Journal of Protozoology* Vol.30, pp. 308–313.

Waridel, P., Frank, A., Thomas, H., Surendranath, V., Sunyaev, S., Pevzner, P. & Shevchenko, A. (2007). Sequence similarity-driven proteomics in organisms with unknown genomes by LC-MS/MS and automated de novo sequencing. *Proteomics.* Vol. 7, pp. 2318-2329.

Wilder, C.L., Park, K., Keegan, P.M. & Platt, M.O. (2011). Manipulating substrate and pH in zymography protocols selectively distinguishes cathepsins K, L, S, and V activity in cells and tissues. *Archives of Biochemistry and Biophysics* Vol. 516, pp. 52-77.

Williams, A.G. & Coombs, G.H. (1995). Multiple protease activities in *Giardia intestinalis* trophozoites. *The International Journal of Parasitology* Vol. 25, pp. 771-778.

Wilkesman, J.G. & Schröder, H.C. (2007). Analysis of serine proteases from marine sponges by 2-D zymography. *Electrophoresis* Vol. 28, pp. 429-436.

Yao, C. (2010). Major surface protease of trypanosomatids: one size fits all? *Infection and Immunity* Vol.78, pp. 22-31.

Zhao, Z. & Russell, P.J. (2003). Trypsin activity assay in substrate-specific one- and two-dimensional gels: a powerful method to separate and characterize novel proteases in active form in biological samples. *Electrophoresis* Vol. 24, pp. 3284-3248.

Zhao, Z., Raftery, M.J., Niu, X.M., Daja, M.M. & Russell, P.J. (2004). Application of in-gel protease assay in a biological sample: characterization and identification of urokinase-type plasminogen activator (uPA) in secreted proteins from a prostate cancer cell line PC-3. *Electrophoresis* Vol. 25, pp. 1142-1148.

Polyacrylamide Gel Electrophoresis an Important Tool for the Detection and Analysis of Enzymatic Activities by Electrophoretic Zymograms

Reyna Lucero Camacho Morales, Vanesa Zazueta-Novoa,
Carlos A. Leal-Morales, Alberto Flores Martínez,
Patricia Ponce Noyola and Roberto Zazueta-Sandoval
University of Guanajuato
México

1. Introduction

Gel electrophoresis of enzymes is a very useful and powerful analytical method, which is at present widely used in many distinct fields of both biological and medical sciences and successfully applied in many different fields of human activity. The tremendous expansion of this methodology is mainly due to its simplicity and its high ability to separate both isoenzymes and alloenzymes, which have proven to be very useful genetic markers. The most important step of enzyme electrophoresis is the detection of native enzymes on electrophoretic gels; it means the procedure of obtaining electropherograms, or zymograms. Detection of enzymes on electrophoresis gels means the visualization of gel areas occupied by specific enzyme molecules after their electrophoretic separation. From this point of view, sometimes the testing of an enzyme-coding DNA sequence for expression of catalytically active enzyme is performed by zymograms, where the use of this technique for this purpose is very effective, cheap, and time saving. The number of applications of zymogram techniques for testing cloning enzyme-coding genes for their expression at the protein level is growing (Pfeiffeer-Guglielmi et al., 2000; Lim et al., 2001; Okwumabua et al., 2001). An absolute prerequisite for this is the specific and sensitive zymogram technique suitable for detection of the enzyme inside the gel and the use of the appropriate substrates.

The zymograms has been used to detect a variety of oxidoreductases (Bergmeyer, 1983), including isoenzymes (Jeng & Wayman, 1987) and to classify various genera of yeast based upon the relative mobility of the activity bands produced by selected enzymes (Goto & Takami, 1986; Yamasaki & Komagata, 1983). Electrophorezed gels are placed in a staining solution containing a reduced substrate such as an alcohol, oxidized cofactor such as NAD^+ or $NADP^+$, a dye such as nitroblue tetrazolium, and an electron acceptor-donor such as phenazine methosulphate. At the location of the appropriate enzyme catalyzing oxidation of substrate and reduction of cofactor, a dark-purple band appears as a result of the precipitate that forms upon reduction of the dye.

Previously we have investigated the use of zymogram staining of native electrophoretic gels as an initial approach to the identification of carbonyl reductase activities against both aliphatic (Silva et al., 2009; Zazueta et al., 2008) and aromatic hydrocarbons (Durón et al., 2005; Zazueta et al., 2003) in *Mucor circinelloides* YR1, an indigenous fungus isolated from petroleum contaminated soil.

Oil spills sometimes occur during routine operations associated with the exploration and production of crude oil. Crude oils vary widening in composition depending on factors such as source bed type and generation temperatures (Hunt, 1979). Biodegradation rates for crude oils will vary due to differences in composition, as reflected by hydrocarbon class distribution: saturates, aromatics, and polars, and the amount *n*-alkanes *versus* branched and cyclic alkanes within the saturated hydrocarbon class (Cook et al., 1974). In nature exist many types of microorganisms useful in the biodegradation processes of contaminant compounds (Atlas, 1995), such as the polycyclic aromatic hydrocarbons (PAH's) that are persistent soil contaminants and many of which have toxic and carcinogenic properties (Hyötyläinen and Oikari, 1999; Cerniglia, 1997).

In bacterial aerobic degradation of aromatic compounds, reactions of metabolic pathways generally lead to the formation of aromatic intermediates containing two hydroxyl constituents, which are subsequently ring-cleaved by excision dioxygenases (Neidle et al., 1992). In many catabolic pathways the formation of such intermediates is carried out by two successive enzymatic steps namely dihydroxylation of the polyaromatic substrate to produce *cis*-diols followed by dehydrogenation (Harayama & Timmis, 1989). The ring hydroxylation is catalyzed by multi-component dioxygenases, while the dehydrogenation is catalyzed by *cis*-diol-dehydrogenases. In mammalian tissues the enzyme dihydro-diol dehydrogenase (DD, EC 1.3.1.20) exists in multiple forms (Hara et al., 1990; Higaki et al., 2002) and catalyses the NADP$^+$-linked oxidation of *trans*-dihydro-diols of aromatic hydrocarbons to the corresponding catechols (Penning et al., 1999). Studies on the metabolism of aromatic hydrocarbons by fungi are limited, nevertheless have been shown to posses the ability to metabolize aromatic compounds (Auret et al., 1971; Ferris et al., 1976) and the aryl oxidative enzymes of fungi appear to be similar to monooxygenases of hepatic microsomes (Cerniglia & Gibson, 1977; Ferris et al., 1976). Smith & Rosazza (1974) have also presented evidence that naphthalene is metabolized to 1-naphtol by six different genera of fungi.

In this work we analyze the cytosolic fraction of YR-1 strain by electrophoretic zymograms, methodology that there is not described in the literature for the NADP$^+$-dependent dihydrodiol dehydrogenase (DD) activities. We analyze all the activity bands corresponding to proteins with DD activity present in an enzymatic extract in only one lane of the electrophoretic gel. Our results show eleven different DD activity bands, five of them are constitutive, DD1-5, since they appears when the strain is growth on glucose, and the others six are induced by different compound added to the culture media as a sole carbon source. Some biochemical-enzyme characteristics as pH, optimal temperature, cofactor dependence, substrate specificity and the effect of cations, EDTA and pyrazole were investigated for DD activities when YR-1 strain was grown in naphthalene as sole carbon source.

2. Materials and methods

2.1 Organisms used and culture conditions

Mucor circinelloides strain YR-1 originally isolated from petroleum-contaminated soil in Salamanca, Guanajuato, Mexico was used as enzymatic source. A defined media containing yeast-peptone-glucose-agar (YPGA) (Bartnicki-García & Nickerson, 1962) was used for strain maintenance, spore collection and mycelium growth. Aerobic mycelium growth was also carried out on salt minimal medium (Alvarado et al. 2002) added with 0.1% (w/v) peptone (sMMP). As a carbon source we added D-glucose (1% w/v) or glycerol (1.0% v/v) or ethanol (2.0% v/v) or n-decanol (1.0% v/v) or n-pentane (1.0% v/v) or n-decane, (1.0% v/v) or n-hexadecane (1.0% w/v) or naphthalene (0.5% w/v) or anthracene (0.5% v/v) or phenanthrene (0.5% w/v) or pyrene (0.5% w/v). Liquid cultures (600 ml) inoculated with spores at a final cell density of 5×10^5/ml were propagated in 2-l Erlenmeyer flasks and incubated in a reciprocating water bath at 28°C for 22 h at 125 rpm for all substrates except glucose that was incubated for 12 h at same other conditions.

2.2 Preparation of purified fractions

Mycelium of 22 h of incubation was harvested by filtration and exhaustively washed with cold sterile-distilled water; mycelial mass was suspended in 15 ml of 20 mM Tris-HCl pH 8.5 buffer containing 1 mM phenylmethanesulphonyl fluoride (previously dissolved in ethanol). Approximately 20 ml of cells was mixed with an equal volume of glass beads (0.45-0.50 mm diameter) and disrupted in a Braun model MSK cell homogenizer (Braun, Melsungen, Germany) for four periods of 30 sec each under a CO_2 stream. The homogenate (crude extract) was centrifuged at 4,300g for 10 min in a J2-21 Beckman rotor in a Beckman JA-20 centrifuge to remove cell walls and unbroken cells, a 1 ml sample of the supernatant was saved. The rest of the supernatant (low speed supernatant) was centrifuged at 31,000g for 20 min in a 70Ti Beckman rotor in a Beckman L8-80 ultracentrifuge and samples of 1 ml of the supernatant was saved; the resulting pellet (mitochondrion rich sample) was resuspended in 2 ml of buffer and saved. The rest of the supernatant was high-speed centrifuged at 164,500g for 45 min in a 70Ti Beckman rotor at 4°C in a Beckman L8-80 ultracentrifuge; the supernatant (cytosolic fraction) was put aside, and the pellet, the mixed membrane fraction (MMF), was resuspended in 2 ml and saved. In all cases samples of different fractions were kept at -70 °C for further studies.

2.3 Gel electrophoresis

The slab gels were 1.5 mm-thick contained 6% (w/v) acrylamide/4% (w/v) bisacrilamide, loaded with the cytosolic fraction of each culture and run in the mini-gel system manufactured by Bio-Rad. The continuous buffer system described by Laemmli (1970) without SDS (native conditions) was used to run for 2.5 h at 80 V. The Rm values were calculated as the ratio of the distance migrated by the stained band divided by the distance migrated by tracking dye; standard deviation was calculated with Excel from three independent experiments and each experiment was made by triplicate on each substrate.

2.4 Enzymatic assays

All enzyme assays were carried out in a final volume of 1 ml and incubated for different times at 25 °C. NAD^+-dependent ADH activity was assayed in the oxidative direction according to Bergmeyer (1983). The enzymatic assays contained 25 mM Tris-HCl (pH 8.5), 2 mM NAD^+ or $NADP^+$, cell-free extract (100-200 μg protein), and 100 mM of the substrate (1R, 2S)-cis-1,2-di-hydro-1,2-naphthalene-diol. The reaction was started by dihydrodiol addition, and reduction of NAD^+ or $NADP^+$ was monitored by the increase in absorbance at

340 nm in a Beckman DU-650 spectrophotometer. One unit of enzyme activity was defined as the amount required reducing 1 μmol of NAD$^+$ or NADP$^+$ per minute at 25°C. Specific dihydrodiol dehydrogenase (DD) activity was expressed as units per milligram of protein.

For DD activity in gels we developed an appropriate methodology because there is not any report in the literature about the detection of these enzymes by means of electrophoretic zymograms, so for we modified the method described for Nikolova & Ward (1991) for alcohol dehydrogenase. Briefly, after non-denaturing 6% (w/v) PAGE, described above, the activity was revealed as follows. The gel was submerged for 120 min in 4 ml of 0.5 M Tris-HCl buffer pH 8.5 containing 0.5 mg phenazine methosulphate (PMS), 7.5 mg p-nitro-blue tetrazolium (PNBT), 14.34 mg NADP$^+$ or NAD$^+$, 1 mM EDTA, 1 mM DTT and 100 mM of (1R, 2S)-cis-1, 2-dihydro-1, 2-naphthalene-diol as substrate. After incubating at 25 °C for 30 min (in dark) with gentle shaking at 80 rpm, the dihydrodiol dehydrogenases or ADH electro-morphs were observed as blue-dark bands.

When substrate specificity of DD was tested, different single alcohols were added to the mixture reaction at a final concentration of 100 mM. The following substrates were tested: N-decanol, n-hexadecanol, n-octadecanol, hexane-1,2,3,4,5,6-hexaol, benzyl alcohol, cholesterol, cis-naphthalene-diol, ethylene-glycol, poly-ethylene-glycol 3350, and sorbitol, were previously dissolved in dioxan and others were prepared in water: methanol, ethanol, propane-1-ol, propane-2-ol, butane-1-ol, pentane-1-ol, propane-1,2,3-triol and methyl propane-1-ol.

The pH, optimal temperature, substrate specificity, and effect of cations, EDTA and pyrazole were performed after a non-denaturing gel, 6% acrylamide, loading 300 μg of protein. The pH determination was performed from 3 to 9 with citrate buffer for 3 to 5, phosphate buffer for 5 to 7 and Tris/HCl buffer for 7 to 9. The temperature effect was tested in a range of 4 to 45 °C, using a freezer or metabolic bath at the desired temperature. The cation effect was tested using 1 mM of CaCl$_2$, MgSO$_4$, ZnSO$_4$ and FeSO$_4$, and for the EDTA, 1mM was also used. The assays were performed in the presence of cis-naphthalene-diol as substrate and NADP$^+$ as electron acceptor; the enzymatic activity was measured over a range of pH values in the forward reaction dihydrodiol → diol.

2.5 Miscellaneous

Protein concentration was measured according Lowry et al. (1951), using bovine serum albumin as standard. Phenylmethanesulphonyl fluoride and cis-naphthalene-diol were purchased from Sigma (St. Louis, MO, USA), the alcohol used as substrates were from J.T. Baker (Phillipsburg, NJ, USA). All reagents were analytical grade.

Densitometric analysis was performed in a Gene Genius Bio-Imaging System V. 6.05.01, SYNGENE, Synoptics Systems. Software used was Gene Tools V. 3.06.02, Syn. Ltd.

3. Results

3.1 Sub-cellular distribution of dihydrodiol dehydrogenase activity

The first approach was to know the sub-cellular distribution of the dihydrodiol dehydrogenase (DD) activity by means of a differential-centrifugation procedure and the spectrophotometer detection of the DD activity from M. circinelloides YR-1 grown in different carbon sources, using a variation of the method described by Bergmeyer (1983). For this purpose we use the commercial substrate cis-naphthalene-diol. If the low speed supernatant is compared, the enzymatic activity was almost 8 times higher when naphthalene rather than glucose was the carbon source and $NADP^+$ was used as electron acceptor (Table 1).

Sample	DD activity (x 10^{-2})					
	$NADP^+$			NAD^+		
	Carbon source					
	Glucose	Ethanol	Naphthalene	Glucose	Ethanol	Naphthalene
4,300 x g						
Supernatant	42	4.0	270	131	12	39
31,000 x g						
Pellet	1.7	ND^a	ND	23	ND	12
Supernatant	4.5	1.3	91	57	23	61
164,500 x g						
Pellet (MMF)	0.7	0.1	0.1	5.2	1.2	ND
Supernatant (Cytosol)	21	0.5	178	59	2.8	1.4

Table 1. $NADP^+$ or NAD^+-dependent dihydrodiol dehydrogenase activities present in sub-cellular fractions of *Mucor circinelloides* YR-1 grown on different carbon sources. Mycelial cells, grown in the indicated carbon sources, were broken (Braun) and fractions obtained by differential-centrifugation. DD activity of the different fractions was measured with cis-naphthalene-diol as substrate and $NADP^+$ or NAD^+ as electron acceptor. The values are the means of three independent experiments with triplicate determinations.
[a] ND, no detected.

This suggests that at least some of the detected activity could be inducible, and as can be seen, the major enzymatic activity is present in the soluble fractions. When NAD+ was used as electron acceptor, the activity found in the low speed supernatant when the fungus was grown in glucose as a carbon source is more than 3 times higher than the one present when naphthalene was used, and more than ten times higher if compared with the activity obtained with ethanol as a carbon source.

These results enhance the interest to investigate how many different activities will be revealed by electrophoretic zymograms in the cytosolic fraction of the fungus when it grown on different carbon sources.

3.2 Use of zymograms to reveal the presence of several dihydrodiol dehydrogenase activities in cytosolic fraction of *M. circinelloides* YR-1 grown on different carbon sources

Aerobically mycelium grown in different carbon sources (see Materials and Methods) was used to obtain the corresponding cytosolic fraction and each one was run on no-denaturing polyacrylamide gels and stained for $NADP^+$-dependent dihydrodiol dehydrogenase activity with cis-naphtalen-diol as substrate.

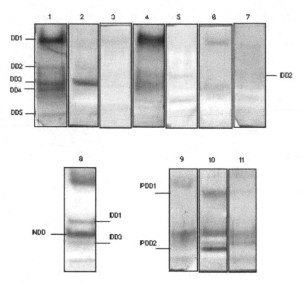

Fig. 1. Dihydrodiol dehydrogenase enzymatic activity present in cytosolic fraction of *M. circinelloides* YR-1 grown in different carbon sources. Mycelia in lane 1 was grown for 12 h and lanes 2-11 were grown for 22 h at 28°C on sMMP medium added with the carbon source indicated. *Lanes*: *1*, D-glucose; *2*, glycerol; *3*, ethanol; *4*, *n*-decanol; *5*, *n*-pentane; *6*, *n*-decane; *7*, *n*-hexadecane; *8*, naphthalene; *9*, anthracene; *10*, phenanthrene; *11*, pyrene, at the concentrations described in Material and Methods. The extracts were electrophoresed and stained as described in Materials and Methods. In all cases 300 µg of protein were loaded in each lane. In this gel 100 mM *cis*-naphthalene-diol was the substrate and NADP⁺ the electron acceptor. These results are representative gels and mycelia were grown up and run on the gels at least three times. The induction patterns were always reproducible.

Under the conditions tested, five bands were seen in glucose as carbon source (Fig. 1, lane 1) which were considered as constitutive dihydrodiol dehydrogenases and identified with a number (1-5) considering their decreasing Rm (DD1-5) (Table 2), and six inducible bands of activity were detected, depending of the carbon source in the culture media used for growth (Fig. 1; Table 3). One of the inducible bands (iDD1) was seen when *n*-decanol or *n*-pentane or *n*-hexadecane or naphthalene was the carbon source (Fig. 1, lanes 4, 5, 7 and 8 respectively, Table 3).

A second inducible NADP⁺-dependent dihydrodiol dehydrogenase activity was seen when *n*-decanol or *n*-pentane was the carbon source (iDD2) (Fig. 1, lanes 4 and 5; Table 3). The third inducible enzymes (iNDD) was seen only when aromatic hydrocarbons were used as sole carbon source in the growth media (Fig. 1, lanes 8 to 11, Table 3). A fourth inducible naphthalene-diol dehydrogenase (iDD3) was induced by some of the alcohols, alkanes and aromatic polycyclic compounds tested (Fig. 1, lanes 3, 4, 6, 8, 10 and 11; Table 3). When phenanthrene was used as the carbon source, two new bands with different relative motilities were revealed, iPDD1 and iPDD2, (Fig. 1, lane 10; Table 3). The iPDD1 was also observed when pyrene was the carbon source (Fig. 1, lane 10 and 11; Table 3).

Polyacrylamide Gel Electrophoresis an Important Tool for the Detection and Analysis of Enzymatic Activities by
Electrophoretic Zymograms

33

Carbon source	Rm^a of DD constitutives				
	1	2	3	4	5
D-Glucose	0.22±0.04	0.4±0.03	0.62±0.01	0.69±0.02	0.90±0.01
Glycerol	-	-	-	-	-
Ethanol	-	-	-	-	-
n-Decanol	0.21±0.01	-	0.61±0.02	0.69±0.03	0.89±0.03
n-Pentane	-	-	-	0.7±0.03	-
n-Decane	0.21±0.01	-	-	-	-
n-Hexadecane	-	-	-	-	-
Naphthalene	0.21±0.01	-	-	-	0.89±0.006
Anthracene	0.22±0.006	-	-	-	0.89±0.02
Phenanthrene	-	-	-	-	0.89±0.01
Pyrene	-	-	-	-	0.89±0.01

Table 2. Constitutive NADP+-dependent dihydrodiol dehydrogenase activities in cytosolic fraction of *Mucor circinelloides* YR-1 grown in different carbon sources.
[a]*Rm*; was calculated as described in Materials and Methods section as its standard deviation.

Carbon source	Rm^a of DD inducibles					
	iDD1	iDD2	iDD3	iPDD1	iPDD2	iNDD
D-Glucose	-	-	-	-	-	-
Glycerol	-	-	-	-	-	-
Ethanol	-	-	-	-	-	-
n-Decanol	0.55±0.002	0.66±0.005		-	-	-
n-Pentane	0.55±0.001	0.66±0.003	0.74±0.002	-	-	-
n-Decane	-	-	-	-	-	-
n-Hexadecane	-	-	0.73±0.010	-	-	0.66±0.005
Naphthalene	0.56±0.003	-	0.73±0.020	-	-	0.67±0.008
Anthracene	-	-	-	-	-	0.67±0.008
Phenanthrene	-	-	0.74±0.010	0.23±0.006	0.78±0.003	0.67±0.008
Pyrene	-	-	0.74±0.004	0.22±0.020	-	0.67±0.008

Table 3. Inducible NADP+-dependent dihydrodiol dehydrogenase activities in cytosolic fraction of *Mucor circinelloides* YR-1 grown in different carbon sources.
[a]*Rm*; was calculated as described in Materials and Methods section as its standard deviation.

3.3 Bands intensity of dihydrodiol dehydrogenase activities in cytosolic fraction of *M. circinelloides* YR-1

With the comparing purpose the activity showed for the denominated iNDD enzyme that is induced when naphthalene was used as a carbon source was taking as a 100% and the others enzymes where referred to this value (Table 4).

In all cases cis-naphthalene-diol and NADP+ were used in the enzymatic assay. When phenanthrene was used as a carbon source there are four different inducible enzymes, iDD3, iPDD1, iPDD2 and iNDD, being iPDD1 and iPDD2 best induced by this carbon source, and the latest is the one that showed the highest induction value of all inducible enzymes (103.5%) (Table 4). The iPDD2 enzyme is the only one that it is induced by only one carbon source (Table 3). In the case of the iDD1, the highest induction value obtained was when naphthalene was used as a carbon source as iDD3 and iNDD enzymes (Table 4). The iDD2 showed its best induction value when n-pentane was used as a carbon source (Table 4). It is noticeable that glycerol, ethanol and n-decane do not induce any of the DD activities.

Carbon source	Band intensity[a] (relative units)					
	iDD1	iDD2	iDD3	iPDD1	iPDD2	iNDD
D-Glucose	0.0	0.0	0.0	0.0	0.0	0.0
Glycerol	0.0	0.0	0.0	0.0	0.0	0.0
Ethanol	0.0	0.0	0.0	0.0	0.0	0.0
n-Decanol	10.2	5.3	0.0	0.0	0.0	0.0
n-Pentane	8.7	18.2	14.6	0.0	0.0	0.0
n-Decane	0.0	0.0	0.0	0.0	0.0	0.0
n-Hexadecane	0.0	0.0	6.3	0.0	0.0	4.5
Naphthalene	56.8	0.0	16.3	0.0	0.0	100.0
Anthracene	0.0	0.0	0.0	0.0	0.0	90.5
Phenanthrene	0.0	0.0	8.1	82.9	103.5	84.5
Pyrene	0.0	0.0	10.2	8.0	0.0	38.4

Table 4. Relative inducibility of NADP[+]-dependent dihydrodiol dehydrogenase activities in cytosolic fraction of *Mucor circinelloides* YR-1 grown in different carbon sources.
[a]Relative units were obtained by densitometry, using the value from iNDD as 100% when the fungus was growth on naphthalene.

3.4 Effect of ethanol as substrate and NAD[+] as electron acceptor on induced dihydrodiol dehydrogenase activities in cytosolic fraction of *M. circinelloides* YR-1 grown on different carbon sources

It is interesting to compare if the inducible DD enzymes are able to use NAD[+] as electron acceptor and/or ethanol as substrate because some dehydrogenases are able to use both of them. In the presence of NAD[+] as electron acceptor and *cis*-naphthalene-diol as substrate, there was not any staining in the region of either constitutive or inducible dihydrodiol dehydrogenase activities (Fig. 2A, lanes 1-5; Table 5). In the presence of ethanol as substrate and NADP[+] as electron acceptor, two ADH activity in cytosolic fraction from mycelium grown on glucose were revealed (Fig. 2B, lane 1) one with a Rm of 0.42 ± 0.008 (denominated ADH1) and the other with a Rm of 0.84 ± 0.003 (denominated ADH2). Under these conditions we also observed the inducible dihydrodiol dehydrogenase enzymes denominated DD3, suggesting that this enzyme also possesses an ADH activity NADP[+]-dependent (Fig. 2B, lane 1). Also, under these assay conditions, two bands were observed when phenanthrene was used as a carbon source to growth the mycelia (Fig. 2B, lane 4). The bands correspond to the DD5 (Rm of 0.90 ± 0.010) and a new ADH, denominated ADH3 with a Rm of 0.94 ± 0.010 (Fig. 2B, lane 4).

When we used NAD⁺ as electron acceptor and ethanol as substrate the denominated ADH1, ADH3 and DD5 activity bands were revealed when the cytosolic fraction from mycelium grown on glucose (Fig. 2C, lane 1).

As a control, a sample of the culture media lacking carbon source (Fig. 2A to C, lane 6) and an assay lacking substrate in the reaction mixture for the activity in zymograms (Fig. 2A to C, lane 7) did not showed any enzymatic activity.

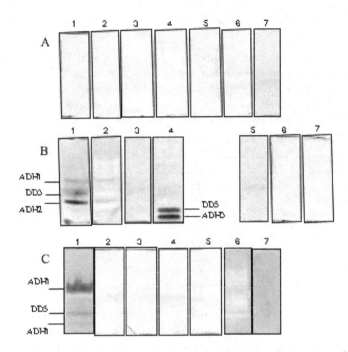

Fig. 2. Detection of dihydrodiol dehydrogenase activities in cytosolic fraction of *M. circinelloides* YR-1 by activity-stained gels. All mycelia were obtained after grown for 22 h at 28°C on the carbon source indicated. A. *Lanes 1*, 1.0% glucose; *2*, 0.5% naphthalene; *3*, 0.5% anthracene; *4*, 0.5% phenanthrene; *5*, 0.5% pyrene; *6*, a sample of the culture media without carbon source; *7*, without substrate in the activity-reaction mixture. The reaction was performed with NAD⁺ and 100 mM cis-naphthalene-diol as substrate, described in Materials and Methods section. B. identical samples as A, but the activity was developed with NADP⁺ and 100 mM ethanol as substrate. C. identical samples as A, but the activity was developed with 100 mM ethanol as substrate. The amount of protein loaded per track was equalized to 300 µg. These results are representative gels and mycelia were grown up and run on the gels at least three times. The activity patterns were always reproducible.

In order to compare the observed activities we performed a densitometric analysis to the bands intensity and the Table 5 shows the obtained values under the condition where the activity have its highest value, taking the iNDD activity as a 100%. As show, the denominated DD1 enzymes, a constitute one, has the highest activity of all, in contrast the activity denominated iDD3 and DD5 measured with ethanol and NAD⁺ are the lowest

(Table 5). It is interesting that only the denominated ADH 1 to 3 have activity with ethanol as a substrate, being the ADH3 the enzyme with the highest activity (Table 5). Surprisingly not a single one activity was revealed when *cis*-naphthalene-diol and NAD+ were used as a substrate (Table 5).

3.5 General properties of NDD activities

DD activities were assayed only in crude cell-free extracts of aerobically-naphtahalene grown mycelia cells because at these moments, we were strongly interested in the NDD activity.

In all cases, the DD activities were tested in cell-free extracts of *M. circinelloides* YR-1.grown in 0.5% of naphthalene as sole carbon source for 22 h at 28 °C. Cytosolic fraction was separated in a native electrophoresis and the amount of protein loaded per track equalized and was equivalent to 300 μg, NADP+ as electron acceptor and 100 mM *cis*-naphthalene-diol as enzyme substrate were used to reveal the zymograms.

| | Band intensity (relative units)[a] | | | | |
| | *cis*-naphthalene-diol | | Ethanol | | |
Enzyme	NADP+	NAD+	NADP+	NAD+	Rm^b
DD 1	104.2	0.0	0.0	0.0	0.21±0.006
DD 2	1.2	0.0	0.0	0.0	0.40±0.010
DD 3	3.4	0.0	0.0	0.0	0.61±0.020
DD 4	3.2	0.0	0.0	0.0	0.69±0.030
DD 5	1.2	0.0	67.6	0.4	0.90±0.010
iDD1	56.8	0.0	0.0	0.0	0.56±0.010
iDD2	18.2	0.0	0.0	0.0	0.66±0.050
iDD3	16.3	0.0	3.6	0.4	0.73±0.010
iPDD1	82.9	0.0	0.0	0.0	0.23±0.006
iPDD2	103.5	0.0	0.0	0.0	0.78±0.003
iNDD	100.0	0.0	0.0	0.0	0.67±0.008
ADH1	0.0	0.0	1.2	3.1	0.42±0.008
ADH2	0.0	0.0	45.2	0.0	0.84±0.003
ADH3	0.0	0.0	74.8	0.0	0.94±0.010

Table 5. Activities of *cis*-naphthalene-diol and alcohol dehydrogenase of cytosolic fraction of *M. circinelloides* YR-1 grown in the best inducer for each one. Densitometric analysis was carried out as described in Materials and Methods. The enzymatic determination was on the gel with *cis*-naphthalene-diol or ethanol as the substrate and NADP+ or NAD+ as the electron acceptor.
[a] Relative units were obtained by densitometry, using the value from iNDD as 100% when naphthalene was the carbon source.

3.5.1 pH dependence

The Fig. 3 shows that the optima pH value for all five activities expressed with n-naphthalene as carbon source and NADP+ and naphthalene-diol in the enzymatic reaction was 8.5. It is noticeable that only the iNDD show activity at pH 3 and little DD activities were showed at pH values below 8.5. It is important to say that the background in the lane for activity revealed at pH 9, was darken because of pH.

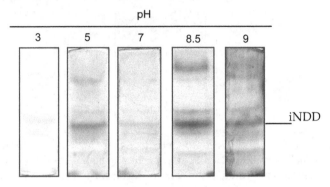

Fig. 3. Effect of pH on dihydrodiol dehydrogenase activities present in cytosolic fraction of *M. circinelloides* YR-1grown in naphthalene. Each track was cut and stained at the indicated pH value.

3.5.2 Temperature

The effect of the temperature on DD activities was tested on cytosolic fraction in a range of temperatures oscillating between 4 and 45 °C. The optimum temperature was 37 °C, notice that even at 45 °C the activity band corresponding to the iNDD can be seen in the zymogram (Fig. 4). It is important to specify that the background in the lanes for activity revealed at 37 or 45 °C were darker because of the incubation temperature.

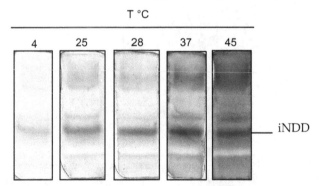

Fig. 4. Temperature effect on dihydrodiol dehydrogenase activity present in cytosolic fraction of *M. circinelloides* YR-1. Each track was cut and activity developed at the indicated temperature.

3.5.3 Requirement of different divalent ions

Different divalent ions were used to prove if some of them were required for DD activities. The Fig. 5 shows that only Ca^{2+} had an enhancing effect on DD activities meanwhile the other divalent metals tested and also EDTA were inhibitory Fe^{2+} > Zn^{2+} > EDTA> Mg^{2+}.

Ion added [1 mM]

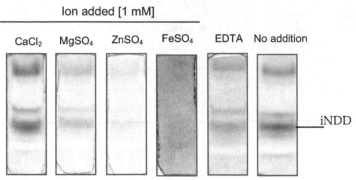

CaCl₂   MgSO₄   ZnSO₄   FeSO₄   EDTA   No addition

iNDD

Fig. 5. Divalent ions effect on dihydrodiol dehydrogenase activity present in cytosolic fraction of *M. circinelloides* YR-1. Each track was cut and stained adding to the reaction mixture the divalent ion indicated at the concentrations described in Materials and Methods.

3.5.4 Pyrazole effect

Pyrazole is a well known ADH competitive inhibitor (Pereira et al., 1992) and this is the principal reason we decide to prove its effect on the different DD activities present in crude cell-free extracts obtained from *M. circinelloides* YR-1 mycelia grown in naphthalene as the sole carbon source. As can be seen in Fig. 6, pyrazole has a little inhibitory effect on the different DD's. In addition, iNDD showed a mild decrease in the level of its activity when measured by staining for activity in gels in presence of pyrazole (Fig. 6)

Pyrazole1 mM

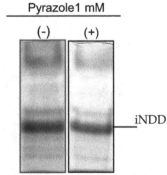

(-)   (+)

iNDD

Fig. 6. Pyrazole effect on dihydrodiol dehydrogenase activity present in the cytosolic fraction of *M. circinelloides* YR-1 grown in naphthalene. Each track was cut and stained adding or not to the reaction mixture 1.0 mM pyrazole.

3.5.5 Substrate specificity of the different inducible and constitutive dihydrodiol dehydrogenase activities

To test the substrate specificity we chose naphthalene as the carbon source in the culture media since we can observe two constitutive band of activity (DD1 and DD5) and three inducible bands (iDD1, iDD3 and iNDD) (Fig. 1, Table 2 and 3). A variety of substrates were tested in the gel making the assay with NADP⁺ as electron acceptor. The constitutive DD1

Polyacrylamide Gel Electrophoresis an Important Tool for the Detection and Analysis of Enzymatic Activities by
Electrophoretic Zymograms

39

enzyme it is the one that shows in a majority of substrates, 14 out of 18, being ethanol, propane-1-ol, benzyl alcohol and sorbitol the substrates where the activity did not show, but when show its intensity is really low (Fig. 7). In contrast, the band with the highest intensity is the DD2 enzyme when propane-1,2,3-triol was the substrate (Fig. 7, lane 10), even when this enzyme cannot be seeing when cis-naphtalene-diol is used as a substrate (Fig. 7, lane 18). As this DD2 enzyme there are others enzymes that did not show with cis-naphthalene-diol and can be seen with others substrates, as it is DD3 and iPDD2, that shows with three and four different substrates respectively (as an example see Fig. 7, lane 10). It is surprising that the iPDD2 enzyme that only showed when phenanthrene was the carbon source to growth the fungus, it is present here depending on the substrate used, propane-1,2,3-triol, 2-methyl propane-1-ol, ethylene-glycol and benzyl alcohol (Fig. 7, lanes 10, 11, 13 and 15).

In the particular case of NDD1, it was present only when naphthalene was the carbon source (Fig. 7) but it was absent in all other aromatic hydrocarbons used as carbon source (Fig. 1) and this enzyme practically only uses cis-naphthalene-diol as substrate (Fig. 7, lane 18). In the case of iDD3 and DD5, both present broad substrate specificity, showing a special preference for short-chain alcohols, including 1-decanol (Fig. 7). There are five bands that show with different substrates, but its intensity is really low and they were not taken in account (Fig. 7).

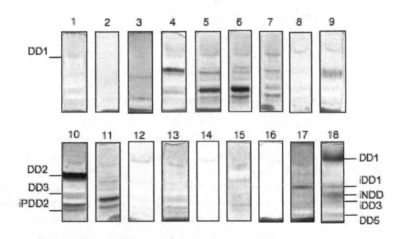

Fig. 7. Substrate specificity of constitutive and inducible DD activities of cytosolic fraction of M. circinelloides YR-1 grown on naphthalene, revealed by activity-stained gels. A variety of substrates were used with NADP+ as electron acceptor. All substrates were tested at 100 mM of final concentration. Lane 1 methanol; 2 ethanol; 3 propane-1-ol; 4 propane-2-ol; 5 butane-1-ol; 6 pentane-1-ol; 7 1-decanol; 8 1-hexadecanol; 9 hexane-1,2,3,4,5,6- hexaol; 10 propane-1,2,3-triol; 11 2-methyl propane-1-ol; 12 1-octadecanol; 13 ethylene-glycol; 14 poly-ethylene-glycol 3350; 15 benzyl alcohol; 16 cholesterol; 17 sorbitol; 18 cis-naphthalen-diol.

4. Discussion and conclusion

Many studies have been done on NAD+-dependent cis-dihydrodiol dehydrogenases (DD) in bacteria (Jouanneau & Meyer, 2006; Van Herwijnen et al., 2003). In the case of NADP+-

dependent *trans*-dihydrodiol dehydrogenases, almost all investigations have been done in mammalian tissues (Carbone et al., 2008; Chang et al, 2009; Chen et al., 2008) but only a few reports about these important enzymes have been done in fungi (Bezalel et al., 1997; Hammel, 1995; Sutherland et al., 1993) particularly in *Phanerochaete chrisosporium* (Bogan & Lamar, 1996; Muheim et al., 1991). At date, there is no any report about the detection of dihydrodiol dehydrogenase activities by means of electrophoretic zymograms in any organism. This methodology represents an interesting approach because in this way it is possible to detect, study and compare the different isoenzymes present in the cell-free extracts of the organism used as enzymatic source. In our own work, YR-1 strain possesses extraordinary metabolic machinery that premises it to survive in a very dangerous place how is a petroleum-contaminated soil.

The results about the localization of DD activities in a differential centrifugation procedure from YR-1 grown in different carbon sources (Table 1), revealed that the activity measured with *cis*-naphthalene-diol as substrate and $NADP^+$ as electron acceptor was only present in the supernatant fractions of each centrifugation speed, suggesting that all DD activity observed must be a soluble enzyme. At date, we cannot discard the possibility that the DD activity could be located in the lumen of some kind of microsomal bodies, because of the drastic ballistic treatment used to homogenize the cells. Actually, we are conducting different experiments employing density gradients and electron transmission microscopy to resolve this question.

Complementary analysis of DD activities by electrophoretic zymograms led us to detect eleven different activities and all of them were $NADP^+$-dependent (Fig. 2) this represents the first report about the detection of DD activities by electrophoretic zymograms, a non-denaturing gel electrophoresis stained with a colored product of the enzymatic reaction.

Of the eleven bands detected, we described five different constitutive DD activities, DD 1-5, since them were observed when D-glucose was the carbon source and only DD-2 was solely induced by this sugar, since the others are induced at least for another carbon source. When *n*-decanol was used as a carbon source, we observed four out of five of the constitutive bands, lacking only the DD-2 band. Its noticeable that only when glycerol, ethanol, n-pentane and n-hexadecane were the carbon source to grow the fungus, not a single constitutive band was observed, may be due to the fact that these compounds only can be metabolized specifically by the induced enzymes. In glucose grown mycelium, all inducible dihydrodiol dehydrogenase activities were absent suggesting that they could be subject to carbon-catabolite repression (Fig. 2).

Surprisingly all the activities described here as DD are able to use *cis*-naphthalene-diol, since this substrate has been describes as bacterial specific (Cerniglia & Gibson 1977). The substrate reported for eukaryotic cells is the *trans*-naphthalene-diol (Cerniglia 1977).

Phenanthrene was the best inducer since when used as a carbon source four out of six inducible bands were observed, *n*-decanol and naphthalene were the second best inducers since each one led the induction of three different enzymes, sharing the bands denominated iDD1 and iDD2. Also in the case of the inducible enzymes glycerol, n-pentane and n-hexadecane were unable to induce any activity. The specific induction of an activity must be due to substrate specificity.

We have shown that on naphthalene, anthracene, phenanthrene or pyrene used as sole carbon source, there exist three different inducible NADP+-dependent dihydrodiol dehydrogenase activities. One of them iNDD, was the only isoenzyme inducible by all aromatic hydrocarbons which presumably is involved in the aromatic hydrocarbon biodegradation pathway in YR-1 strain (Fig. 1). In particular iNDD was capable to use only cis-naphthalene-diol as substrate (Fig. 7) suggesting that this enzyme is specific part of the metabolic pathway of the naphthalene; all other activity bands are ADH substrate unspecific

It is interesting that iPDD2 has broad substrate specificity when NADP+ was the electron acceptor, suggesting that it could be one of the different dehydrogenases belonging to the microsomal system of alcohol (ethanol) oxidation [MEOS (Krauzova et al., 1985)]. No one of all inducible DD's activities showed to be able to use NAD+ as electron acceptor.

For iNDD, naphthalene was the best inducer and pyrene the worst. In the case of both iPDD1 and 2, phenanthrene was the only inducer of these enzymes, however pyrene shows a very low inducer effect on iPDD1 (Table 3). The finding of two inducible (iNDD and iPDD1) and one constitutive (DD1) enzymes that uses specifically a dihydrodiol as substrate is in agreement with the number of three possible DD's of predicted function, reported in database of *Mucor circinelloides* (Torres-Martínez et al., 2009).

With regard to the constitutive dihydrodiol dehydrogenase activities present in YR-1 strain, four of them use only cis-naphthalene-diol as substrate: DD1, 2, 3 and 7; DD2, use both cis-naphthalene-diol and with high efficiency propane-1,2,3-triol, indicating that it can be the glycerol dehydrogenase-1 (iGlcDH1 inducible by 1-decanol) described previously by ourselves in YR-1 (Camacho et al., 2010).

Our above-mentioned findings with *M. circinelloides* YR-1 dihydrodiol dehydrogenase activities are indicative of developmental regulation of the different DD's enzymes; this interpretation is supported by following observations: in zymograms for DD's activities when YR-1 was grown in different carbon sources is showed a differential pattern of the activity bands depending of the carbon source used in the culture media. The present results suggest the existence of eleven enzymes with dihydrodiol dehydrogenase activity. Particularly important the DD1 that could be the constitutive DD, and iNDD iPDD1, that could be part of the aromatic hydrocarbon biodegradation pathway in YR-1 strain for naphthalene or the others aromatic hydrocarbon, respectively. Future genomic analysis after isolation of the respective genes should prove the existence of one gene for each constitutive or inducible activity in agreement with the *M. circinelloides* data base prediction. The details of the possible interaction between alcohols or hydrocarbons metabolism remain to be determined.

5. Acknowledgments

Support for this research by Universidad de Guanajuato (México), is gratefully acknowledged.

6. References

Alvarado, C. Y., Gutiérrez-Corona, F. & Zazueta-Sandoval, R. (2002). Presence and physiologic regulation of alcohol oxydase activity in an indigenous fungus isolated from petroleum-contaminated soils. *Applied Biochemistry and Biotechnology*. Vol. 98-100, No. 1-9 (March 2002), pp. 243-255. ISSN: 0273-2289.

Atlas, R. M. (1995). Bioremediation. *Chemical and Engineering News*. Vol. 73, No. 14 (April 1995), pp. 32-42. ISSN 0009-2347. DOI: 10.1021/cen-v073n014.p032

Auret, B. J., Boyd, D. R., Robinson, P. M. & Watson, C.G. (1971). The NIH Shift During the Hydroxylation of Aromatic Substrates by Fungi. *Journal of the Chemical Society D: Chemical Communications* No. 24, (December 1971), pp. 1585-1587. ISSN 1364-548X.

Bartnicki-García, S. & Nickerson, W. J. (1962). Nutrition, Growth and Morphogenesis of *Mucor rouxii*. *Journal of Bacteriology*. Vol. 84, No. 4 (October, 1962), pp. 841-858. ISSN 0021-9193.

Bergmeyer, H. U. (1983). Alkoholdehydrogenase. In: Bergmeyer, H.U. (ed) *Methods of enzymatic analysis*, Verlag Chemie, Weinheim, Vol. 11, 3rd edn. pp-139-145.

Bezalel, L., Hadar, Y. & Cerniglia, C.E. (1997). Enzymatic Mechanisms Involved in Phenanthrene Degradation by the White-Rot Fungus *Pleurotus ostreatus*. *Applied and Environmental Microbiology*. Vol. 63, No. 7, (July 1997) pp. 2495-2501. ISSN 0099-2240.

Bogan, B. W. & Lamar, R. T. (1996). Polycyclic Aromatic Hydrocarbon-Degrading Capabilities of *Phanerochaete laevis* HHB-1625 and Its Extracellular Ligninolytic enzymes. *Applied and Environmental Microbiology*. Vol. 62, No. 5, (May 1996), pp. 1597-1603. ISSN 0099-2240.

Camacho, M. R. L., Durón, C. A. & Zazueta-Sandoval, R. (2010). Analysis of Glycerol Dehydrogenase Activities Present in *Mucor circinelloides* YR-1. *Antonie Van Leeuwenhoek*. Vol. 98, No. 4, (November 2010), pp. 437-445, ISSN 0003-6072.

Carbone, V., Hara, A. & El-Kabbani, O. (2008). Structural and Functional Features of Dimeric Dihydrodiol Dehydrogenases. *Cellular and Molecular Life Science*. Vol. 65, No. 10, (May 2008), pp. 1464-1474, ISSN 1420-682X.

Cerniglia, C. E. (1997). Fungal Metabolism of Polycyclic Aromatic Hydrocarbons: Past, Present and Future Applications in Bioremediation. *Journal of Industrial Microbiology and Biotechnology*. Vol. 19, No. 5-6 (November 1997), pp. 324-333. ISSN 1367-5435.

Cerniglia, C. E. & Gibson, D. T. (1977). Metabolism of Naphthalene by *Cunninghamella elegans*. *Applied and Environmental Microbiology*. Vol. 34, No. 4, (October 1977), pp. 363-370. ISSN 0099-2240.

Chang, C. H., Chen, L. Y., Chan, P. C., Yeh, T. K., Kuo, J. S., Ko, J. C. & Fang, Y. H. (2009). Overexpression of Dihydrodiol Dehydrogenase as a Prognostic Marker in Resected Gastric Cancer Patients. *Digestive Diseases and Sciences*. Vol. 54, No. 2 (February 2009), pp. 342-347. ISSN 0163-2116.

Chen, J., Adikari, M., Pallai, R., Parckh, K. H. & Simpkins, H. (2008). Dihydrodiol Dehydrogenases Regulate the Generation of Reactive Oxygen Species and the Development of Cisplatin Resistance in Human Ovarian Carcinoma Cells. *Cancer Chemotherapy and Pharmacology*. Vol. 61, No. 6 (May 2008), pp. 979-987. ISSN 0344-5704.

Cook, F. D., Jobson, A., Phillippe, R. & Westlake, D. W. S. (1974). Biodegradability and crude oil composition. *Canadian Journal of Microbiology*. Vol. 20, No. 7 (July 1974), pp. 915-928. ISSN 0008-4166.

Durón-Castellanos, A., Zazueta-Novoa, V., Silva-Jiménez, H., Alvarado-Caudillo, Y., Peña Cabrera, E. & Zazueta-Sandoval, R. (2005). Detection of NAD+-Dependent Alcohol Dehydrogenase Activities in YR-1 Strain of *Mucor circinelloides*, a Potential Bioremediator of Petroleum-Contaminated Soils. *Applied Biochemistry and Biotechnology*. Vol. 121-124, No. 1-3 (March 2005), pp. 279-288. ISSN 0273-2289.

Ferris, J. P., MacDonald, L. H., Patrie, M. A. & Martin, M. A. (1976). Aryl Hydrocarbon Hydroxylase Activity in the Fungus *Cunninghamella bainieri*: Evidence for the Presence of Cytochrome P-450. *Archives of Biochemistry and Biophysics*. Vol. 175, No. 2 (August 1976), pp. 443-452. ISSN 0003-9861.

Goto, S. & Takami, H. (1986). Classification of Ascoideaceous Yeasts Based on the Electrophoretic Comparison of Enzymes and Coenzymes Q Systems. *The Journal of General and Applied Microbiology*, Vol. 32, No. 4 (June 1986), pp. 271-282. ISSN 1349-8037.

Hammel, K. E. (1995). Mechanisms for Polycyclic Aromatic Hydrocarbon Degradation by Ligninolytic Fungi. *Environmental Health Perspectives*, Vol. 103. Suppl. 5 (June 1995), pp. 41-43.

Hara, A., Taniguchi, H., Nakayama, T. & Sawada, H. (1990). Purification and Properties of Multiple Forms of Dihydrodiol Dehydrogenases from Human Liver. *Journal of Biochemistry*. Vol. 108, No. 2 (August 1990), pp. 250-254. ISSN 0021-924X.

Harayama, S. & Timmis, K. N. (1989). *In: Genetics of bacterial diversity*. Hopwood, D.A., and Chater K.E.. eds. Academic Press, pp. 151-174. New York, USA.

Higaki, Y., Kamiya, T., Usami, N., Shintani, S., Shiraishi, H., Ishikura, S., Yamamoto, I. & Hara, A. (2002). Molecular Characterization of Two Monkey Dihydrodiol Dehydrogenases. *Drug Metabolism and Pharmacokinetics*. Vol. 17, No. 4 (September 2002), pp. 348-356, ISSN 1347-4367.

Hunt, J. M. (1979). *Petroleum geochemistryand geology*. W. M. Freeman, San Francisco, CA. pp. 221. ISBN 0 7167 1005 6.

Hyötyläinen, T. & Olkari, A. (1999). The Toxicity and Concentrations of PAHs in Creosote-Contaminated Lake Sediment. *Chemosphere*. Vol. 38, No. 5 (February 1999), pp. 1135-1144. ISSN 0045-6535.

Jeng, R. S., Shiyuan, Y. & Wayman M. (1987). Isoenzyme and Protein Patterns of Pentose-Fermenting Yeasts. *Canadian Journal of Microbiology*. Vol. 33, No. 11 (November 1987), pp. 1017-1023. ISSN 0008-4166.

Jouanneau, Y. & Meyer, C. (2006). Purification and Characterization of an Arene *Cis*-Dihydrodiol Dehydrogenase Endowed with Broad Substrate Specificity Toward Polycyclic Hydrocarbon Hydrodiols. *Applied and Environmental Microbiology*. Vol. 72, No. 7 (July 2006), pp. 426-434. ISSN 0099-2240.

Krauzova, V. I., Il'chenko, A. P., Sharyshev, A. A. & Lozinov, A. B. (1985). Possible Pathways of the Oxidation of Higher Alcohols by Membrane Fractions of Yeasts Cultured on Hexadecane and Hexadecanol. *Biochemistry*. Vol. 50, No. 5 (May 1985), pp. 609-615. ISSN 0006-2960.

Laemmli, U. K. (1970). Cleavage of Structural Proteins During the Assembly of Head of Bacteriophage T6. *Nature*. Vol. 227, No. 5259 (August 1970), pp. 680-685. ISSN 0028-0836.

Lim, W. J., Park, S. R., Cho, S. J. Kim, M. K., Ryu, S. K., Hong, S. Y., Seo, W. T., Kim, H. & Yun, H. D. (2001). Cloning and characterization of an intracellular isoamylase gene from *Pectobcterium chrysantemi* PY35. *Biochemical and Biophysical Chemical Communications*. Vol. 287, No. 2 (September 2001), pp.348-354. ISSN 0006-921X.

Lowry, O. H., Rosebrough, N. J., Farr, A. L. & Randall, R. J. (1951). Protein Measurement with the Folin Phenol Reagent. *Journal of Biological Chemistry*, Vol. 193, No. 1 (November 1951), pp. 265-275. ISSN 0021-9258.

Muheim, A., Waldner, R., Sanglard, D., Reiser, J., Schoemaker, H. E. & Leisola M. S. (1991). Purification and Properties of an Aryl-Alcohol Dehydrogenase from the White-Rot Fungus *Phanerochaete chrysosporium*. *European Journal of Biochemistry*. Vol. 195, No. 2 (January 1991), pp. 369-375. ISSN 0014-2956.

Neidle, E., Hartnett, C., Ornston, L. N., Bairoch, A., Rekik, M. & Harayama, S. (1992). *Cis*-Diol Dehydrogenases Encoded by the TOL pWWO Plasmid *xylL* Gene and the *Acinetobacter calcoaceticus* Chromosomal *benD* Gene Are Members of the Short –

Chain Alcohol Dehydrogenase Superfamily. *European Journal of Biochemistry*. Vol. 204, No. 1, (February 1992), pp. 113-120. ISSN 0014-2956.

Nikolova, P. & Ward, O. P. (1991). Production of L-phenylacetyl Carbinol by Biotransformation: Product and By-Product Formation and Activities of the Key Enzymes in Wild Type and ADH Isoenzyme Mutants of *Saccharomyces cerevisiae*. *Biotechnology and Bioengineering*. Vol. 38, No. 5 (August 1991), pp. 493-498. ISSN 1097-0290.

Okwumabua, O., Persaud, J. S., & Reddy, P. G. (2001). Cloning and characterization of the gene encoding the glutamato dehydrogenase of *Streptococcus suis* serotype 2. *Clinical and Diagnostic Laboratory Immunology*. Vol. 8, No. 2 (March 2001), pp. 251-257. ISSN 556-6811.

Penning, T. M., Burczynski, M. E., Hung, C. F., McCoull, K. D. Palackal, N. T. & Tsuruda, L. S. (1999). Dihydrodiol dehydrogenases and polycyclic aromatic hydricarbon activation: Generation of reactive and redox *o*-quinones. *Chemical Research in Toxicology*. Vol. 12, No. 1 (December 1998), pp. 1-18. ISSN 0893-228X.

Pereira, E. F., Aracava, Y., Aronstam, R. S., Barreiro, E. J. & Alburquerque, E. S. (1992). Pyrazole, an Alcohol Dehydrogenase Inhibitor, Has a Dual Effects on N-methyl-D-asp. *Journal of Pharmacology and Experimental Theory*. Vol. 261, No.1 (April 1991), pp. 331-340. ISSN 0022-365.

Pfeiffeer-Guglielmi, B., Broer, S., Broer, A. & Hamprecht, B. (2000). Isoenzyme pattern of glycogen phosphorylase in the rat nervous system and rat astroglia-rich primary cultures: electrophoretic and polymerase chain reaction studies. *Neurochemical Research*, Vol. 25, No. 11 (November 2000), pp. 1485-1491. ISSN 0364-3190.

Silva, J. H., Zazueta-Novoa, V., Durón, C. A., Rodríguez, R. C., Leal-Morales, C. A. & Zazueta-Sandoval, R. (2009). Intracellular Distribution of Fatty Alcohol Oxidase Activity in *Mucor circinelloides* YR-1 Isolated from Petroleum Contaminated Soils. *Antonie Van Leeuwenhoewk*. Vol. 96, No. 4 (November 2009), pp. .527-535. ISSN 0003-6072.

Smith, R. V. & Rosazza, J. P. (1974). Microbial Models of Mammalian Metabolism. Aromatic Hydroxylation. *Archives of Biochemistry and Biophysics*. Vol. 161, No. 2 (April 1974), pp. 551-558. ISSN 003-9861

Sutherland, J. B., Fu, P. P., Yang, S. K., Von Tungeln, L. S., Casillas, R. P., Crow, S. A. & Cerniglia, C. E. (1993). Enantiomeric Composition of the *Trans*-Dihydrodiols Produced by Phenanthrene by Fungi. *Applied and Environmental Microbiology*. Vol. 59, No. 7 (July 1993), pp. 2145-2149. ISSN 0099-2240.

Torres-Martínez, S., Garre, V., Corrochano, L., Eslava, A. P., Baker, S.E. & others. (2009). Project CBS277.49, v1.0 http//genome.jgi-PSF.org.

Van Herwijnen, R., Springael, D., Slot, P., Govers A. J. H. & Parsons, R. J. (2003). Degradation of Anthracene by *Mycobacterium* sp. Strain LB501T Proceeds Via a Novel Pathway Throught o-Phtalic Acid. *Applied and Environmental Microbiology*. Vol. 69, No. 1 (January 2003), pp. 186-190. ISSN 0099-2240.

Yamazaki, M. & Komagata, K. (1983). An Electrophoretic Comparison of Enzymes of Ballistosporogenous Yeasts. *The Journal of General Applied Microbiolgy*. Vol. 29, No. 2 (February 1983), pp. 115-143. ISSN 0022-1260.

Zazueta-Sandoval, R., Durón, C. A. & Silva, J. H. (2008). Peroxidases in YR-1 strain of *Mucor circinelloides* a Potential Bioremediator of Petroleum-Contaminated Soils. *Annals of Microbiology*. Vol. 58, No. 3 (September 2008), pp. 421-426. ISSN 1590-4261.

Zazueta-Sandoval, R., Zazueta-Novoa, V., Silva-Jiménez, H. & Ortíz, R. C. (2003). A Different Method of Measuring and Detecting Mono and Di-Oxygenase Activities: Key Enzymes in Hidrocarbon Biodegradation. *Applied Biochememistry and Biotechnology*. Vol. 108, No. 1-3 (March 2003), pp. 725-736. ISSN 0273-2289.

Part 2

Temporal Temperature Gel Electrophoresis

Temporal Temperature Gel Electrophoresis to Survey Pathogenic Bacterial Communities: The Case of Surgical Site Infections

Romano-Bertrand Sara[1,2], Parer Sylvie[1,2], Lotthé Anne[1,2], Colson Pascal[3],
Albat Bernard[4] and Jumas-Bilak Estelle[1,2]
[1]*University Montpellier 1, Equipe pathogènes et environnements, UMR 5119 ECOSYM*
[2]*University Hospital of Montpellier, hospital hygiene and infection control team*
[3]*University Hospital of Montpellier, Cardio-thoracic intensive care unit*
[4]*University Hospital of Montpellier, Cardio-thoracic surgery unit*
France

1. Introduction

The main objective of this chapter is to review 16S rRNA gene-based PCR-Temporal Temperature Gel Electrophoresis (TTGE) methods with emphasis on its use in medical microbiology and infectious diseases. As an example of application, we describe optimization, validation and results of an original approach for exploring the microbiology of surgical site infections with a focus on particular constraints related to low microbial load found in this setting. Finally, PCR-TTGE will be situated in the evolution of medical microbiology and infectious disease medicine toward the analysis of complex microbial communities.

2. State of the art

2.1 From human microbiome to pathogenic bacterial communities

At the early 21th century, studies on human-associated bacteria showed that there are at least 10 times as many bacterial cells as human cells, i.e. 10^{14} bacterial cells in the human gut (Turnbaugh et al., 2007). The current estimation of the number of genes in the human genome is about 23000 (Wei & Brent, 2006). Based on the diversity of gut microbes and the average number of genes contained in one bacterial genome, the diversity of bacterial genes in human gut was guessed to be 100 times greater than that of our human genome (Bäckhed et al., 2005). This number seems to be underestimated, since a more recent publication estimates to more than 9.000.000 the number of unique genes in human gut bacterial community (Yang et al., 2009). This huge community is named the human microbiome, a term coined by Joshua Lederberg in 2001 "to signify the ecological community of commensal, symbiotic, and pathogenic microorganisms that literally share our body space" (Lederberg & McCray, 2001). A new concept is to consider human organism as an assemblage of human and bacterial cells organized into organs, tissues, and cellular communities amounting to a super- or a hetero-organism. Metagenomics show clear differences between microbiomes in various body sites and together with metatranscriptomics and metaproteomics reveal how microbiomes contribute to organ and tissue functions. Consequently, human biology can no longer concern itself only with human cells: microbiomes at different body sites and functional metagenomics must be considered part of systems biology (Pflughoeft & Versalovic, 2011). The biological concept of hetero-organism further evolves into new medical conception on health and disease, involving not only human genes and cells but also related genomes from environment around and inside human body. For instance, the change in gut microbiome is associated with local pathology such as Crohn's disease but also with systemic diseases such as obesity (Turnbaugh et al., 2009), diabetes (Cani et al., 2007) and hypertension (Holmes et al., 2008), among other chronic diseases (Proal et al., 2009; Lampe, 2008). The microbiome of various human body sites has already been described for the digestive tract, the mouth, the skin, the genital tract and it will continue with the « Human Microbiome Project » (Turnbaugh et al., 2007).

Microbiome and its disturbances are now a major area of microbiology research, as attested by the leading publications produced in this field. However, these advances in 'microbiomology' have not yet change the patterns of medical reflection. In the field of infectious diseases, bacteriologists still strive to obtain a pure culture preferentially isolating known pathogens and focus on the morphology, physiology, and genetics of the small subset of the microorganisms in the body that are known to cause disease. Hence, most bacteria present in a community are neglected and so are the complex relationships among members of the community. It is well-known that a large range of endogenous bacteria can cause opportunistic infections such healthcare associated infections (HAI), consecutive to microbiome disequilibrium, and that strict pathogens must cross the cutaneous or mucosal barriers covered with local microbiota before developing their virulence. Therefore, considering the whole human microbiota is required to better understand the physiopathology of infections and improve their prevention and treatment of bacterial infectious diseases that remain major challenges in public health. Confronting microbiomes analyses to clinical issues will warrant these analyses to be good predictors and markers in infectious disease. This implies that microbiomes analyses become routine, which is not yet the case.

2.2 How to explore bacterial communities?

Bacterial communities are classically assessed through culture-dependent methods based on colony isolation on solid medium, sometimes after enrichment by growth in liquid medium. This way, the description of complex communities inevitably involved time-consuming steps of growth on multiple media under varying conditions, while the full description of the community is not insured. Indeed, it is now obvious that the real microbial diversity is poorly represented by the cultured fraction, and conventional culture techniques have been shown to explore less than 1% of the whole bacterial diversity in environment samples, such as soil samples (Riesenfeld et al., 2004). It is also estimated that as much as 20% to 60% of the human-associated microbiome, depending on body site, is unculturable (NIH HMP Working Group et al., 2009). A renewal of studies on complex bacterial communities has been made possible by molecular methods that allow direct analysis of bacteria present in a

Temporal Temperature Gel Electrophoresis to Survey Pathogenic Bacterial Communities:
The Case of Surgical Site Infections

49

sample, while avoiding the bias of cultivability. The emergence of molecular ecology and metagenomics offers the potential of determining microbial diversity in an ecosystem without prior laboratory enrichment, isolation and growth on artificial media thanks to sophisticated methodological and computational tools. Generally, culture-independent approaches allow a precise description of an ecosystem by assessing its genetic diversity. As an example of molecular ecology efficiency, Dekio et al (2005) show that the total skin microflora explored by culture-independent molecular profiling is greater than previously believed, finding 22 potentially novel members, comprising 9 species and 13 phylotypes not yet described as members of skin microbiota (Dekio et al., 2005).

Metagenomics examine the complexity of the community by sequencing genomic libraries made from DNA extracted directly from the sample containing a complex mix of different bacteria. The complete metagenomic approach will give the total gene content of a community, thus providing data about biodiversity but also function and interactions (Tyson et al., 2004). For the purpose of biodiversity studies, metagenomics can focus on one common gene shared by all members of the community. The most commonly used culture-independent method relies on amplification and analysis of the 16S rRNA genes in a microbiome (Nossa et al., 2010).

16S rRNA genes are widely used for documentation of the evolutionary history and taxonomic assignment of individual organisms because they have highly conserved regions for construction of universal primers and highly variable regions for identification of individual species (Woese, 1987). The 16S rRNA gene, in spite of some recognized pitfalls (von Wintzingerode et al., 1997), remains today the most popular marker for studying the specific diversity in a bacterial community. The different 16S rRNA genes representative of the community are amplified by PCR and then separated and identified either by cloning and Sanger sequencing or by direct pyro-sequencing (Nossa et al., 2010). Tools for sequence-specific separation after bulk PCR amplification, such as T-RFLP (Terminal-Restriction Fragment Lenght Polymorphism) (Kitts, 2001), D-HPLC (Denaturing High Performance Liquid Chromatography) (Penny et al., 2010), CDCE (Constant Denaturing Capillary Electrophoresis) (Thompson et al., 2004), SSCP (Single Stand Conformation Polymorphism) (Ege et al., 2011), DGGE (Denaturing Gradient Gel Electrophoresis) (Muyzer et al. 1993), TGGE (Temperature Gradient Gel Electrophoresis) (Zoetendal et al., 1998) and TTGE (Temporal Temperature Gradient Gel Electrophoresis) (Ogier et al., 2002), can also be used. Methods based upon separation in denaturing electrophoresis appear particularly suitable for the routine follow-up of microbiomes with low or medium diversity (Roudière et al., 2007). They provide a "fingerprint" of the community diversity.

2.3 16S rRNA gene PCR-TTGE: Advantages and limitations

PCR-TTGE is a PCR-denaturing gradient gel electrophoresis that allows separation of DNA fragments in a temporal gradient of temperature (Yoshino et al., 1991; Ogier et al., 2002). PCR amplicons of the same size but with different sequences are separated in the gel. In a denaturing acrylamide gel, DNA denatures in discrete regions called melting domains, each of them displaying a sequence specific melting temperature. When the melting temperature (Tm) of the whole amplicon is reached, the DNA is denatured creating branched molecules. This branching reduces DNA mobility in the gel. Therefore, amplicons of the same size but with different nucleotide compositions can be separated based on differences in the behavior of their melting domains. When DNA is extracted and amplified from a complex community, TTGE lead to the separation of the different amplicons and produce a banding pattern characteristic of the community. The gradient obtained by varying the temperature over time in TTGE generally produces more clear and reproducible profiles than does the chemical gradient in DGGE.

Direct observation and counting bands on the TTGE profile provides a diversity score that roughly corresponds to the number of molecular species in the sample. However, it must be remembered that two amplicons with different sequences can give identical migration distances when their Tm are identical. The banding profile can be further analyzed by the affiliation of each band to a species or other taxon. Affiliation can be realized by comparing the migration distance of each band to a molecular ladder, named diversity ladder (Roudière et al., 2009; Ogier et al., 2002), constructed by using amplicons corresponding to species known to be representative of the community under study. A more accurate analysis of each band can be achieved by cutting bands from the gel, extracting DNA from bands and sequencing. This way the diversity ladder will be completed and updated. A method associating the use of diversity ladder with sequencing has shown its efficiency in describing bacterial communities of low complexity such as the gut microflora of neonates (Roudière et al., 2009). Such an approach is simple enough to survey dynamics of bacterial communities on a wide range of samples, particularly in health and disease (Jacquot et al., 2011).

PCR-TTGE appears not suitable for investigating highly diverse communities. This limit is due to number of bands that can be separated within the length of the gel. Optimization of TTGE conditions allows separation of bands by a minimum of 0.1 mm over all the gel length. We showed that the number of specific bands separable by TTGE could not exceed about 50 for the migration of an artificial diversity ladder. For instance, a ladder contained 53 different bacterial species found in stool of neonates can be efficiently separated in the optimized TTGE conditions (Roudière et al., 2009). For samples obtained from natural ecosystem, TTGE would be difficult to interprete if the diversity exceeds 25 to 30 bands.

Prior to TTGE migration, other technical steps are limitative and should be carefully considered and optimized. Particularly, DNA should be recovered and amplified from all the genotypes in the community, i.e. extraction and PCR should be as universal as possible. Special attention should be given to *Firmicutes* and *Actinobacteria* because they display thick and resistant cell wall. The extraction efficiency should be tested on a wide panel of bacteria to scan a large range of bacterial types. Extraction is generally improved by the use of large-spectrum lytic enzymes and/or by a mechanical grinding (Roudière et al., 2009; Le Bourhis et al., 2007).

The PCR itself is a cause of limitations in the PCR-TTGE approach. Molecular methods are often praised for their sensitivity. However, this detection sensitivity can fail when complex samples are analyzed. For example, detection thresholds of 10^3-10^4 CFU/mL are currently described for universal PCR-TTGE or PCR-DGGE (Le Bourhis et al., 2007; Temmerman et al., 2003; Roudière at al., 2009). The detection limit in PCR-TTGE cannot be easily assessed as it depends on both CFU/g count of each species and the relative representation of species in the community. Minor populations of less than 1% of total population are undetectable by PCR-TTGE. This breakpoint is commonly reported for denaturing-gel-based methods used in microbial ecology (Ogier et al., 2002; Zoetendal et al., 1998; Roudière et al, 2009).

Temporal Temperature Gel Electrophoresis to Survey Pathogenic Bacterial Communities:
The Case of Surgical Site Infections

51

PCR can also be affected by preferential PCR amplification that may hinder the detection of some genotypes when a complex mix of DNA molecules is used as template. Preferential PCR amplification can be caused by primer mismatches at the annealing sites for some genotypes or by a lower rate of primer hybridization to certain templates due to a low local denaturation (Kanagawa, 2003). PCR carried out on complex bulk DNA can produce heteroduplexes particularly in later cycles when primer concentration decreases and the concentration of PCR products is high (Kanagawa, 2003). Chimeric amplicons can also be formed in later PCR cycles when template concentration is high enough to allow the re-annealing of templates before primer extension (von Wintzergerode et al., 1997; Kanagawa, 2003). All these artifacts can generate additional signals that do not correspond to genotypes in the sample. Heteroduplexes and chimera produce additional bands in the TTGE pattern that lead to an overestimation of the diversity. These artifactual bands can be detected either on the basis of their very short migration distance or by sequencing. Consequently, the crude diversity index determined by simple band count should be optimized after exclusion of heteroduplexes and chimeric bands.

Last but not least, a major parameter in community studies is the choice of a molecular marker allowing the genotyping of the whole community. The notion developed by Woese that rRNA genes could identify living organisms by reconstructing phylogenies resulted in the adoption of 16S rRNA gene in microbiology (Woese, 1987). Its universality and the huge number of sequences stored in databases have established 16S rRNA gene as the "gold standard" not only in microbial phylogeny, systematics, and identification but also microbial ecology (Case et al., 2007).

For the purpose of identifying an isolated bacterial strain, the complete 16S rRNA gene (1500 bp) is generally used, giving accurate affiliation to a species in most cases. In PCR-TTGE experiments, the amplified fragments are short (200 to 400 bp) to allow migration in polyacrylamide gel. The fragment amplified should contain hypervariable regions of the 16S rRNA gene in order to compensate for the lack of information due to the small sequence size by a high rate of mutation. Bacterial 16S rRNA genes comprise nine hypervariable regions, V1-V9, exhibiting sequence diversity among species (Van de Peer et al., 1996). In most studies, the V3 region located in the 5' part of the gene is chosen (Jany & Barbier, 2008). However, the phylogenetic information is sometimes insufficient to achieve species identification. Depending on the bacterium, sequences provide identification to the genus or family level only. Consequently, the diversity of the community is not described by a list of bacterial species but by a list of operational taxonomic units (OTUs) corresponding to the lower taxonomic level being accurately identified.

At the genomic level, rRNA genes are generally organized in multigene families (Acinas et al., 2004). The members of a rRNA multigene family are subject to a homogenization process allowing the multiple gene copies to evolve in concert. In a concerted evolution mode, mutation occurring in one copy will be fixed in all of them or lost from all. Thus, rRNA sequences show low variability within species, subspecies or genome (Liao, 2000). However, intra-genomic heterogeneity in the form of nucleotide differences between 16S rRNA gene copies are often described. For examples micro-heterogeneity has been identified in *Escherichia coli*, *Mycobacterium terrae*, *Paenibacillus polymyxa*, members of the classes *Mollicutes*, and *Actinomycetales* (Teyssier et al., 2003). Analyzing of complete genome sequences has recently assessed intra-genomic heterogeneity. For genomes with more than one rRNA operon, 62% display some degree of sequence divergence between 16S rRNA loci in a same genome (Case et al., 2007). In PCR-TTGE, the intra-genomic 16S rDNA heterogeneity can lead to multiple bands for a single OTU and then to an overestimation of OTU diversity. This pitfall inherent to the 16S rRNA gene marker will be avoided by band sequencing. However, as a pre-requisite of diversity analysis by PCR-TTGE, the major known species expected in a particular ecosystem should be individually studied by TTGE in order to explore heterogeneity in 16S rRNA gene copies (Roudière et al, 2007; Michon et al., 2011).

Alternative markers can also be proposed such as rpoB (Case et al., 2007) but universal rpoB PCR primers allowing the exploration of the whole bacterial diversity can not be designed (personal data) and the databases remain poor in rpo sequences.

Several authors remarked that culture-independent methods regularly fail to identify species obtained using culture-dependent methods (Jany & Barbier, 2008). By contrast, culture-dependent methods have yielded information on the structure of microbial populations but they are limited by the in vitro growth capacity of most bacteria in a community. Such a discrepancy is not observed for all the communities studied. For instance, culture dependant and independent approaches of the premature neonate gut microbiome give globally congruent results (Roudière et al., 2009). However, it is accepted that culture-independent methods remain the only approach for monitoring the rapid dynamics of microbial communities. Nevertheless, the two types of methods reveal different images of the same community and combining culture-dependent and culture-independent methods may be worthwhile to obtain a more accurate view of the structure of the microbial community (Case et al., 2007).

In spite of the limited growth capacity of most bacteria, culture-dependent methods remain the sole approach available for monitoring sub-populations selected on the basis of phenotypic traits such as dependence to metabolites or resistance to antimicrobial agents. In this context, PCR-TTGE can be used after culture in specific conditions in order to describe the diversity of cultivable population. The colonies growing in diverse conditions can be bulked and further analyzed by PCR-TTGE as described before. This culture- and genetic-based mixed approach is particularly suitable to describe dynamics of populations according to their level of resistance to antimicrobial drugs in natural environments (Vanhove et al., 2011).

Culture-independent approaches have previously shown their interest in cardiology to detected new or atypical infectious agents (Marchandin et al., 2009; Daïen et al., 2010). Considering the interest and limitations of 16S rRNA PCR-TTGE, we will proposed a protocol for describing the diversity and following the dynamics of the bacterial community that colonize surgical wound of the patient during hospitalization for cardiac surgery. We will show how 16S rRNA PCR-TTGE is particularly suited to the low bacterial diversity encountered in aseptic surgical settings, where antibiotic prophylaxis and cutaneous antisepsis effectively reduce the bacterial load of patients.

In addition, one example of the use of 16S PCR-TTGE in a culture-dependent analysis will be detailed. This approach associates determination of Minimal Inhibitory Concentration (MIC) at the community level and determination of the diversity by 16S PCR-TTGE in the resistant sub-population at each concentration of antimicrobial agent.

Temporal Temperature Gel Electrophoresis to Survey Pathogenic Bacterial Communities:
The Case of Surgical Site Infections

53

3. An original method of PCR-TTGE to learn more about the physiopathology of surgical site infection in cardiac surgery

3.1 Surgical site infections: State of the art

3.1.1 Epidemiology of surgical site infection

Surgical site infections (SSIs) are among the most frequent healthcare associated infections (HAIs), along with urinary tract and pulmonary infections, and remain an unresolved problem for modern medicine, their occurrence having significant impact on patient morbidity, length of stay and cost of care. Data from longitudinal surveillance studies show SSI rates of 1 to 5% (Klevens et al., 2007; Astagneau et al., 2009; de Lissovoy et al., 2009), whereas higher rates are reported from interventional studies, where control groups can have up to 8.5% infection rates (Bode et al., 2010; Perl et al., 2002), owing to different population case mixes. Large-scale epidemiological studies have identified risk factors for SSI that can be grossly classified as related to patient condition, surgical procedure and environment. The American National Nosocomial Infection Surveillance system (NNIS) developed an easily calculated risk index that combines the patient-related risk assessment of the American Society of Anesthesiologists (ASA) score, and 2 surgical procedure-related factors: type of surgery as defined by pre- or per-operative microbial contamination (Altemeier classification, from I – clean surgery- to IV – septic surgery), and duration of operation exceeding 75th percentile for a given procedure. For all categories of surgery, there is a linear increase in the incidence of SSI when the NNIS risk index increases (Coello et al., 2005). Beside these surveillance-derived risk assessment scores, prospective controlled studies identified many more factors associated with a higher risk of SSI. Most important are poorly controlled diabetes mellitus, malignant diseases, smoking, advanced age, per operative hypothermia, emergency surgery (Coello et al., 2005). Specific risk factors have been identified for cardiac surgery: obesity, pre operative myocardial infarction, chronic obstructive broncho-pulmonary disease, duration of extra corporeal circulation, early post operative bleeding, combined valve and coronary bypass procedures... (Filsoufi et al., 2009).

In the mid 1980s, 30 to 35% of nosocomial (i.e. hospital-acquired) infections were deemed evitable (Haley et al., 1985). A recent analysis of infection control interventional studies estimates that as many as 26 to 54% of SSIs could be avoided by comprehensive implementation of evidence-based prevention strategies, foremost of which are pre operative cutaneous antisepsis, no pre operative shaving of surgical site, timely antibiotic prophylaxis and strict per operative glycemic control (Umsheid et al., 2011).

Still, even when all known preventive measures are implemented, even in low-risk (i.e. clean, non urgent) surgery for low risk patients, SSI can occur. This seemingly irreducible rate of "inevitable" infections raises the problem of how surgical site infections develop. Success of preventive measures based on antisepsis and optimization of patient status compounds the hypothesis that infection results from disequilibrium between host defense mechanisms and microbial infectiveness. In a 10-to-1 inequity between human cells and colonizing microbes, the balance of power in ensured by integrity of skin and mucous membranes, both obviously disrupted by surgery (Wenzel, 2010). However, the intimate mechanisms of infection are not known, starting with the origin of germs involved.

3.1.2 Are surgical site infections related to endogenous or exogenous bacteria?

SSIs are 4.5 times more frequent in patients with nasal carriage of *Staphylococcus aureus* compared to non-carriers. In carrier patients who developed SSI, *S. aureus* isolated from infection site was identical to the one isolated from anterior nares in 84% of cases (Perl et al., 2002). The case for an endogenous bacterial origin is indirectly made by the 60% reduction rate in *S. aureus* SSIs obtained by thorough pre operative decontamination (Bode et al., 2010). Nasal carriage of *S. aureus* was found to be associated with a higher rate of SSIs in orthopedic surgery in a French multicentric study (Berthelot et al., 2010). However, in this study, only 27% of *S. aureus* infections were molecularly linked to an endogenous strain. This can be due to insufficient sensitivity of carriage detection, or to the fact that infections don't necessarily have an endogenous origin.

Indeed, bacteria involved in SSIs can also originate from an exogenous source in the per operative environment. Several studies report cases of cross contamination of surgical sites with bacteria molecularly linked to health care professionals or other patients (Perl et al., 2002). Contamination of surgical instruments and devices increases over time inside the operating room (Dalstrom et al., 2008). Under different air treatment devices, air contamination can vary from 8 to 34% (Knobben et al., 2006). In the air above the surgical site, *S. aureus* and coagulase negative staphylococci can be found, mostly molecularly linked to nasal and pharyngeal carriage by operating team personnel (Edminston et al., 2005).

3.1.3 From contamination to infection

The origin of germs notwithstanding, what induces ordinarily commensal germs to become pathogenic is mostly unexplained. Indeed, in spite of the frequent presence of germs around and in the surgical site (up to 4.1% deep tissue samples found positive in clean orthopedic surgery (Byrne et al., 2007)), only very few infections actually develop.

Antibiotic prophylaxis certainly thwarts the development of most SSIs, at least insofar as bacterial load remains low and antibiotic spectrum and pharmacokinetics are adapted to the germs and tissues involved (Classen et al., 1992). However, the occurrence of SSIs in spite of adequate prophylaxis leads to consider the mechanism of infection from a bacterial point of view. Infections are often the consequences of disequilibrium in the relationships inside the microbiota or between the microbiota and the host. SSIs are typically infections related to microbiota disequilibrium, their major etiologic agents being members of the skin microbiota such as *S. aureus*, coagulase-negative Staphylococci and *Propionibacterium acnes*. The mechanism of selection of these bacteria from the skin community remains unknown and could be highlighted by dynamic survey of the microbial re colonization by a suitable method. Exploring the microbiota disequilibrium and its dynamics should provide insights into the mechanisms involved in SWI infections.

We hereafter present the use of TTGE in culture-independent and culture-based approaches to study the dynamics of bacterial communities involved in SSIs. The study was undertaken in the setting of clean cardiothoracic surgery, because SSI rates remain unacceptably high in coronary artery bypass grafts (CABG), in spite of ongoing efforts to minimize patient- and procedure-related risk factors (Filsoufi et al., 2009).

Temporal Temperature Gel Electrophoresis to Survey Pathogenic Bacterial Communities:
The Case of Surgical Site Infections

55

3.2 Culture-independent approach: Use of TTGE to survey the dynamics of bacterial communities involved in surgical wound colonization and infection

3.2.1 Patients and samples

Forty cotton swabs were collected from 5 patients who underwent CABG surgery at the Service of Thoracic and Cardiovascular Surgery of the Montpellier University Hospital (France). For each patient, swabs were sampled during intervention at the surgical site, superficially and deeply. The first sample was taken on the skin after cutaneous antisepsis just before incision and the second one in sub-cutaneous tissue once incision made. Thirdly, sternum edges were sampled after sawing. The fourth sample was mediastinal tissue after positioning of sternal retractors. At the end of the operation, mediastinum, sternum edges, sub-cutaneous tissue and skin were sampled again.

3.2.2 Cell lysis, extraction of bacterial DNA and PCR amplification

Bacterial genomic DNA was directly extracted from bulk cells present on cotton swabs using an enzymatic method (MasterPure Gram positive DNA purification kit, EPICENTRE Biotechnologies ®) according to the manufacturer's recommendations with modifications as described by (Roudière et al., 2009). This method has been previously described as efficient on a wide range of bacteria including Gram-positive bacteria (Roudière et al., 2009; Jacquot et al., 2011). A fragment about 1465 bp of the 16SrRNA gene was amplified using the primers 27f (5'-GTGCTGCAGAGAGTTTGATCCTGGCTCAG-3') and 1492r (5'-CACGGATCCTACGGGTACCTTGTTACGACTT-3'). The PCRs were carried out in 50 µL of reaction mixture containing 200 nM of each primer (Sigma Genosys), 200 nM each dNTP (Fermentas), 1U of *Taq* polymerase (Promega) in the appropriate reaction buffer, and 1 µL of crude DNA extract as the template. PCR conditions were 30 cycles of 1 min at 94°C, 1 min at 65°C, and 2 min at 72°C.

The 199-bp fragment (from position 338 to position 536, *Escherichia coli* numbering) overlapping the 16S rDNA V2-V3 variable region (Neefs et al., 1993; Sundquist et al., 2007) was amplified using the primers HDA1-GC (primer HDA1 with a fragment rich in GC – the 'GC clamp' – added to the 59 extremity) and HDA2 (Ogier et al., 2002). The reaction mixture (50 µl) consisted of 200 nM of each primer (Sigma Genosys), 200 mM each dNTP (Fermentas), 2.5 U FastStart Taq DNA polymerase (Roche, France) in the appropriate reaction buffer, with 1.8 mM MgCl$_2$. One µL of DNA previously amplified was added to the reaction buffer and the thermal cycling was as follows: 95°C for 2 min; 35 cycles of 95°C for 1 min, 62°C for 30 s, 72°C for 1 min ; and 72°C for 7 min. PCR products were checked by electrophoresis in a 1.5 % agarose gel before TTGE migration.

3.2.3 TTGE migration

TTGE migration was performed in the DCode Universal Mutation Detection System (Bio-Rad Laboratories). Gels were composed of 8 % (w/v) bisacrylamide (37.5: 1), 7 M urea, 40 ml N,N,N9,N9-tetramethylethylenediamine, and 0.1 % (w/v) ammonium persulfate, and were run in 16 Tris/acetate/EDTA buffer at pH 8.3. DNA was loaded on the gel with in-house dye marker (50% sucrose, 0.1% bromophenol blue) using capillary tips. The electrophoresis conditions were 46 V for 16 h with an initial temperature of 63°C and a final temperature of 70°C corresponding to an increase of 0.4°C h^{-1}. In order to obtain thin discrete bands, a pre-migration for 15 min at 63°C and 20 V was done after loading. Pre-migration and migration were performed with additional magnetic shaking in the electrophoresis chamber. Gels were stained for 15 min with 0.5 mg ethidium bromide ml[-1] in 1x TAE buffer, washed for 45 min in 1x TAE buffer, and photographed under UV illumination.

3.2.4 TTGE band sequencing

Each TTGE band for further analysis was cut out of the gel with a disposable sterile scalpel to avoid contamination between bands. Gel slices were washed twice in molecular biology grade water and incubated overnight at 37 °C in 10 mM Tris buffer (pH 8.5) to allow DNA diffusion. Amplification of a single 16S rRNA gene V3 region copy was performed using 1 µl band eluate and the primers HDA1 without a GC-clamp and HDA2. The PCR was carried out in 50 ml reaction mixture containing 200 nM of each primer, 200 mM each dNTP, 2.5 mM $MgCl_2$ and 2.5 U Taq DNA polymerase (Promega) in the appropriate buffer. PCR conditions were 94 °C for 2 min; 35 cycles of 45 s at 95 °C, 30 s at 62 °C, 1min at 72 °C; and 10 min at 72 °C. PCR products were checked by electrophoresis in a 1.5 % agarose gel and sequenced on an ABI 3730xl sequencer (Cogenics). Each sequencing chromatograph was visually inspected and corrected. The sequences were analyzed by comparison with Genbank (http://www.ncbi.nlm.nih.gov/) and RDPII databases (http://rdp.cme.msu.edu/) using Basic Local Alignment Search Tool (BLAST) and Seqmatch programs, respectively and affiliated to an OTU as recommended (Drancourt et al., 2000; Stackebrandt & Goebel, 1994).

3.2.5 Optimization of DNA amplification in samples with low load of template

In most assays, the bacterial load is low because samples originated from quasi-sterile sites, i.e. the mediastinum during the surgical intervention. When applied on DNA samples extracted from a low number of bacteria ($<10^3$/tube) (Roudière et al., 2009), a single PCR produced a faint or no signal in agarose gel electrophoresis (data not shown). We showed that the signal was equivalent to the positive control for DNA extracted from skin, a site with a resident microbiota but faint for the samples taken from sub-cutaneous and mediastinal tissues, sites expected to be almost sterile (data not shown). To gain in sensitivity, we performed a nested-PCR approach with pre-amplification of the almost complete 16S rRNA gene, further used as template for a second PCR focused on the V2-V3 hypervariable regions of the gene. As expected, nested-PCR resulted in an improved signal in agarose gel for all samples including samples with low bacterial load (Figure 1A and 1B). However, nested-PCR could induce biases such as amplification of contaminant DNA, preferential amplification of a sub-population of template and heteroduplex formation, among others (Park & Crowley, 2010. Consequently, the enhanced quantity of amplified DNA in agarose gel had to be confirmed in TTGE in order to assess the quality of amplified DNA and qualitatively compare single and nested PCR. Figure 2 shows the TTGE profiles after single- and nested-PCR obtained for 4 samples. We observed that nested-PCR led to an enrichment of the profiles while all the bands detected by single PCR-TTGE were also observed in nested PCR-TTGE. The PCR-TTGE profiles presented in Figure 2 displayed neither obvious heteroduplex bands nor smears due to amplification of miscellaneous contaminant DNA. However, preferential amplification was observed particularly for

Temporal Temperature Gel Electrophoresis to Survey Pathogenic Bacterial Communities:
The Case of Surgical Site Infections

57

sample AD6 In the corresponding TTGE profile, the faint bands seen in single PCR-TTGE profile were very faint in the nested PCR-TTGE profile while original and intense bands appeared in the latter profile. The reproducibility is also satisfactory (data not shown).

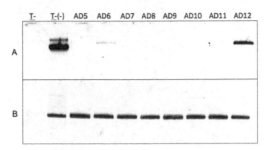

Fig. 1. Migration in agarose gel after single (A) and nested PCR (B) for samples from the patient AD. AD5, skin before incision; AD6, sub-cutaneous tissue after incision; AD7, sternum edges; AD8, mediastinum at the beginning of the operation; AD9, mediastinum at the end of intervention; AD10, edges of sternum before closing; AD11, sub-cutaneous tissue before suture; AD12, skin at the end of the procedure.

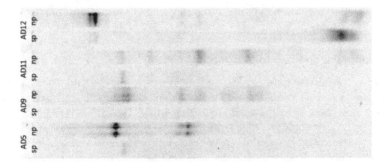

Fig. 2. TTGE migration after single PCR (sp) and nested PCR (np) on samples from the patient AD. AD6 sub-cutaneous tissue after incision; AD9, mediastinum before suture; AD11, sub-cutaneous tissue before suture; AD12, skin at the end of the surgery.

3.2.6 PCR-TTGE profiles and diversity index

About fifteen samples from the sternal region were recovered during hospital stay of each patient undergoing programmed CABG surgery. TTGE profiles were obtained for each sample after nested-PCR and showed clear bands easily separated from each other (data not shown). This result confirmed that 16S rRNA gene PCR-TTGE was a suitable method to survey the bacterial community from skin and wound tissues in cardiac surgery. By numbering the bands in each profile, crude diversities index (DI) were determined. Crude DI varied from 2 to 12 (mean value = 6). The bacterial communities presenting the lower mean value of crude DI (DI=4.2) were sampled on the skin just after preoperative antisepsis. Subcutaneous tissues sampled just before the closure of the wound at the end of the operation displayed the higher crude DI (DI=8.2). The dynamics of evolution of mean crude DI over the surgical operation showed that the bacterial diversity globally increased from the start to the end of the surgery, from 4.2 to 7.6. The increase of crude DI was also observed at each site of sampling: 4.2 to 7.6 for skin, 7.4 to 8.2 for subcutaneous tissues, 4.4 to 5.2 for sternum banks and 5.4 to 6.8 for mediastin. Figure 3 showed a representative DI dynamics for a patient.

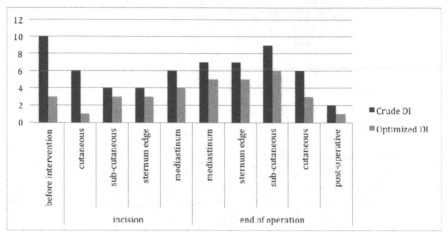

Fig. 3. Representative evolution of crude and optimized DI during the course of the surgical operation of the patient AD.

3.2.7 Operational Taxonomic Units diversity

Each band with specific migration distance was cut from the gel and the eluted DNA was reamplified, sequenced and compared to genetic databases for OTU affiliation. For the five patients presented here, we detected 16 different OTUs (Table 1). Four major OTU, *Staphylococcus* sp., *Flavobacteriaceae*, *Comamonadaceae* and *Propionibacterium* sp., were detected in all patients. *Acinetobacter* sp. and *Corynebacterium* sp. are also frequently detected. In contrast, OTUs such as *Bosea* sp. and *Bacillus* sp. were specific of one single patient. Most major OTUs were present in all patient at all step of the surgery operation and in all sampling sites. *Flavobacteriaceae* and *Comamonadaceae* were atypical Gram-negative bacilli recently described in the skin metagenome (Grice et al., 2009). These atypical Gram-negative bacilli were particularly represented in subcutaneous and in deep tissues (Figure 4). Band identification by sequencing allowed refining the crude DI by the elimination of multiple bands corresponding to a single OTU, as observed for bacteria that displayed 16S rRNA genes heterogeneity. In some cases, the DI dramatically decreased as shown in Figure 3.

3.3 Culture-dependent approach: Use of PCR-TTGE to assess the impact of antimicrobial prophylaxis on the whole skin microbiome

3.3.1 Methods

Sternal skin microbiome was sampled with cotton swabs, before hospitalization, in four patients requiring CABG surgery. This sample reflects the normal resident skin microbiome of the patients. Sternal skin samples were streaked on 5% blood and nutrient agar plates supplemented with different concentrations of cefamandole (0, 0.125, 0.25, 0.5, 2, 4, 8 and 32 µg/mL) and incubated for 24h at 37°C in anaerobiosis and aerobiosis. The resulting colonies

Temporal Temperature Gel Electrophoresis to Survey Pathogenic Bacterial Communities:
The Case of Surgical Site Infections

59

OTU	patient				
	AB	AD	AE	AF	AG
Staphylococcus sp.	X	X	X	X	X
Clostridiaceae				X	
Enterococcus sp.		X			
Acinetobacter sp.		X		X	X
Lactococcus sp.	X				
Streptococcus sp.				X	
Flavobacteriaceae	X	X	X	X	X
Comamonadaceae	X	X	X	X	X
Bosea sp.	X				
Corynebacterium sp.	X	X		X	X
Anaerococcus sp.				X	
Bacillus sp.			X		
Micrococcaceae	X	X	X		
Rhizobiaceae		X			X
Parococcus sp.	X	X		X	X
Propionobacterium sp.	X	X	X	X	X

Table 1. Repartition of the OTUs detected in the skin microbiota of five patients.

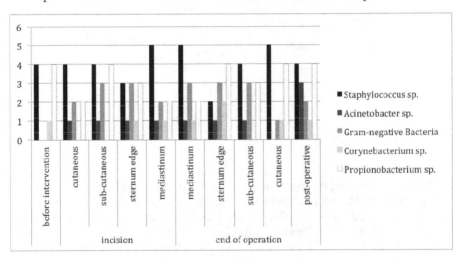

Fig. 4. Number of patients carrying the major OTUs according to the step of the surgery procedure.

were counted and the whole bacterial cultures were then harvested for DNA extraction. DNA extraction, single PCR HDA, TTGE migration and band affiliation to an OTU were performed as described in previous section.

3.3.2 Results: Impact of antimicrobial prophylaxis on skin microbiota

Most samples displayed cultivable microbiota susceptible to cefamandole, i.e. no culture was observed on agar plates with 8mg/L, which was the cefamandole critical concentration used to define susceptibility versus resistance. At lower concentrations (from 0.125 mg/L to 4 mg/L), we observed a decrease of colonies in number and in diversity, comparatively to the control plate. The mean colony counts went from 80 CFU/plate without antibiotics to 30 CFU/plate with 0.125 mg/L cefamandole and to 6 CFU/plate with 1mg/L cefamandole. All the colonies were bulked and analyzed by 16S rRNA gene PCR-TTGE. Profiles comparison allowed to determine the concentration of antimicrobial agents that inhibited 25%, 50%, 75% and 90% of the microbial diversity and to identify the resistant bacteria to the species level. These data will be related to dynamics of wound re-colonization after surgery.

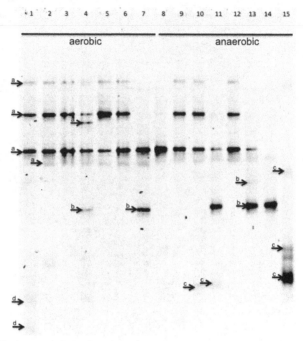

Fig. 5. TTGE gel illustrating the culture-dependent approach on skin sample of the patient CB3. The gel shows differences in banding patterns according to the concentration of cefamandole and the culture atmosphere. Lanes 1 and 8, culture without cefamandole; Lanes 2 and 9, cefamandole 0.125mg/L; Lanes 3 and 10, cefamandole 0.25mg/L; Lanes 4 and 11, cefamandole 0.5mg/L; Lanes 5 and 12, cefamandole 1 mg/L; Lanes 6 and 13, cefamandole 2mg/L; Lanes 7 and 14, cefamandole 4mg/L; Lane 15, cefamandole 8mg/L. Bands corresponding to OTUs are shown with arrow: a for *Staphylococcus* sp., b for *Bacillus* sp., c for *Corynebacterium* sp. and d for *Propionibacterium* sp.

The TTGE profiles from colonies harvested on control plates without antibiotics showed the predominant cultivable population. For the four patients studied herein, the corresponding bacteria belonged to the genera *Staphylococcus* (4 to 6 species) and *Propionibacterium*. For two patients, the cultivable skin microbiota was susceptible to cefamandole because no colonies were observed on plates with more than 8 mg/L, which was the cefamandole critical concentration used to define susceptibility versus resistance. For the two other patients, colonies were observed for 8 mg/L plates incubated in aerobiosis and even for 32 mg/L plates incubated in anaerobiosis. Resistant bacteria belonged to the genera *Bacillus* and *Corynebacterium*. For lower concentrations, which could correspond to the tissue

Temporal Temperature Gel Electrophoresis to Survey Pathogenic Bacterial Communities:
The Case of Surgical Site Infections

61

concentrations in the surgical wound (0.125mg/L to 4mg/L), TTGE profiles showed a modification of the microbiota with loss of bands affiliated to the genus *Staphylococcus* and appearance of other bands affiliated to the genera *Corynebacterium* and *Micrococcus* or to *Bacillus cereus* group. These bands corresponding to OTUs with a decreased susceptibility to cefamandole revealed minor OTUs that were undetected in the total cultivable microbiota, i.e. on plates without antibiotics. For instance in the patient CB (Figure 5), a band corresponding to *Bacillus* sp. was observed from 0,5 à 4mg/L cefamandole and a band corresponding to *Corynebacterium* sp. was observed in at least 8mg/L cefamandole.

4. Conclusion: Why should 'microbiomology' be on the benchtop of the medical microbiologist?

In this experiment we applied PCR TTGE to the study of microbiome dynamics in skin and tissues during a cardiac surgery procedure. We present the results obtained for 5 patients in a culture-independent approach and in 4 patients in a culture-dependent approach. Our results enable us to describe the bacteria involved in surgical site colonization during the procedure. The bacteria detected belonged to OTUs previously described in the skin microbiome (Grice et al., 2009). Gram-positive bacteria belonged to *Bacilli* and *Clostridia* in the phylum *Firmicutes*, and to *Corynebacterium*, *Propionibacterium* and *Micrococcaceae* in the phylum *Actinobacteria*. Gram-negative bacteria mainly belonged to *Proteobacteria* but also to *Flavobacteria*. This result confirmed that the nested-PCR TTGE approach gave a good representation of the skin microbiome even after antisepsis when the bacterial load is very low.

We have showed for the first time that the skin microbiota contaminated the surgical wound at all steps of the surgery in quantities that can be detected by molecular methods. The PCR-TTGE appeared as a suitable method to follow-up the microbiology of a clean wound such as a surgical wound. The approach could also give insights about the activity of antibiotics, e.g. agents used for antibioprophylaxis, at the community level and not only on isolated strains as performed currently.

In order to find relationships between the dynamics of microbiome and the clinical evolution of patients, a cohort of 120 patients will be studied for their surgical wound microbiota and their clinical outcome. Such clinical studies are necessary to understand to physio-pathological mechanism of wound re-colonization and SSI. PCR-TTGE appeared as easy and cost-effective enough to provide microbiological arguments at the community level in large clinical studies. Other very efficient methods such as pyrosequencing are until now not convenient for clinical studies on large cohorts of patients.

More generally the study presented here underlines that is now time for microbiome studies to exit from research labs and to take place on the benchtop of the medical microbiologist.

Current medical microbiology still rests on the shoulders of Leeuwenhoek, Koch and Pasteur. Although every microbiologist knows that a bacterium does not live alone but in complex communities, routine practice still consists in watching a small subset of isolated microbes under microscope and on Petri dishes. It is now clear that the human microbiome has profound effects on health and disease. Ambitious project are in progress such as the international Human Microbiome Consortium and the Human Microbiome Project. High-throughput sequencing and metagenomics allow us to explore new fields of the relationships between human and bacteria, considering these relationships at the community level. Facing the huge volume of data generated by metagenomics, microbiologists and even more so physicians are perplexed. For instance, Grice et al. admit that "...hairy, moist underarms lie a short distance from smooth dry forearms, but these two niches are likely as ecologically dissimilar as rainforests are to deserts" (Grice et al., 2009). A conceptual revolution is occurring while in medical microbiology practice, skin is still shown as a sheet covered by *S. epidermidis*.

How can we deal with this new world we discover within and around ourselves with our PCR and sequencers? To change our understanding of health and disease, it is now time to develop ways of seeing patterns and interpret them among the staggering biodiversity of microbes. Extending microbial ecology to healthy and diseased microbiota as previously done for mouth (Bik et al., 2010), stomach (Bik et al., 2006), gut (Yang et al., 2009) and skin (Grice et al., 2009) could elucidate the physio-pathology of diseases. For instance, description of the baseline skin microbiome is a step toward testing the therapeutic potential of manipulating the microbiome in skin disorders (Grice et al., 2009). Studies on psoriasis (Gao et al., 2008) describe selective microbial shifts associated with diseases and suggest that therapies might require not only inhibiting the growth of pathogenic bacteria, but also promoting the growth of mutualistic bacteria. Moreover, antibiotic exposure and hygienic practices such as antisepsis modify the skin microbiome. Methods that help understand naturally occurring mutualistic microbial communities will provide insights into the conditions that favor the emergence of antibiotic-resistant organisms.

It is now urgent to confront 'microbiomology' data to clinical data in order to define new indicators for re-evaluating the risk of disease and for adapting therapeutics. Before the generalization of high-throughput metagenomics on the microbiologist's benchtop, alternative methods such as PCR-TTGE or other methods based on denaturant electrophoresis should be considered. Although they are less exhaustive and powerful than pyrosequencing, these approaches are easy to handle in all laboratories and are cost-effective. They can be used to study large cohorts of patients, thus deriving clinical benefits from the conceptual revolution brought by contemporary researches on human microbial ecology.

5. References

Acinas Sylvia G, Marcelino LA, Klepac-Ceraj V & Polz MF. (2004). Divergence and Redundancy of 16S rRNA Sequences in Genomes with Multiple rrn Operons. *Journal of Bacteriology*, Vol.186, No.9 (May 2004), pp 2629–2635, ISSN 1098-5530

Astagneau Pascal, L'Hériteau F, Daniel F, Parneix P, Venier AG, Malavaud S, Jarno P, Lejeune B, Savey A, Metzger HM, Bernet C, Fabry J, Rabaud C, Tronel H, Thiolet JM, Coignard B, for the ISO-RAISIN Steering Group. (2009). Reducing surgical site infection incidence through a network: results from the French ISO-RAISIN surveillance system. *Journal of Hospital Infection*, Vol.72, No.2 (June 2009), pp 127-134, ISSN 0195-6701

Bäckhed Fredrik, Ley RE, Sonnenburg JL, Peterson DA, Gordon JI. (2005). Host-bacterial mutualism in the human intestine. *Science*, Vol.307, No.5717 (March 2005), pp. 1915–1920, ISSN 1095-9203

Temporal Temperature Gel Electrophoresis to Survey Pathogenic Bacterial Communities:
The Case of Surgical Site Infections

63

Berthelot Philippe, Grattard F, Cazorla C, Passot JP, Fayard JP, Meley R, Bejuy J, Farizon F, Pozzetto B & Lucht F. (2010). Is nasal carriage of Staphylococcus aureus the main acquisition pathway for surgical-site infection in orthopaedic surgery? *European Journal of Clinical Microbiology and Infectious Diseases*, Vol.29, No.4 (January 2010), pp 373-382, ISSN 1435-4373

Bik Elisabeth M, Eckburg PB, Gill SR, Nelson KE, Purdom EA, Francois F, Perez-Perez G, Blaser MJ & Relman DA. (2006). Molecular analysis of the bacterial microbiota in the human stomach. *Proceedings of the National Academy of Sciences of the United States of America*, Vol.103, No.3 (January 2006), pp 732–737, ISSN 1091-6490

Bik Elisabeth M, Long CD, Armitage GC, Loomer P, Emerson J, Mongodin EF, Nelson KE, Gill SR, Fraser-Liggett CM & Relman DA. (2010). Bacterial diversity in the oral cavity of 10 healthy individuals. *The ISME Journal*, Vol.4, No.8 (August 2010), pp 962–974, ISSN 1751-7370

Bode Lonneke GM, Kluytmans JAJW, Wertheim HFL, Bogaers C, Vandenbroucke-Grauls CMJE, Roosendaal R, Troelstra A, Box ATA, Voss A, van der Tweel I, van Belkum A, Verbrugh HA & Vos MC. (2010). Preventing surgical-site infections in nasal carriers of Staphylococcus aureus. (2010). *The New England Journal of Medicine*, Vol.362, No.1 (January 2010), pp 9-17, ISSN 1533-4406

Byrne Ann Maria, Morris S, McCarthy T, Quinlan W & O'Byrne JM. (2007). Outcome following deep wound contamination in cemented arthroplasty. *International Orthopaedics*, Vol.31, No.1 (February 2007), pp 27-31, ISSN 1432-5195

Cani Patrice D, Amar J, Iglesias MA, Poggi M, Knauf C, Bastelica D, Neyrinck AM, Fava F, Tuohy KM, Chabo C, Waget A, Delmée E, Cousin B, Sulpice T, Chamontin B, Ferrières J, Tanti J-F, Gibson GR, Casteilla L, Delzenne NM, Alessi MC & Burcelin R. (2007). Metabolic endotoxemia initiates obesity and insulin resistance. *Diabetes*, Vol.56, No.7 (July 2007), pp 1761–1772, ISSN 1939-327X

Case Rebecca J, Boucher Y, Dahllöf I, Holmström C, Doolittle WF & Kjelleberc S. (2007). Use of 16S rRNA and rpoB genes as molecular markers for microbial ecology studies. *Applied and Environmental Microbiology*, Vol.73, No.1 (January 2007), pp 278-288, ISSN 1098-5336

Classen David C, Evans RS, Pestotnik SL, Horn SD, Menlove RL & Burke JP. (1992). The timing of prophylactic administration of antibiotics and the risk of surgical-wound infection. *The New Englang Journal of Medecine*, Vol.326, No.5 (January 1992), pp 281-286, ISSN 1533-4406

Coello Rosa, Charlett A, Wilson J, Ward V, Pearson A & Borriello P. (2005). Adverse impact of surgical site infections in English hospitals. *Journal of Hospital Infection*, Vol.60, No.2 (June 2005), pp 93-103, ISSN 0195-6701

Daïen Claire I, Cohen JD, Makinson A, Battistella P, Jumas-Bilak E, Jorgensen C, Reynes J & Raoult D. (2010). Whipple's endocarditis as a complication of tumour necrosis factor-alpha antagonist treatment in a man with ankylosing spondylitis. *Rheumatology*, Vol.49, No.8 (February 2010), pp 1600-1602, ISSN 1462-0332

Dalstrom Daivd J, Venkatarayappa I, Manternach AL, Palcic MS, Heyse BA, & Prayson MJ. (2008). Time-dependent contamination of opened sterile operating-room trays. *The Journal of Bone and Joint Surgery*, Vol.90, No.5 (May 2008), pp 1022-1025, ISSN 1535-1386

Dekio Itaru, Hayashi H, Sakamoto M, Kitahara M, Nishikawa T, Suematsu M & Benno M. (2005). Detection of potentially novel bacterial components of the human skin microbiota using culture- independent molecular profiling. *Journal of Medical Microbiology*, Vol.54, No.12 (December 2005), pp 1231–1238, ISSN 0022-2615

Drancourt Michel, Bollet C, Carlioz A, Martelin R, Gayral JP & Raoult D. (2000). 16S ribosomal DNA sequence analysis of a large collection of environmental and clinical unidentifiable bacterial isolates. *Journal of Clinical Microbiology*, Vol.38, No.10 (October 2000), pp 3623–4630, ISSN 0095-1137

Edminston Charles E, Seabrook GR, Cambria RA, Brown KR, Lewis BD, Sommers JR, Krepel CJ, Wilson PJ, Sinski S & Towne JB. (2005). Molecular epidemiology of microbial contamination in the operating room environment: is there a risk for infection? *Surgery*, Vol.138, No.4 (October 2005), pp 573-582, ISSN 1532-950X

Ege Markus J, Mayer M, Normand AC, Genuneit J, Cookson WO, Braun-Fahrländer C, Heederik D, Piarroux R, von Mutius E & GABRIELA Transregio 22 Study Group. (2011). Exposure to environmental microorganisms and childhood asthma. *The New England Journal of Medicine*, Vol.364, No.8 (February 2011), pp 701-709, ISSN 1533-4406

Filsoufi Farzan, Castillo JG, Rahmanian PB, Broumand SR, Silvay G, Carpentier A & Adams DH. (2009). Epidemiology of deep sternal wound infection in cardiac surgery. *Journal of Cardiothoracic and Vascular Anesthesia*, Vol.23, No.4 (August 2009), pp 488-494, ISSN 1053-0770

Gao Zhan, Tseng CH, Strober BE, Pei Z & Blaser MJ. (2008). Substantial alterations of the cutaneous bacterial biota in psoriatic lesions. *Public Library of Science one*, Vol.3, No.7 (July 2008), e2719, ISSN 1932-6203

Grice Elisabeth A, Kong HH, Conlan S, Deming CB, Davis J, Young AC; NISC Comparative Sequencing Program, Bouffard GG, Blakesley RW, Murray PR, Green ED, Turner ML, Segre JA. (2009). Topographical and temporal diversity of the human skin microbiome. *Science*, Vol.324, No 5931 (May 2009), pp 1190-1192, ISSN 1095-9203

Haley Robert W, Culver DH, White JW, Morgan WM, Emori TG Munn VP & Hooton TM. (1985). The efficacy of infection surveillance and control programs in preventing nosocomial infections in US hospitals. *American Journal of Epidemiology*, Vol.121, No.2 (February 1985), pp 182-205 ISSN 1476-6256

Holmes Elaine, Loo RL, Stamler J, Bictash M, Yap IK, Chan Q, Ebbels T, De Iorio M, Brown IJ, Veselkov KA, Daviglus ML, Kesteloot H, Ueshima H, Zhao L, Nicholson JK & Elliott P. (2008). Human metabolic phenotype diversity and its association with diet and blood pressure. *Nature*, Vol.453, No.7193 (May 2008), pp 396–400, ISSN 0028-0836

Jacquot Aurelien, Neveu D, Aujoulat F, Mercier G, Marchandin H, Jumas-Bilak E & Picaud JC. (2011). Dynamics and clinical evolution of bacterial gut microflora in extremely premature patients. *The Journal of Pediatrics*, Vol.158, No.3 (March 2011), pp 390-396, ISSN 1432-1076

Jany Jean-Luc & Barbier G. (2008). Culture-independent methods for identifying microbial communities in cheese. *Food Microbiology*, Vol.25, No.7 (October 2008), pp 839-848, ISSN 0740-0020

Temporal Temperature Gel Electrophoresis to Survey Pathogenic Bacterial Communities:
The Case of Surgical Site Infections

65

Kanagawa Takahiro. (2003). Bias and artifacts in multitemplate polymerase chain reaction (PCR). *Journal of Bioscience and Bioengineering*, Vol.96, No.4 (December 2003), pp 317–323, ISSN 1347-4421

Kitts Christopher L. (2001). Terminal restriction fragment patterns: a tool for comparing microbial communities and assessing community dynamics. *Current Issues in Intestinal Microbiology*, Vol.2, No.1 (March 2001), pp 17-25, ISSN 1466-531X

Klevens Monia R, Edwards JR, Richards CL, Horan TC, Gaynes RP, Pollock DA & Cardo DM. (2007). Estimating health care-associated infections and deaths in U.S. hospitals, 2002. *Public Health Reports*, Vol.122, No.2 (March-April 2007), pp 160-166, ISSN 0094-6214

Knobben Bas AS, van Horn JR, van der Mei HC & Busscher HJ. (2006). Evaluation of measures to decrease intra-operative bacterial contamination in orthopaedic implant surgery. *Journal of Hospital Infection*, Vol.62, No.2 (February 2006), pp 174-180, ISSN 1532-2939

Lampe Johanna W. (2008). The Human Microbiome Project: getting to the guts of the matter in cancer epidemiology. *Cancer Epidemiology, Biomarkers & Prevention*, Vol.17, No.10 (October 2008), pp 2523-2524, ISSN 1538-7755

Le Bourhis Anne-Gaëlle, Doré J, Carlier J, Chamba J, Popoff M & Tholozan J. (2007). Contribution of C. beijerinckii and C. sporogenes in association to C. tyrobutyricum to the butyric fermentation in Emmental type cheese. *International Journal of Food Microbiology*, Vol.113, No.2 (2007), pp 154–163, ISSN 1879-3460

Lederberg Joshua & McCray AT. (2001). 'Ome Sweet 'Omics - a genealogical treasury of words. *The Scientist*, Vol.15, No.7 (April 2001), pp 8-8, ISSN 1547-0806

Liao Daiqing. (2000). Gene conversion drives within genic sequences: concerted evolution of ribosomal RNA genes in bacteria and archaea. *Journal of Molecular Evolution*, Vol.51, No.4 (June 2000), pp 305-317, ISSN 1432-1432

de Lissovoy Gregory, Fraeman K, Hutchins V, Murphy D, Song D & Vaughn BB. (2009). Surgical site infection: incidence and impact on hospital utilization and treatment costs. *American Journal of Infection Control*, Vol.37, No.5, (June 2009), pp 387-397, ISSN 1527-3296

Marchandin Hélène, Battistella P, Calvet B, Darbas H, Frapier JM, Jean-Pierre H, Parer S, Jumas-Bilak E, Van de Perre P & Godreuil S. (2009). Pacemaker surgical site infection caused by Mycobacterium goodii. *Journal of Medical Microbiology*, Vol.58, No.4 (April 2009), pp 517-520, ISSN 1473-5644

Michon Anne-Laure, Aujoulat F, Roudière L, Soulier O, Zorgniotti I, Jumas-Bilak E & Marchandin H. (2010). Intragenomic and intraspecific heterogeneity in rrs may surpass interspecific variability in a natural population of Veillonella. *Microbiology*, Vol.156, No.7 (July 2010), pp 2080–2091, ISSN 1465-2080

Michon Anne-Laure, Jumas-Bilak E, Imbert A, Aleyrangues L, Counil F, Chiron R & Marchandin H. (2011). Intragenomic and intraspecific heterogeneity of the 16S rRNA gene in seven bacterial species from the respiratory tract of cystic fibrosis patients assessed by PCR-Temporal Temperature Gel Electrophoresis. *Pathologie Biologie*, Vol. 2956, (May 2011), pp 1-6, ISSN 1768-3114

Muyzer Gerard, De Waal E.C & Uittierlinden A.G. (1993). Profiling of Complex Microbial Populations by Denaturing Gradient Gel Electrophoresis Analysis of Polymerase

Chain Reaction-Amplified Genes Coding for 16S rRNA. *Applied And Environmental Microbiology*. Vol.59, No.3 (March 1993), pp 695-700, ISSN 0099-2240

Neefs Jean-Marc, Van de Peer Y, De Rijk P, Chapelle S & De Wachter R. (1993). Compilation of small ribosomal subunit RNA structures. *Nucleic Acids Research*, Vol.21, No.13 (July 1993), pp 3025–3049, ISSN 1362-4962

NIH HMP Working Group, Peterson J, Garges S, Giovanni M, McInnes P, Wang L, Schloss JA, Bonazzi V, McEwen JE, Wetterstrand KA, Deal C, Baker CC, Di FrancescoV, Howcroft TK, Karp RW, Lunsford RD, Wellington CR, Belachew T, Wright M, Giblin C, David H, Mills M, Salomon R, Mullins C, Akolkar B, Begg L, Davis C, Grandison L, Humble M, Khalsa J, Little AR, Peavy H, Pontzer C, Portnoy M, Sayre MH, Starke-Reed P, Zakhari S, Read J, Watson B & Guyer M. (2009). The NIH Human Microbiome Project. *Genome Research*, Vol.19, No.12 (December 2009), pp 2317-2323, ISSN 1424-859X

Nossa Carlos W, Oberdorf WE, Yang L, Aas JA, Paster BJ, Desantis TZ, Brodie EL, Malamud D, Poles MA & Pei Z. Design of 16S rRNA gene primers for 454 pyrosequencing of the human foregut microbiome. *World Journal og Gastroenterology*, Vol.16, No.33 (September 2010), pp 4135-4144, ISSN 1007-9327

Ogier Jean-Claude, Son O, Gruss A, Tailliez P & Delacroix-Buchet A. (2002). Identification of the bacterial microflora in dairy products by temporal temperature gradient gel electrophoresis. *Applied and Environmental Microbiology*, Vol.68, No. 8 (August 2002), pp 3691-701, ISSN 1098-5336

Park Joong-Wook & Crowley DE. (2010). Nested PCR bias: a case study of Pseudomonas spp. in soil microcosms. *Journal of environmental monitoring*, Vol.12, No.4 (April 2010), pp 985-988, ISSN 1464-0333

Penny Christian, Nadalig T, Alioua M, Gruffaz C, Vuilleumier S & Bringel F. (2010). Coupling of denaturing high-performance liquid chromatography and terminal restriction fragment length polymorphism with precise fragment sizing for microbial community profiling and characterization. *Applied and Environmental Microbiology*, Vol.76, No.3 (February 2010), pp 648-651, ISSN 1098-5336

Perl Trish M, Cullen JJ, Wenzel RP, Zimmerman B, Pfaller MA, Sheppard D, Twombley J, French PP, Herwaldt LA, & The Mupirocin and the Risk of Staphylococcus aureus Study Tean. (2002). Intranasal mupirocin to prevent postoperative Staphylococcus aureus infections. *The New England Journal of Medicine*, Vol.346, No.24 (June 2002), pp 1871-1877, ISSN 1533-4406

Pflughoeft Kathryn J & Versalovic J. (2011). Human Microbiome in Health and Disease. *Annual Review of Pathology: Mechanisms of Disease*, Vol.7, No.1 (September 2011) ISSN 1553-4014 [Epub ahead of print]

Proal Amy D, Albert PJ & Marshall T. (2009). Autoimmune disease in the era of the metagenome. *Autoimmunity Reviews*, Vol.8, No.8 (July 2009), pp 677-681, ISSN 1568-9972

Riesenfeld Christian S, Schloss PD & Handelsman J. (2004). METAGENOMICS: Genomic Analysis of Microbial Communities. *Annual Review of Genetics*, Vol.38 (July 2004), pp 525–552, ISSN 0066-4197

Roudière Laurent, Lorto S, Tallagrand E, Marchandin H, Jeannot JL & Jumas-Bilak E. (2007). Molecular fingerprint of bacterial communities and 16S rDNA intra-species

Temporal Temperature Gel Electrophoresis to Survey Pathogenic Bacterial Communities:
The Case of Surgical Site Infections

67

heterogeneity: a pitfall that should be considered. *Pathologie Biologie*, Vol.55, No.8-9 (November 2007), pp 434-440, ISSN 1768-3114

Roudière Laurent, Jacquot A, Marchandin H, Aujoulat F, Devine R, Zorgniotti I, Jean-Pierre H, Picaud JC & Jumas-Bilak E. (2009). Optimized PCR-Temporal Temperature Gel Electrophoresis compared to cultivation to assess diversity of gut microbiota in neonates. *Journal of Microbiological Methods*, Vol.79, No.2 (November 2009), pp 156-165, ISSN 0167-7012

Stackebrandt Erko & Goebel BM. (1994). Taxonomic note: a place for DNA-DNA reassociation and 16S rRNA sequence analysis in the present species definition in bacteriology. *International Journal of Systematic Bacteriolgy*, Vol.44, No.4 (October 1994), pp 846–849, ISSN 0020-7713

Sundquist Andreas, Bigdeli S, Jalili R, Druzin ML, Waller S, Pullen KM, El-Sayed YY, Taslimi MM, Batzoglou S & Ronaghi M. (2007). Bacterial flora-typing with targeted, chip-based Pyrosequencing. *BMC Microbiology*, Vol.7, No.108 (November 2007) pp 1-11, ISSN 1471-2180

Temmerman Robin, Masco L, Vanhoutte T, Huys G & Swings J. Development and Validation of a Nested-PCR-Denaturing Gradient Gel Electrophoresis Method for Taxonomic Characterization of Bifidobacterial Communities. *Applied and Environmental Microbiology*, Vol.69, No.11 (November 2003), pp 6380–6385, ISSN 1098-5336

Teyssier Corine, Marchandin H Siméon De Buochberg, Ramuz M & Jumas-Bilak E. (2003). Atypical 16S rRNA gene copies in Ochrobactrum intermedium strains reveal a large genomic rearrangement by recombination between rrn copies. *Journal of Bacteriology*, Vol.185, No.9 (May 2003), pp 2901–2909, ISSN 1098-5530

Thompson Janelle R, Randa MA, Marcelino LA, Tomita-Mitchell A, Lim E & Polz MF. (2004). Diversity and dynamics of a north atlantic coastal Vibrio community. *Applied and Environmental Microbiology*, Vol70, No.7 (July 2004), pp 4103-4110, ISSN 1098-5336

Turnbaugh Peter J, Ley RE, Hamady M, Fraser-Liggett CM, Knight R & Gordon JI. (2007). The human microbiome project. *Nature*, Vol.449, No.7164 (October 2007), pp 804–810, ISSN 1476-4687

Turnbaugh Peter J, Hamady M, Yatsunenko T, Cantarel BL, Duncan A, Ley RE, Sogin ML, Jones WJ, Roe BA, Affourtit JP, Egholm M, Henrissat B, Heath AC, Knight R & Gordon JI. (2009). A core gut microbiome in obese and lean twins. *Nature*, Vol.457, No.7228 (January 2009), pp 480–484, ISSN 1476-4687

Tyson Gene W, Chapman J, Hugenholtz P, Allen EE, Ram RJ, Richardson PM, Solovyev VV, Rubin EM, Rokhsar DS & Banfield JF. (2004). Community structure and metabolism through reconstruction of microbial genomes from the environment. *Nature*, Vol.428, No.6978 (March 2004) pp 37–43, ISSN 1476-4687

Umsheid Craig A, Mitchell MD, Doshi JA, Agarwal R, Williams K & Brennan PJ. (2011). Estimating the proportion of healthcare-associated infections that are reasonably preventable and the related mortality and costs. *Infection Control and Hospital Epidemiology*, Vol.32, No.2 (February 2011), pp 101-114, ISSN 1559-6834

Van de Peer Yves, Chapelle S & De Wachter R. (1996). A quantitative map of nucleotide substitution rates in bacterial rRNA. *Nucleic Acids Research*, Vol.24, No.17 (September 1996), pp 3381–3391, ISSN 1362-4962

Vanhove Audrey, Jumas-Bilak E, Aujoulat F, Roger F, Vallaeys T, Marchandin H & Monfort P. An original approach for studying the bacterial resistance to antimicrobial agents at the community level: application to natural brackish lagoon water. FEMS, Geneva, Switzerland, June 2011

Wei Chaochun & Brent MR. (2006). Using ESTs to Improve the Accuracy of de novo Gene Prediction. *BMC Bioinformatics*, Vol.7, No.327 (July 2006), pp 1-10, ISSN 1471-2105

Wenzel Richard P. (2010). Minimizing surgical-site infections. The New England Journal of Medicine, Vol.362, No.1 (January 2010), pp 75-77, ISSN 1533-4406

von Wintzingerode Friedrich, Gobel UB & Stackebrandt E. (1997). Determination of microbial diversity in environmental samples: pitfalls of PCR-based rRNA analysis. *FEMS Microbiology Reviews*, Vol.21, No.3 (November 1997), pp 213-229, ISSN 1574-6976

Woese Carl R. (1987). Bacterial evolution. *Microbiological Reviews*, Vol.51, No.2 (June 1987), pp 221-271, ISSN 1740-1534

Yang Xing, Xie L, Li Y & Wei C. (2009). More than 9,000,000 unique genes in human gut bacterial community: estimating gene numbers inside a human body. *Public Library of Science One*, Vol.4, No.6 (June 2009), e6074, ISSN 1932-6203

Yoshino Kenji, Nishigaki K & Husimi Y. (1991). Temperature sweep gel electrophoresis: a simple method to detect point mutations. *Nucleic Acids Research*, Vol.19, No.11 (June 1991), pp 3153, ISSN 1362-4962

Zoetendal Erwin G, Akkermans AD & de Vos WM. (1998). Temperature gradient gel electrophoresis analysis of 16S rRNA from human faecal samples reveals stable and host-specific communities of active bacteria. *Applied and Environmental Microbiology*, Vol.64, No.10 (October 1998), pp 3854-3859, ISSN 0099-2240

Part 3

Two-Dimensional Gel Electrophoresis (2-DE)

4

Two-Dimensional Gel Electrophoresis Reveals Differential Protein Expression Between Individual *Daphnia*

Darren J. Bauer, Gary B. Smejkal and W. Kelley Thomas
Hubbard Center for Genome Studies, University of New Hampshire
USA

1. Introduction

Analysis of individual genetic variation is paramount to understanding how organisms and communities respond to changes in the environment and requires a model system with well-developed molecular resources and a solid foundation of ecological knowledge. Traditional genetic model systems (*E. coli*, yeast, fly, worm, and mouse) have served as workhorses in elucidating virtually all of the knowledge in modern molecular biology. While these systems were chosen for their robustness in laboratory studies, virtually nothing is known about their life histories in their native environment. By contrast, new-model systems, which have typically been studied in depth, from an ecological perspective are severely limited in regards to their molecular resources.

The model organism *Daphnia* has been utilized as an ecological model for centuries, and now with the sequencing of the genome complete and the development of the associated molecular resources, it is poised as one of the few model systems with the necessary molecular and ecological tools to answer the questions of response to the environment (Colborne et al., 2011). Long recognized as a model for ecological research, the freshwater crustacean *Daphnia* is rapidly maturing into a powerful model for understanding basic biological processes, within an ecological context. A common resident of lakes and ponds, *Daphnia* has been the subject of over a century of study in the areas of rapid environmental response, physiology, nutrition, predation, parasitology, toxicology and behavior (Edmondson, 1987). The reproductive cycle of *Daphnia* is ideal for experimental genetics. Generation time in the laboratory rivals that of almost all other model eukaryotic systems, reaching maturity within 5-10 days. Under favorable environmental conditions, *Daphnia* reproduce through parthenogenesis, allowing the conservation of genetic lines. Sexual reproduction is induced by environmental changes allowing the production of inbred or outbred lineages. The sexually produced diapausing eggs, termed ephippia, can be stored viably for considerable periods. Moreover, they have been hatched from lake sediments up to a century old (Hairston et al., 2001; Limburg & Weider, 2002) allowing tracking of genetic changes over ecological and evolutionary time scales. *Daphnia* are transparent throughout life, allowing for studies of tissue-specific gene expression at any life stage and direct observation of parasites and pathogens. As a result, there is a growing body of work in

Daphnia related to regulation of developmental genes, the genetic basis of evolutionary ecology, and parasite resistance and immunity.

Understanding and predicting how individual organisms respond at the molecular level to environmental change will provide new insight into the evolution of complex biological systems. This insight will lead to the development of new predictive models of host-pathogen interactions, environmental stress and community dynamics as a function of environment and genotype/phenotype (National Science Board, 2000) advancing the field of individualized molecular medicine. As the number of organisms with complete genome sequence increases and technological improvements allow more sequence to be generated at a lower cost, the ability to look at genetic variation in a variety of organisms is greater than ever. However, to understand the role of genetic variation in the context of the natural environment, a model system with two critical components, (1) well-developed molecular resources and (2) a well-understood ecological knowledge base, is essential. Until recently, model systems typically possessed one of these components in depth while the other was nominal or lacking altogether. The recent sequencing of the entire *Daphnia pulex* genome and the establishment of the still growing molecular toolbox (ESTs, genetic map, arrays, etc.) represents the first model system with both components in place.

Organisms respond to environmental change through relatively quick changes in gene expression or through evolutionary response over multiple generations. To better comprehend the effect of gene expression on phenotype, an understanding of genetic variation for gene expression is necessary. A comprehensive understanding of genetic variation is obtained by sampling between and within populations, including individual organisms, directly from their natural environment. The well-documented ecological understanding of *Daphnia* makes the system uniquely suited to this and allows researchers to collect and sample individuals of wild populations directly from their native environment.

Our goal was to demonstrate that it is possible to detect biologically relevant variation in protein expression from an individual *Daphnia*. Using pressure cycling technology (PCT) for sample preparation and two-dimensional gel electrophoresis (2-DE), we have demonstrated that differences in protein expression between individual *Daphnia* with distinct genotypes and exhibiting biologically relevant phenotypic differences are detectable. The ability to detect and analyze individual differences for a large number of proteins represents an important step towards understanding the connection between genotype/phenotype and the environment.

2. Materials and methods

2.1 *Daphnia* and algae cultures

2.1.1 Algae cultures

Starter cultures of the green algae *Ankistrodesmus falcatus* were obtained from UTEX, The Culture Collection of Algae at The University of Texas (Austin, TX, USA). *A. falcatus* was grown in 2 L aerated, air-filtered culture vessels containing GTk media at 25°C under continuous illumination. GTk contains the following macronutrients, 0.2 mM $CaCl_2$, 2.5 mM KNO_3, 0.3 mM $MgSO_4$, 0.4 mM Na_2HPO_4 and the following micronutrients, 150 μM EDTA,

Na_2, 20 µM $FeSO_4$, 2 µM $ZnSO_4$, 1 µM $NaMoO_3$, 0.6 µM $CuSO_4$, $CoCl_2$ and 14 µM $MnCl_2$ (Leland Jahnke, Personal Communication).

2.1.2 *D. magna* with unique phenotype

D. magna starter cultures were obtained from Sachs Systems Aquaculture (St. Augustine, FL, USA). Stabilized cultures were maintained in 8 L of 25% mineralized water (Vermont Spring Water Company, Brattleboro, VT, USA) at a density of 60-120 individuals/L. *D. magna* were cultured at 22° ± 1°C under constant illumination with standard fluorescent bulbs. Cultures were maintained at pH 7.0-7.4 by the addition of 100 g/L crushed coral (Tideline Aquatics, Hanahan, SC, USA) supplied in nylon bags. Starter cultures were fed daily with 1 mL/L of *Nanochloropsis* microalgae liquid concentrate (Reed Maricultures, Campbell, CA, USA) for the first four weeks, followed by 0.1 mL/L thereafter.

2.1.3 *D. magna* with unique genotype and *D. pulex*

D. magna clones Iinb1 and Xinb3 were isolated from Munich, Germany and Tvärminne, Finland, respectively (Rottu et al., 2010). *D. pulex* clone Log50 was obtained from the *Daphnia* Genomics Consortium stock (www.wfleabase.org/stocks). Xinb3 and Log50 are the clones for the respective, *D. magna* and *D. pulex* genome projects. Cultures were maintained in 8 L of COMBO media (Killham et al., 1998) at a density of 30 individuals/L at 20° ± 1°C under a 16:8 hours, light:dark, low intensity photoperiod, and fed 1mg Carbon/L of *A. falcatus*.

2.2 Harvesting of *Daphnia*

Daphnia gut contents were minimized by allowing the microcrustaceans to feed on copolymer microspheres of 4.3-micron mean diameter (Duke Scientific, Fremont, CA, USA) for one hour prior to harvesting. Microspheres were fed at a concentration equal to the number of algal cells previously supplied. *Daphnia* were harvested by filtration through 250 um Nitex mesh (Sefar America, Depew, NY, USA) and flash frozen. Average mass of adult *Daphnia pulex* was 0.1158 ± 8.3 mg fully hydrated and 0.05285 ± 10.60 mg dehydrated (n = 50). Average mass of adult *D. magna* was 1.37± 0.46 mg fully hydrated and 0.23 ± 0.06 mg when dehydrated (n = 64).

2.3 Pressure Cycling Technology (PCT)

PCT has been shown to be an effective means for isolating proteins from a variety of microorganisms, as well as many difficult-to-lyse samples such as *Caenorhabditis elegans* (Geiser et al., 2002; Smejkal et al., 2006b; Smejkal et al., 2007). PCT, which subjects samples to rapid cycles of pressure, facilitated the extraction of proteins from single *Daphnia magna* with and without ephippia and from single *Daphnia pulex*.

Daphnia were transferred to tared PULSE Tubes (Pressure BioSciences, Inc., South Easton, MA, USA) and suspended in 500 uL of 7M urea, 2M thiourea, and 4% CHAPS supplemented with 100 mM dithiothreitol (DTT) and protease inhibitor cocktail P-2714 (Sigma Aldrich Chemicals, St. Louis, MO). An additional 900 uL of mineral oil was added to accommodate the necessary volume for the PULSE Tubes. The tubes were placed in the Barocycler NEP-3229 (Pressure BioSciences, Inc., South Easton, MA, USA) for 60 pressure cycles, each cycle consisting of 10 seconds at 35,000 psi followed by rapid depressurization and held for 2 seconds at atmospheric pressure. Following PCT, each PULSE Tube was coupled to an Ultrafree-CL centrifugal filtration device with a 5-micron pore size (Millipore Corporation, Danvers, MA, USA) and evacuated by centrifugation for 1 minute at 1000 RCF. The PULSE Tube was removed and centrifugation continued for 4 minutes at 4000 RCF, followed by the removal of the mineral oil.

2.4 Reduction, alkylation, and ultrafiltration

Samples were transferred to ULTRA-4 ultrafiltration devices with 10 kDa molecular weight cut-off (Millipore Corporation, Danvers, MA, USA). Proteins larger than 10 kDa are retained in the ultrafiltration device. Centrifugation assisted the ultrafiltration, and the samples were exchanged with fresh UTC until the final DTT concentration was 10 mM. Reduction and alkylation of the samples were performed directly in the ultrafiltration devices using 5 mM tributylphosphine and 50 mM acrylamide as described (Smejkal et al., 2006a). The alkylation reaction was terminated by resuming centrifugation and ultrafiltrative exchange. Bradford Reagent (Sigma-Aldrich Chemicals, St. Louis, MO, USA) was used to measure the protein concentration in each sample.

2.5 IEF and 2-DE

Two-hundred uL of each sample was placed onto individual wells in IPG rehydration trays from Proteome Systems (Woburn, MA, USA). Bio-Rad ReadyStrip IPG strips with a pH range of 3-10, 4-7, or 7-10 (Hercules, CA, USA) were placed onto each sample, and the tray was placed into a humidifying Ziploc bag. Rehydration occurred over six hours until all the sample was visibly absorbed by the strip. At the termination of rehydration, strips were placed into isoelectric focusing trays and ran at 10,000 volts (maximum voltage) for 110,000 accumulative volt-hours. Strips were equilibrated twice for 10 minutes in 375 mM Tris-HCl containing 2.5% SDS, 3M urea, 0.01% and phenol red, then placed onto Criterion Tris-HCl 8-16% IPG+1 gels (Bio-Rad Laboratories, Hercules, CA) and ran at 120 V and 60 mA/gel. 2-DE gels were ran for all IPGs, only 4-7 gels are shown.

2.6 Digital image analyses

The 24-bit images were analyzed using PDQuest™ software (Bio-Rad, v.7.1). Background was subtracted, and protein spot density peaks were detected and counted. A reference pattern was constructed from one of the individual gels to which each of the gels in the matchset was matched. Numerous proteins that were uniformly expressed in all patterns were used as landmarks to facilitate rapid gel matching. After matching, the total spot count was determined for each gel.

3. Results and discussion

3.1 Protein variation between individual *D. magna* of distinct genotypes

Our first goal was to demonstrate that differences in protein expression could be detected between individual *Daphnia* of distinct genotypes. Individual *D. magna* from Iinb1 and Xinb3 genotypes were isolated, proteins extracted and analyzed in quadruplicate by 2-DE as describe above. Silver staining detected an average of 687 ± 11 protein spots from the Xinb3 gels and 692 ± 14 protein spots from the Iinb1 gels (figure 1). After normalization of the gel images based on total intensity, 679 spots were matched between the two gel images. One Hundred thirty six spots showed a two-fold or greater difference in spot intensity. Seventy-nine of these were more abundant in Xinb3 and 57 were more abundant in the Iinb1. To illustrate these differences, the silver stained gels were digitally colored. The Iinb1 gels were red and the Xinb3 gels were green. The gels were then superimposed. (Figure 2).

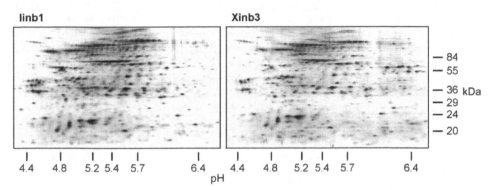

Fig. 1. Representative silver-stained 2D gels of individual *Daphnia magna* with distinct genotypes. Quadruplicate gels revealed 692 ± 14 spots in Iinb1 clone (left) and 687 ± 11 spots in Xinb3 clone (right).

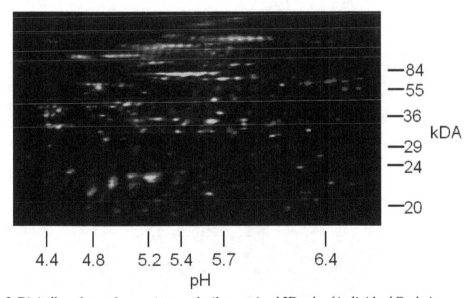

Fig. 2. Digitally enhanced, superimposed, silver-stained 2D gels of individual *Daphnia magna* with distinct genotypes. Red indicates spots unique to Iinb1, green indicates spots unique to Xinb3, yellow indicate spots shared by both genotypes.

3.2 Protein variation between individual *D. magna* of distinct phenotypes

We were also able to detect differences between individual *Daphnia* with distinct phenotypic differences. Individual *D. magna*, with and without ephippia, were isolated, proteins extracted and analyzed by 2-DE as described above. Silver staining detected an average of 524.5 ± 7.8 protein spots in 2D gels produced from single *D. magna*, with and without ephippia (figure 3). After normalization of the gel images based on total intensity, 386 spots were matched between the two gel images. Eighty-four spots showed a three-fold or greater difference in spot intensity. Fifty-five of these were more abundant in the parthenogenic (no ephippia) animal, while 29 were more abundant in the sexual animal. In addition, eleven protein spots were unique to the parthenogenic phenotype, while 49 protein spots were unique to the sexual phenotype. This demonstrates the feasibility of 2-DE and image analysis for the differentiation of *Daphnia* phenotypes isolated in the field as indicators of environmental variables. Other studies with parthenogenic and sexual *Daphnia carinata* were able to identify several proteins that were differentially expressed between the two phenotypes by 2-DE; however, 100's of animals were used (Zhang et al., 2006). It is interesting to note that using single animals, we discovered similar patterns of up-regulation in the parthenogenic individual in comparison to the sexual phenotype. While Zhang et al.'s goal was only to gain insight into the genes involved in the switch to sexual reproduction, our method, using single *Daphnia magna* with and without ephippia, allows the sampling of these candidate genes within a population.

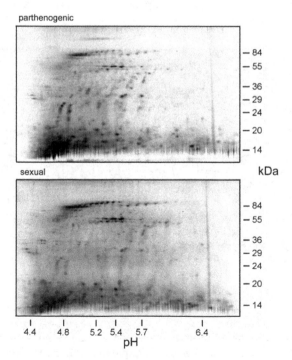

Fig. 3. Silver stained 2D gels of individual *Daphnia magna* with unique phenotypes.

3.3 Extending approach to single *D. pulex*

As *D. magna* is the largest of the *Daphnia* genus, we used PCT to extract protein from single *D. pulex* to demonstrate that our technique would be feasible with smaller individuals. Even when using the much smaller *Daphnia pulex,* we were able to detect approximately 900 spots from a single individual. It is reasonable to expect that differences in 2-DE between single *D. pulex* with phenotypic differences would also be detectable. As an indication of reproducibility of our method, we ran duplicate 2-DE of PCT- extracted proteins from 1, 2, 3, 4 and 5 *Daphnia pulex*. Figure 4 shows representative gels of 1, 2 or 3 *Daphnia pulex*. Figure 5 shows the number of protein spots detected and the standard deviation for *Daphnia pulex* gels of 1, 2, 3, 4 and 5 individuals. The low standard deviation indicates that PCT and 2-DE is an efficient and highly reproducible method of sample preparation and protein comparison.

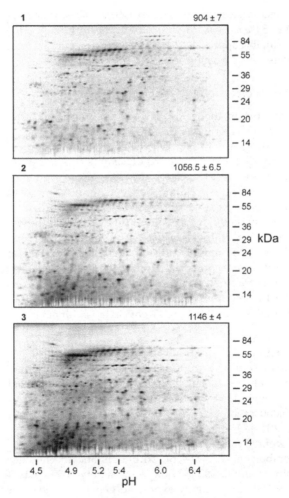

Fig. 4. Representative, silver stained 2D gels of 1, 2 & 3 *Daphnia pulex*.

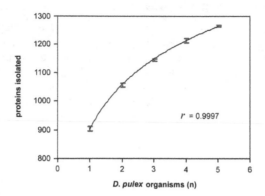

Fig. 5. Graph of detected protein spots from duplicate silver stained 2D gels of proteins extracted 1, 2, 3, 4 & 5 *Daphnia pulex*. The shape of the curve suggests that our method is an efficient and reproducible approach to protein extraction and the majority of the unique proteins are recovered from a single animal.

3.4 Protein functional diversity

To properly evaluate our method, it is necessary to understand the functional diversity of the proteins sampled. The proteins detected on any 2D gel will be biased towards the most abundant and the goal is to sample gene expression for proteins of diverse functions. To evaluate the likely diversity of proteins detectable and comparable by this method, we first generated a theoretical 2D gel for the 2000 most highly expressed genes in the *Daphnia pulex* genome. The correlation between mRNA expression levels and protein abundance is a debated topic; (for a brief review, see Greenbaum et al., 2003) however, recent studies have found a high correlation (Lu et al., 2007; Tuller et al., 2007). Using the recently completed draft of the *Daphnia pulex* genome, we found the top 2000 gene prediction models with the most Expressed Sequence Tag (EST) evidence using BLAST (Altschul et al., 1990). The theoretical pI and MW of these genes were calculated using the Compute pI/MW tool from ExPASy (Gasteiger et al., 2003) and the results graphed using Excel to create the theoretical 2D gel (Figure 6). Importantly, the pI range used on the actual gels (indicated by the rectangle) covers a significant portion of the predicted proteome.

To assay the functional diversity of these theoretical proteins, we utilized the 25 eukaryotic orthologous groups (KOGs) (Totusov et al., 2003) that were assigned to the *Daphnia pulex* genes as part of the genome sequence annotation project. Many *D. pulex* genes have no homology to any entries at NCBI; therefore, it is not surprising that of the approximately 30,000 predicted genes only 18,371 have been assigned to a KOG class (Colbourne et al., 2011) Of the 2000 most highly expressed genes, 298 had been assigned to a KOG class. Twenty four of the 25 KOG classes were represented, and only "coenzyme transport and metabolism" was absent. To understand the functional diversity that may be present on a single animal 2D gel of approximately 1000 spots, we looked at the KOG assignments of the 1000 most highly expressed genes. Ninety-seven have been assigned to a KOG class, with 21 of the 25 KOG classes being represented. Not represented were "coenzyme transport and metabolism", "secondary metabolites, biosynthesis, transport and catabolism", "nuclear structure" and "chromatin structure and dynamics" (Figure 7). In general, KOG classes that are well represented in the *Daphnia* genome are also well represented in the top 2000 of most highly expressed genes. As 84% of the KOG classes are represented in the top 1000 most highly expressed genes, we feel that a single animal 2D gel of approximately 1000 spots should represent a diverse sampling of biologically relevant proteins.

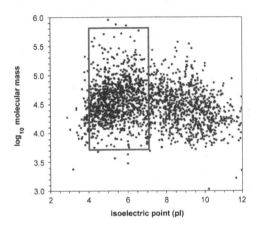

Fig. 6. Theoretical 2D gel of the top 2000 highly expressed *D. pulex* genes. The rectangle represents the separation range (MW and pI) of actual 2D gels (pH range 4 – 7).

pI range	number of theoretical proteins [a]	percent of total predicted	number of observed proteins	coefficient of variation
3-4	17	0.9	-	-
4-7	1027	51.4	904.0 ± 7.0 [b]	0.008
7-10	749	37.5	381.5 ± 26.5 [c]	0.069
> 10	207	10.4	-	-
total	2000	-	1285.5 ± 33.5	0.026

[a] Proteins predicted from sequence does not account for charge
[b] Spots detected in 2D gels from single D. pulex organisms.
[c] Spots detected in 2D gels of a multiple D. magna organisms.
(184.0 ± 2.6 mg or approximately 130 daphnids)

Table 1. Distribution of 2000 most abundant *Daphnia* proteins predicted from genome sequence

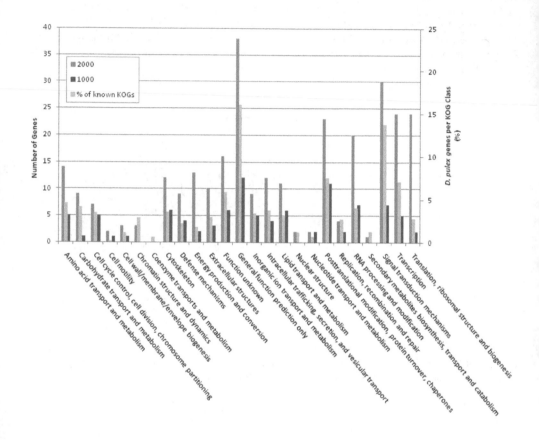

Fig. 7. Distribution of highly expressed *D. pulex* genes across 25 KOG classes. Blue and red bars indicate the number of genes in each class from the top 2000 and 1000, respectively, most highly expressed genes (Left-hand axis). Green bars indicate the total genes assigned to each KOG class as a percentage of the total number of genes (18,371) assigned to a KOG class (Right-hand axis). To further characterize the predicted protein gel, we compared it to our observed spot counts. Table 1 summarizes the number of predicted proteins and the number of actual proteins within specific pI ranges. Through the generation of several *D. magna* (both single animal and multiple-animal) 2D gels (not shown), we were able to visualize a total of 1285 protein spots. Seventy percent of these were observed in the 4-7 pI range, while the theoretical gel predicted 51% in this same range. It is important to note that the theoretical 2D does not account for post-translational modifications and was generated from *D. pulex* genes. Both of these factors will contribute to differences between the predicted (*D. pulex*) and observed (*D. magna*) number of proteins.

3.5 Basic proteins constituents of the *D. magna* proteome

Initially, single Daphnid extracts were analyzed on IPG strips with a pH range of 3-10. Since more than 80% of the proteins visualized by silver staining were in the pH 4-7 range, subsequent analyses were performed using IPGs pH 4-7. However, the theoretical 2D does predict a significant number of proteins (47%) in the basic range (7-10). Due to their relative low abundance, the visualization of very basic proteins (pH 7-10) in Daphnids required many more organisms. For this, 184 ± 3 mg of *D. magna* (approximately 135 organisms) were cultured under either normal or hypoxic conditions and processed in 1.3 mL of ProteoSOLVE IEF Reagent. The samples were concentrated two-fold by ultrafiltration, and IEF was performed on IPGs pH 7-10, followed by silver staining and image analysis. Silver staining detected 355 and 408 spots (gels not shown) in pH range 7-10 from *D. magna* cultured under normal or hypoxic conditions respectively, further illustrating the utility of 2-DE for detecting phenotypic differences influenced by specific environments.

4. Conclusion

Organisms live in ever changing environments. Understanding how individuals respond and adapt to these environments on a molecular level forms the basis for advances in personalized medicine and requires model systems with both well-developed ecological knowledge and molecular resources. The freshwater crustacean *Daphnia* now has these two requirements in place. We have demonstrated the ability of 2-DE to identify protein differences between single *Daphnia magna* with distinct genotypes (Iinb1 and Xinb3), distinct phenotypic differences (presence or absence of ephippia) and when cultured in different environments (normal or hypoxic conditions).

We predict that the detectable proteins on a single animal 2D gel, while biased towards the most abundant proteins, represent a functionally diverse set of proteins. This technique represents an important step to a greater understanding of individual variation of gene expression and is critical to advancing the field of EEFG. However, as the use of silver stain convolutes downstream mass spectrometry, the number of protein spots that can be confidently identified is significantly curtailed. The full potential of single animal gels will be realized with the development of comprehensive 2D maps of the *Daphnia* proteome.

5. References

Altschul SF, Gish W, Miller W, Myers EW, & Lipman DJ. 1990. Basic local alignment search tool. *Journal of Molecular Biology*, 215(3):403-410.

Colbourne J.K., *et. al.*, 2011. The Ecoresponsive Genome of *Daphnia pulex*. *Science*, Feb 4; 331(6017): 555-61.

Edmondson WT. 1987. *Daphnia* in Experimental Ecology: Notes on Historical Perspectives. In: Peters RH, R. dB, editors. *Daphnia. Pallanza: Memorie dell'Instituto Italiano di Idrobiologia*, 45:11-30.

Gasteiger E, Gattiker A, Hoogland C, Ivanyi I, Appel RD & Bairoch A. 2003. ExPASy: The proteomics server for in-depth protein knowledge and analysis. *Nucleic Acids Research*, 31(13):3784-3788.

Geiser HA, Hanneman AJ, & Reinhold V. 2002. HTP proteome-glycome analysis in Caenorhabditis elegans. *Glycobiology*, 12(10):650.

Greenbaum D, Colangelo C, Williams K, & Gerstein M. 2003. Comparing protein abundance and mRNA expression levels on a genomic scale. *Genome Biology*, 4(9):117.

Hairston NG, Jr., Holtmeier CL, Lampert W, Weider LJ, Post DM, Fischer JM, Caceres CE, Fox JA, & Gaedke U. 2001. Natural selection for grazer resistance to toxic cyanobacteria: Evolution of phenotypic plasticity?, *Evolution*, 55(11): 2203-2214.

Kilham SS, Kreeger DA, Lynn SG, Goulden CE, & Herrera L. 1998. COMBO: a defined freshwater culture medium for algae and zooplankton. *Hydrobiologia*, 377:147-159.

Limburg PA, & Weider LJ. 2002. `Ancient' DNA in the resting egg bank of a microcrustacean can serve as a palaeolimnological database. *Proceedings of the Royal Society of London:Series B*, 269:281-287

Lu P, Vogel C, Wang R, Yao X, & Marcotte EM. 2007. Absolute protein expression profiling estimates the relative contributions of transcriptional and translational regulation. *Nature Biotechnology*, 25(1):117-124.

National Science Board. 2000. Environmental Science and Engineering for the 21st Century: The role of the National Science Foundation. Report number: NSB 00-22.

Rottu J, Jansen B, Colson I, De Meester L, & Ebert D. 2010. The first-generation *Daphnia magna* linkage map. *BMC Genomics*, 11:508.

Smejkal GB, Li C, Robinson MH, Lazarev AV, Lawrence NP, & Chernokalskaya E. 2006a. Simultaneous reduction and alkylation of protein disulfides in a centrifugal ultrafiltration device prior to two-dimensional gel electrophoresis. *Journal of Proteome Research*, 5(4):983-987.

Smejkal GB, Robinson MH, Lawrence NP, Tao F, Saravis CA, & Schumacher RT. 2006b. Increased protein yields from Escherichia coli using pressure-cycling technology. *Journal of Biomolecular Techniques*, 17(2):173-175.

Smejkal GB, Witzmann FA, Ringham H, Small D, Chase SF, Behnke J, & Ting E. 2007. Sample preparation for two-dimensional gel electrophoresis using pressure cycling technology. *Analytical Biochemistry*, 363(2):309-311.

Tatusov RL, Fedorova ND, Jackson JD, Jacobs AR, Kiryutin B, Koonin EV, Krylov DM, Mazumder R, Mekhedov SL, Nikolskaya AN, Rao BS, Smirnov S, Sverdlov AV, Vasudevan S, Wolf YI, Yin JJ, & Natale DA. 2003. The COG database: an updated version includes eukaryotes. *BMC Bioinformatics*, 4:41.

Tuller T, Kupiec M, Ruppin E. 2007. Determinants of Protein Abundance and Translation Efficiency in S. cerevisiae. *PLoS Computational Biology*, 3(12):e248.

Zhang M-F, Zhao Y-L, & Zeng C. 2006. Differential protein expression between parthenogenetic and sexual female of *Daphnia* (*Ctenodaphnia*) *carinata*. *Acta Zoologica Sinica*, 52(5):916 -923.

5

Two Dimensional Gel Electrophoresis in Cancer Proteomics

Soundarapandian Kannan[1], Mohanan V. Sujitha[1],
Shenbagamoorthy Sundarraj[1] and Ramasamy Thirumurugan[2]
[1]Proteomics and Molecular Cell Physiology Lab
Department of Zoology, Bharathiar University, Coimbatore
[2]Department of Animal Science, Bharathidasan University, Tiruchirappalli
India

1. Introduction

Two-dimensional electrophoresis (2-DE) is a powerful and widely used method for the analysis of complex protein mixtures extracted from cells, tissues, or other biological samples. This technique sort's protein according to two independent properties in two discrete steps: the first-dimension step, isoelectric focusing (IEF), separates proteins according to their isoelectric points (pI); the second-dimension step, SDS-polyacrylamide gel electrophoresis (SDS-PAGE), separates proteins according to their molecular weights (Mr, relative molecular weight). Each spot on the resulting two-dimensional array corresponds to a single protein species in the sample. Thousands of proteins can thus be separated, and information such as the protein pI, the apparent molecular weight, and the amount of each protein obtained. The separation of proteins by 2-DE dates back to the 1950s. The first 2-DE technique was developed by Smithies and Poulik in 1956 and O'Farrell, 1975 and Klose, 1975 significantly modified this method to elucidate protein profile. In the original technique, the first-dimension separation was performed in carrier ampholyte-containing polyacrylamide gels cast in narrow tubes.

The power of 2-DE as a biochemical separation technique has been recognized virtually since its introduction. Its application, however, has become significant only in the last few years because of a number of developments. The introduction of immobilized pH gradients and Immobiline™ reagents brought superior resolution and reproducibility to first-dimension IEF. Based on this concept, Görg *et al.*, 1989 and Gorg, 1991 developed the currently employed 2-D technique, where carrier ampholyte-generated pH gradients have been replaced with immobilized pH gradients and tube gels replaced with gels supported by a plastic backing. New mass spectrometry techniques have been developed that allow rapid identification and characterization of very small quantities of peptides and proteins extracted from single 2-D spots. More powerful, less expensive computers and software are now available, rendering thorough computerized evaluations of the highly complex 2-D patterns economically feasible. Data about entire genomes (or substantial fractions thereof) for a number of organisms are now available, allowing rapid identification of the gene encoding a protein separated by 2-DE. The World Wide Web provides simple, direct access to spot pattern databases for the comparison of electrophoresis results and genome sequence databases for assignment of sequence information.

A large and growing application of 2-DE in "proteome analysis." Proteome analysis is "the analysis of the entire Protein complement expressed by a genome". The analysis involves the systematic separation, identification, and quantification of many proteins simultaneously from a single sample. Two-dimensional electrophoresis is used e due to its unparalleled ability to separate thousands of proteins simultaneously. Two-dimensional electrophoresis is also unique in its ability to detect post- and co-translational modifications, which cannot be predicted from the genome sequence. Applications of 2-DEinclude proteome analysis, cell differentiation, and detection of disease markers, monitoring therapies, drug discovery, cancer research, purity checks, and microscale protein purification.

"Proteomics" is the large-scale screening of the proteins of a cell, organism or biological fluid, a process, which requires stringently controlled steps of sample preparation, 2-DE, image detection and analysis, spot identification, and database searches. Moreover, Proteomics studies lead to the molecular characterization of cellular events associated with cancer progression, cellular signaling, developmental stages etc. Proteomics studies of clinical tumor samples have led to the identification of cancer-specific protein markers, which provide a basis for developing new methods for early diagnosis and early detection and clues to understand the molecular characterization of cancer progression. A keystone of conventional proteomics is high-resolution 2D gel electrophoresis followed by protein identification using mass spectrometry.

As a technique with high-flux and high resolution, proteomics has been widely applied in proteome analysis of tumors. The onset and development of the tissues and cells can be detected at the entire protein level through analyzing the differential expression of proteins. The combination of 2-DE and mass spectrometry can be used to identify differential proteins between tumor cells and normal original cells, and these differential proteins imply a large quantity of biological information. Some of the special proteins are special markers of tumors. The most consistently successful proteomic method is the combination of two-dimensional gel electrophoresis (2DE) for protein separation, visualization, and mass spectrometric (MS) identification of proteins using peptide mass fingerprints and tandem MS peptide sequencing.

The experiments form the basis of proteomics, and present significant challenges in data analysis, storage and querying. The core technology of proteomics is 2-DE. At present, there is no other technique that is capable of simultaneously resolving thousands of proteins in one separation procedure. The replacement of classical first-dimension carrier ampholyte pH gradients with well-defined immobilized pH gradients has resulted in higher resolution, improved inter-laboratory reproducibility, higher protein loading capacity, and an extended basic pH limit for 2-DE. With the increased protein capacity, micropreparative 2-DE has accelerated spot identification by mass spectrometry and Edman sequencing. The remarkable improvements in 2-DE resulting from immobilized pH gradient gels, together with convenient new instruments for IPG-IEF, will make critical contributions to advances in proteome analysis.

A comprehensive understanding of protein–protein interactions is an important step in our quest to understand how the information contained in a genome is put into action. Although a number of experimental techniques can report on the existence of a protein–

protein interaction, very few can provide detailed structural information. NMR spectroscopy is one of these, and in recent years several complementary NMR approaches, including residual dipolar couplings and the use of paramagnetic effects, have been developed that can provide insight into the structure of protein–protein complexes.

Two-dimensional gel electrophoresis for separation of complex protein samples coupled with mass spectrometry for protein identification has been used to analyze protein expression patterns for many sample types. Inherent in the use of this technique is information on not only full-length protein expression, but expression of modified, splice variant, cleavage product, and processed proteins. Any protein modification that leads to a change in overall protein charge and/or molecular weight (MW) will generate a different spot on the 2-DE. Modification specific staining can identify whether a specific post-translational modification is responsible for the shift, and mass spectrometry can potentially identify the source of isoelectric point (pI) and/or MW differences. Due to the lack of complete coverage for a protein's amino acid sequence using either matrix-assisted laser desorption/ionization mass spectrometry (MALDI-MS) or high-performance liquid chromatography (HPLC) tandem mass spectrometry (LC-MS/MS), there has been limited success in using MS to identify isoforms and post-translational modifications. While the theoretical MW is often slightly higher than the MW of the fully processed protein due to cleavage of signal and pro-peptides, there can also be post-translational modifications that increase the protein's gel MW. Thus, an exploration into the causes of the difference in the theoretical MW and the MW as seen in the gel can yield information about the state of the protein. When the gel MW of a given protein is significantly lower than the calculated weight, the gel spot represents a protein fragment. The extent to which proteins are present as fragments or variants in tissues and fluids has not been determined, but the combination of 2-DE, Western blotting, and mass spectrometry–based protein identification makes such analyses possible. Two-dimensional gel electrophoresis of human mammary tissue, followed by immune blotting, resulted in multiple spots at significantly differing molecular weights. The function of protein fragments is dependent on activation processes and localization properties. This Chapter will be critically analyzed as per the contents given in the synopsis with up-to-date informations.

2. Overview of experimental design

2.1 Experimental design

2.1.1 Sample preparation

Efficient and reproducible sample preparation methods are a key to successful 2-DE (Rabilloud 1999, Macri et al. 2000, Molloy 2000). Sample preparation methods range from extraction with simple solubilization solutions to complex mixtures of chaotropic agents, detergents, and reducing agents. Sample preparation can include enrichment strategies for separating protein mixtures into reproducible fractions.

An effective sample preparation procedure will:

1. Reproducibly solubilize proteins of all classes, including hydrophobic proteins
2. Prevent protein aggregation and loss of solubility during focusing
3. Prevent postextraction chemical modification, including enzymatic or chemical degradation of the protein sample

4. Remove or thoroughly digest nucleic acids and other interfering molecules
5. Yield proteins of interest at detectable levels, which may require the removal of interfering abundant proteins or nonrelevant classes of proteins

Most protein mixtures will require some experimentation to determine optimum conditions for 2-D PAGE. Variations in the concentrations of chaotropic agents, detergents, ampholytes, and reducing agents can dramatically affect the 2-D pattern.

2.1.2 Solubilization

Solubilization of proteins is achieved by the use of chaotropic agents, detergents, reducing agents, buffers, and ampholytes. These are chosen from a small list of compounds that meet the requirements, both chemically and electrically, for compatibility with the technique of IEF in IPG strips. The compounds chosen must not increase the ionic strength of the solution, to allow high voltages to be applied during focusing without producing high currents. Thorough discussion of solubilization methods, including new variations, can be found in several books (Pennington and Dunn 2001, Rabilloud 2000).

2.1.3 Chaotropic agents

Urea is the most commonly used chaotropic agent in sample preparation for 2-D PAGE. Thiourea can be used to help solubilize many otherwise intractable proteins. Urea and thiourea disrupt hydrogen bonds and are used when hydrogen bonding causes unwanted aggregation or formation of secondary structures that affect protein mobility. Urea is typically used at 8 M. Thiourea is weakly soluble in water, but is more soluble in high concentrations of urea, so a mixture of 2 M thiourea and 5–8 M urea is used when strongly chaotropic conditions are required (Rabilloud 1998).

2.1.4 Detergents

Detergents are added to disrupt hydrophobic interactions and increase solubility of proteins at their pI. Detergents must be nonionic or zwitterionic to allow proteins to migrate according to their own charges. Some proteins, especially membrane proteins, require detergents for solubilization during isolation and to maintain solubility during focusing. Ionic detergents such as SDS are not compatible with IEF, but can be used with concentrated samples in situations where the SDS can be unbound from the proteins by IEF-compatible detergents that compete for binding sites. Nonionic detergents such as octylglucoside, and zwitterionic detergents such as CHAPS and its hydroxyl analog, CHAPSO, can be used. CHAPS, CHAPSO, or octylglucoside concentrations of 1–2% are recommended (Rabilloud 1999). New detergents are emerging that have great potential in proteomics, including SB 3-10 and ASB-14 (Chevallet et al. 1998). Some proteins may require detergent concentrations as high as 4% for solubility (Hermann et al. 2000).

2.1.5 Carrier ampholytes

A fundamental challenge with IEF is that some proteins tend to precipitate at their pI. Even in the presence of detergents, certain samples may have stringent salt requirements to maintain the solubility of some proteins. Salt should be present in a sample only if it is an absolute requirement, and then only at a total concentration less than 40 mM. This is problematic since any salt included will be removed during the initial high-current stage of focusing. Salt limits the voltage that can be achieved without producing high current, increasing the time required for focusing. Proteins that require salt for solubility are subject to precipitation once the salt is removed. Carrier ampholytes sometimes help to counteract insufficient salt in a sample. They are usually included at a concentration of ≤0.2% (w/v) in sample solutions for IPG strips. High concentrations of carrier ampholytes will slow down IEF until they are focused at their pI, since they carry current and hence limit voltage. Some researchers have increased resolution by varying the ampholyte composition.

2.1.6 Reducing agents

Reducing agents such as dithiothreitol (DTT) or tributylphosphine (TBP) are used to disrupt disulfide bonds. Bond disruption is important for analyzing proteins as single subunits. DTT is a thiol reducing agent added in excess to force equilibrium toward reduced cysteines. At 50 mM it is effective in reducing most cystines, but some proteins are not completely reduced by this treatment. If the concentration of DTT is too high it can affect the pH gradient since its pKa is around 8. TBP is a much more effective reducing agent than DTT. It reacts to reduce cystines stoichiometrically at low millimolar concentrations (Herbert *et al.* 1998). It is chemically more difficult to handle than DTT, but Bio-Rad has solved this problem by supplying it in a form safe for shipping and lab use.

2.1.7 Prefractionation

Reducing the complexity of the sample loaded on a 2-D gel can increase the visibility of minor proteins. Techniques such as differential extraction (Molloy *et al.* 1998), subcellular fractionation (Taylor *et al.* 2000, Morel *et al.* 2000), chromatography (Fountoulakis *et al.* 1999), or prefocusing in a preparative IEF device such as the Rotofor® system (Masuoka *et al.* 1998, Nilsson *et al.* 2000) have been used to reduce the complexity of samples.

2.1.8 Removal of albumin and IgG

The isolation of lower-abundance proteins from serum or plasma is often complicated by the presence of albumin and immunoglobulin G (IgG). Albumin is the most abundant protein (~60–70%) in serum and IgG is the second most abundant protein (10–20%). These two proteins effectively act as major contaminants, masking the presence of many co-migrating proteins, as well as limiting the amount of total serum protein that can be resolved on a 2-D gel. In the past, removal of albumin and IgG usually required separate chromatography methodologies for each of the two species. Now, Bio-Rad's Aurum™ serum protein kit allows selective binding and simultaneous removal of both albumin and IgG from serum or plasma samples prior to 2-DE.

2.1.9 Sequential extraction

One method for reducing sample complexity is the basis of the ReadyPrep™ sequential extraction kit. This protocol takes advantage of solubility as a third independent means of protein separation. Proteins are sequentially extracted in increasingly powerful solubilizing solutions. More protein spots are resolved by applying each solubility class to a separate gel, thereby enriching for particular proteins while simplifying the 2-D patterns in each gel. An increase in the total number of proteins is detected using this approach (Molloy *et al.* 1998).

2.1.10 Removal of DNA

The presence of nucleic acids, especially DNA, interferes with separation of proteins by IEF. Under denaturing conditions, DNA complexes are dissociated and markedly increase the viscosity of the solution, which inhibits protein entry and slows migration in the IPG. In addition, DNA binds to proteins in the sample and causes artifactual migration and streaking. The simplest method for removal of DNA is enzymatic digestion. Adding endonuclease to the sample after solubilization at high pH (40 mM Tris) allows efficient digestion of nucleic acids while minimizing the action of contaminating proteases. The advantage of the endonuclease method is that sample preparation can be achieved in a single step, by the addition of the enzyme prior to loading the first-dimension IPG.

2.1.11 Protein load

The amount of protein applied to an IPG strip can range from several micrograms to 1 mg or more (Bjellqvist *et al.* 1993a). Some of the factors affecting the decision of how much protein to load are:

a.   Subsequent analysis. Enough of the protein of interest must be loaded for it to be analyzed. With the Ready Gel® mini system (7 cm IPG), detection of moderately abundant proteins in complex mixtures with Coomassie Brilliant Blue R-250 dye requires on the order of 100 µg total protein. With the same load, many low-abundance proteins can be detected with more sensitive stains such as silver or SYPRO Ruby protein gel stain.

b.   The purpose of the gel. If the gel is being run solely for the sake of getting a good image of well-resolved proteins for comparative studies or for publication, the protein load would be the minimum amount that is stainable.

c.   The abundance of the proteins of interest. If the purpose is to study low-copy-number proteins, a large mass of a protein mixture might be loaded (Wilkins *et al.* 1998).

d.   The complexity of the sample. A highly complex sample containing many proteins of widely varying concentrations might require a compromise load so that high-abundance proteins don't obscure low-abundance proteins. By enriching a sample for specific types of proteins using prefractionation techniques, each individual protein will be at a higher relative concentration, which means that enough material can be loaded for detection of low-abundance constitutents.

e.   pH range of IPG strip. In general, larger amounts of total protein can be loaded on a narrow-range IPG strip. Only the proteins with a pI within the strip pH range will be represented within the second-dimension gel.

2.2 The first dimension: Isoelectric Focusing (IEF)

2.2.1 Isoelectric point (pI)

Differences in proteins' pI are the basis of separation by IEF. The pI is defined as the pH at which a protein will not migrate in an electric field and is determined by the number and types of charged groups in a protein. Proteins are amphoteric molecules. As such, they can carry positive, negative, or zero net charge depending on the pH of their local environment. For every protein there is a specific pH at which its net charge is zero; this is its pI. Proteins show considerable variation in pI, although pI values usually fall in the range of pH 3–12, with the majority falling between pH 4 and pH 7. A protein is positively charged in solution at pH values below its pI and negatively charged at pH values above its pI.

2.2.2 IEF

When a protein is placed in a medium with a pH gradient and subjected to an electric field, it will initially move toward the electrode with the opposite charge. During migration through the pH gradient, the protein will either pick up or lose protons. As it migrates, its net charge and mobility will decrease and the protein will slow down. Eventually, the protein will arrive at the point in the pH gradient equal to its pI. There, being uncharged, it will stop migrating. If this protein should happen to diffuse to a region of lower pH, it will become protonated and be forced back toward the cathode by the electric field. If, on the other hand, it diffuses into a region of pH greater than its pI, the protein will become negatively charged and will be driven toward the anode. In this way, proteins condense, or are focused, into sharp bands in the pH gradient at their individual characteristic pI values. Focusing is a steady-state mechanism with regard to pH. Proteins approach their respective pI values at differing rates but remain relatively fixed at those pH values for extended periods. By contrast, proteins in conventional electrophoresis continue to move through the medium until the electric field is removed. Moreover, in IEF, proteins migrate to their steady state positions from anywhere in the system.

2.2.3 IPG strips

A stable, linear, and reproducible pH gradient is crucial to successful IEF. IPG strips offer the advantage of gradient stability over extended focusing runs (Bjellqvist *et al.* 1982). IPG strips are much more difficult to cast than carrier ampholyte gels (Righetti 1983); however, IPG strips are commercially available, for example as ReadyStrip™ IPG strips. pH gradients for IPG strips are created with sets of acrylamido buffers, which are derivatives of acrylamide containing both reactive double bonds and buffering groups. The general structure is $CH_2=CH–CO–NH–R$, where R contains either a carboxyl [–COOH] or a tertiary amino group (e.g., N (CH3)2). These acrylamide derivatives are covalently incorporated into polyacrylamide gels at the time of casting and can form almost any conceivable pH gradient (Righetti 1990).

2.2.4 Choice of pH gradient ranges

Use of broad-range strips (pH 3–10) allows the display of most proteins in a single gel. With narrow-range and micro-range overlapping gradient strips, resolution is increased by expanding a small pH range across the entire width of a gel. Since many proteins are focused in the middle of the pH range 3–10, some researchers use nonlinear (NL) gradients to better resolve proteins in the middle of the pH range and to compress the width of the extreme pH ranges at the ends of the gradients. However, overlapping narrow-range and micro-range linear IPG strips can outperform a nonlinear gradient and display more spots per sample. This result is due to the extra resolving power from use of a narrower pI range per gel. Use of overlapping gradients also allows the ability to create "cyber" or composite gels by matching spots from the overlapping regions using imaging software.

2.2.5 IPG strip (2-D array) size

The 17 cm IPG strips and large-format gels have a large area to resolve protein spots; however, they take a long time to run. Using a mini system instead of, or as a complement to, a large gel format can provide significant time savings. A mini system is perfect for rapid optimization of sample preparation methods. Switching to a large format then allows thorough assessment of a complex sample and identification of proteins of interest. In many cases, a mini system consisting of narrow-range IPG strips can then be used to focus in on the proteins of interest. Throughput of the 2-D process is a consideration in choosing gel size. The ability to cast or run 12 gels at a time in any of 3 size formats is very useful in gathering proteomic results. In some cases, mini systems (7 cm ReadyStrip IPG strips with Mini-PROTEAN® 3 format gels, or 11 cm ReadyStrip IPG strips with Criterion™ precast gels) can completely replace large 2-D systems, providing speed, convenience, and ease in handling. The availability of narrow and micro overlapping pH-range ReadyStrip IPG strips can increase the effective width of pI resolution more than 5-fold after accounting for overlapping regions. When 3 narrow-range overlapping ReadyStrip IPG strips are used with the Criterion system, the resolution in the first dimension is increased from 11 to 26 cm. When micro-range strips are used, the resolution in the first dimension is expanded from 11 to 44 cm.

2.2.6 Estimation of pI

The pI of a protein can be estimated by comparing the position of the protein spot of interest to the position of known proteins or standards separated across the same pH gradient (Bjellqvist et al. 1993b, Garfin 2000). ReadyStrip IPG strips contain linear gradients, so the pI of an unknown protein can be estimated by linear interpolation relative to proteins of known pI.

2.2.7 Sample application

Commercial IPG strips are dehydrated and must be rehydrated to their original gel thickness (0.5 mm) before use. This allows flexibility in applying sample to the strips. There are 3 methods for sample loading: passive in-gel rehydration with sample, active in-gel rehydration with sample, or cup loading of sample after IPG rehydration. Introducing the sample while the strips are rehydrating is the easiest method. In some specific instances, it is best to rehydrate the strips and then apply sample through sample cups while current is applied.

2.2.8 Sample application during rehydration

For both active and passive rehydration methods, the sample is introduced to the IPG strip at the time of rehydration. As the strips hydrate, proteins in the sample are absorbed and distributed over the entire length of the strip (Sanchez et al. 1997). In the case of active rehydration, a very low voltage is applied during rehydration of the strips. Proteins enter the gel matrix under current as well as by absorption. The PROTEAN IEF cell has preprogrammed methods designed to accommodate active rehydration. Active rehydration is thought to help large proteins enter the strip by applying electrical "pull". Because the voltage is applied before all the solution and proteins are absorbed into the gel, the pH of a protein's environment will be the pH of the rehydration buffer, and the protein will move according to its mass-to charge ratio in that environment. Thus, small proteins with a higher mobility have a higher risk of being lost from the strip. With passive rehydration, proteins enter the gel by absorption only. This method allows efficient use of equipment since strips can be rehydrated in sample rehydration trays while other samples are being focused in the IEF cell.

Whether the strips are hydrated actively or passively, it is very important that they be incubated with sample for at least 11 hr prior to focusing. This allows the high molecular weight proteins time to enter the gel after the gel has become fully hydrated and the pores have attained full size. These sample application methods work because IEF is a steady-state technique, so proteins migrate to their pI independent of their initial positions.

The advantages of this approach are:

a. Sample application is simple (Görg et al. 1999)
b. Sample application during rehydration avoids the problem of sample precipitation, which often occurs with cup loading (Rabilloud 1999)
c. Shorter focusing times can be used because the sample proteins are in the IPG strip prior to IEF
d. Very large amounts of protein can be loaded using this method

2.2.9 Sample application by cup loading

Cup loading can be beneficial in the following cases (Cordwell et al. 1997, Görg et al. 2000):

• When samples contain high levels of DNA, RNA, or other large molecules, such as cellulose
• For analytical serum samples that have not been treated to remove albumin
• When running basic IPG strips; e.g., pH 7–10
• For samples that contain high concentrations of glycoproteins

Because of its relative difficulty and tendency toward artifacts, cup loading should be avoided if possible. When loading the protein sample from a cup, the IPG strips must be rehydrated prior to sample application. The IPG strips can be rehydrated in a variety of ways.

The rehydration tray is recommended although IPG strips are often rehydrated in 1 or 2 ml pipettes that have been sealed at both ends with Para film. Sample volumes of up to 100 µl can be loaded later onto each gel strip using a sample cup.

2.2.10 Power conditions and resolution in IEF

During an IEF run, the electrical conductivity of the gel changes with time, especially during the early phase. When an electrical field is applied to an IPG at the beginning of an IEF run, the current will be relatively high because of the large number of charge carriers present.

As the proteins and ampholytes move toward their pIs, the current will gradually decrease due to the decrease in the charge on individual proteins and carrier ampholytes. The pH gradient, strip length, and the applied electrical field determine the resolution of an IEF run. According to both theory and experiment, the difference in pI between two adjacent IEF-resolved protein bands is directly proportional to the square root of the pH gradient and inversely proportional to the square root of the voltage gradient at the position of the bands (Garfin 2000). Thus, narrow pH ranges and high-applied voltages yield high resolution in IEF. The highest resolution can be achieved using micro-range IPG strips and an electrophoretic cell, such as the PROTEAN IEF cell, capable of applying high voltages. IEF runs should always be carried out at the highest voltage compatible with the IPG strips and electrophoretic cell. However, high voltages in electrophoresis are accompanied by large amounts of generated heat. The magnitude of the electric field that can be applied and the ionic strength of the solutions that can be used in IEF are limited. Thin gels are better able to dissipate heat than thick ones and are therefore capable of withstanding the high voltage that leads to higher resolution. Also, at the completion of focusing, the current drops to nearly zero since the carriers of the current have stopped moving. The PROTEAN IEF cell is designed to provide precise cooling, allowing the highest possible voltages to be applied. (A default current limit of 50 µA per strip is intended to minimize protein carbamylation reactions in urea sample buffers. This limit can be increased to 99 µA per strip.)

Consistent and reproducible focusing requires that the time integral of voltage (volt-hours) be kept consistent. It is usually necessary to program IEF runs to reach final focusing voltages in stages. This approach clears ionic constituents in the sample from the IPG strips while limiting electrical heating of the strips. The PROTEAN IEF cell allows for multistep runs at durations set by time or volt-hours. The number of volt-hours required to complete a run must be determined empirically. A more complex sample in terms of number of proteins or even a different sample buffer might require increased volt-hours. The time needed to achieve the programmed volt-hours depends on the pH range of the IPG strip used as well as sample and buffer characteristics. If different strips are run at the same time, the electrical conditions experienced by individual strips will be different, perhaps exposing some strips to more current than desired, since the total current limit is averaged over all strips in a tray.

2.3 The second dimension: SDS-PAGE

2.3.1 Protein separation by molecular weight (MW)

Second-dimension separation is by protein mass, or MW, using SDS-PAGE. The proteins resolved in IPG strips in the first dimension are applied to second-dimension gels and separated by MW perpendicularly to the first dimension. The pores of the second-dimension gel sieve proteins according to size because dodecyl sulfate coats all proteins essentially in proportion to their mass. The net effect is that proteins migrate as ellipsoids with a uniform negative chargeto- mass ratio, with mobility related logarithmically to mass (Garfin 1995).

2.3.2 Gel composition

Homogeneous (single-percentage acrylamide) gels generally give excellent resolution of sample proteins that fall within a narrow MW range. Gradient gels have two advantages:

they allow proteins with a wide range of MW to be analyzed simultaneously, and the decreasing pore size along the gradient functions to sharpen the spots.

2.3.3 Single-percentage gels

The percentage of acrylamide, often referred to as %T (total percentage of acrylamide plus crosslinker) determines the pore size of a gel. Most protein separations use 37.5 parts acrylamide to 1 part bis-acrylamide (bis). Some researchers substitute piperazine bis-acrylamide (PDA), which can reduce silver staining background and give higher gel strength. If the total percentage of acrylamide plus crosslinker is higher, the smaller is the pore size. A suitable %T can be estimated from charts of mobility for proteins of different MW.

2.3.4 Gradient gels

Gradient gels are cast with acrylamide concentrations that increase from top to bottom so that the pore size decreases as proteins migrate further into the gels. As proteins move through gradient gels from regions of relatively large pores to regions of relatively small pores, their migration rates slow. Small proteins remain in gradient gels much longer than they do in single-percentage gels that have the same average %T, so both large and small molecules may be resolved in the same gel. This makes gradient gels popular for analysis of complex mixtures that span wide MW ranges. A gradient gel, however, cannot match the resolution obtainable with a properly chosen single concentration of acrylamide. A good approach is to use gradient gels for estimates of the complexities of mixtures. A proteomics experiment might start out with an 8–16%T gradient for global comparison. After interesting regions of the 2-D array have been identified, a new set of single-percentage gels may be run to study a particular size range of proteins. It is simplest and often most cost and labor effective to purchase commercially available precast gradient gels.

2.3.5 Precast gels

High-quality precast gels are preferred for high-throughput applications. They provide savings in time and labor, and the precision-poured gradients result in reproducibility among runs. Precast gels differ from handcast gels in that they are cast with a single buffer throughout and without SDS. During storage, different buffers in the stacking and resolving gels would mingle without elaborate means to keep them separate, and thus have no practical value. In addition, because the sample contains SDS, and the dodecyl sulfate ion in the cathode buffer moves faster than the proteins in the gel, keeping them saturated with the detergent, precast gels are made without SDS.

2.3.6 Transition from first to second dimension

The transition from first-dimension to second-dimension gel electrophoresis involves two steps: equilibration of the resolved IPG strips in SDS reducing buffer, and embedding of the strip on the top of the second-dimension gel. Proper equilibration simultaneously ensures that proteins are coated with dodecyl sulfate and that cysteines are reduced and alkylated. The equilibrated IPG strips are placed on top of the gel and fixed with molten agarose solution to ensure good contact between the gel and the strip.

2.3.7 Second dimension and high throughput

Since the first dimension can be run in batches of 12–24 strips at a time, it is desirable to run the same number of samples in the second dimension. Precast gels ensure high reproducibility among samples and help reduce the work involved in running large numbers of samples. Alternatively, gels can be hand cast 12 at a time under identical conditions with multi-casting chambers. The Dodeca cells save time, space, and effort, and help to ensure that gels are run under the same electrical conditions for highest throughput and reproducibility.

2.3.8 MW estimation

The migration rate of a polypeptide in SDS-PAGE is inversely proportional to the logarithm of its MW. The larger the polypeptide, the more slowly it migrates in a gel. MW is determined in SDS-PAGE by comparing the migration of protein spots to the migration of standards. Plots of log MW versus the migration distance are reasonably linear. Gradient SDS-PAGE gels can also be used to estimate MW. In this case, log MW is proportional to log (%T). With linear gradients, %T is proportional to distance migrated, so the data can be plotted as log MW vs. log (migration distance). Standard curves are actually sigmoid. The apparent linearit of a standard curve may not cover the full MW range for a given protein mixture in a particular gel. However, log MW varies sufficiently slowly to allow accurate MW estimates to be made by interpolation, and even extrapolation, over relatively wide ranges (Garfin 1995). Mixtures of standard proteins with known MW are available from Bio-Rad in several formats for calibrating the migration of proteins in electrophoretic gels. Standards are available unstained, prestained, or with tags for development with various secondary reagents (useful when blotting). Standards can be run in a reference well, attached to the end of a focused IPG strip by filter paper, or directly embedded in agarose onto the second-dimension gel

2.4 Detection of proteins in gels

2.4.1 Guidelines for detection of proteins in gels

Gels are run for either analytical or preparative purposes. The intended use of the gel determines the amount of protein to load and the means of detection. It is most common to make proteins in gels visible by staining them with dyes or metals. Each type of protein stain has its own characteristics and limitations with regard to the sensitivity of detection. Sometimes proteins are transferred to membranes by western blotting to be detected by immunoblotting, glycoprotein analysis, or total protein stain. If the purpose of gel electrophoresis is to identify low-abundance proteins (e.g., low-copy-number proteins in a cell extract, or contaminants in a purification scheme), then a high protein load (0.1–1 mg/ml) and a high-sensitivity stain, such as silver or a fluorescent stain, should be used (Corthals et al. 2000). When the intention is to obtain enough protein for use as an antigen or for sequence analysis, then a high protein load should be applied to the gel and the proteins visualized with a staining procedure that does not fix proteins in the gel. Quantitative comparisons require the use of stains with broad linear ranges of detection. The sensitivity that is achievable in staining is determined by: 1) the amount of stain that binds to the proteins; 2) the intensity of the coloration; 3) the difference in coloration between stained proteins and the residual background in the body of the gel (the signal-to-noise ratio). Unbound stain molecules can be washed out of the gels without removing much stain from the proteins. All stains interact differently with different proteins (Carroll *et al*. 2000). No stain will universally stain all proteins in a gel in proportion to their mass. The only observation that seems to hold for most stains is that they interact best with basic amino acids. For critical analysis, replicate gels should be stained with two or more different stains. Of all stains available, colloidal Coomassie Blue (Bio-Safe™ Coomassie) appears to stain the broadest spectrum of proteins. It is instructive, especially with 2-D PAGE gels, to stain a colloidal Coomassie Blue-stained gel with silver or to stain a fluorescently stained gel with colloidal Coomassie Blue or silver. Very often, this double staining procedure will show a few differences between the protein patterns. It is most common to stain gels first with Coomassie Blue or a fluorescent stain, and then restain with silver. However, the order in which the stains are used does not seem to be important, as long as the gels are washed well with high-purity water between stains.

2.4.2 Coomassie blue staining

Coomassie Brilliant Blue R-250 is the most common stain for protein detection in polyacrylamide gels. Coomassie Brilliant Blue R-250 and G-250 are wool dyes that have been adapted to stain proteins in gels. The "R" and "G" designations indicate red and green hues, respectively. Coomassie R-250 requires on the order of 40 ng of protein per spot for detection. Absolute sensitivity and staining linearity depend on the proteins being stained. The staining solution also fixes most proteins in gels. Bio-Safe Coomassie stain is made with Coomassie Brilliant Blue G-250. Bio-Safe Coomassie stain is a ready-to-use, single-reagent protein stain. Sensitivity can be down to 10 ng, and greater contrast is achieved by washing the gel in water after staining. Used stain can be disposed of as nonhazardous waste and the procedure does not fix proteins in the gel.

2.4.3 SYPRO ruby fluorescent staining

SYPRO Ruby protein gel stain has desirable features that make it popular in high-throughput laboratories. It is an endpoint stain with little background staining (high signal-to noise characteristics) and it is sensitive and easy to use. SYPRO Ruby protein stain does not detect nucleic acids. SYPRO Ruby protein stain is sensitive to 1–10 ng and can be linear over 3 orders of magnitude. It is compatible with high through put protocols and downstream analysis, including mass spectrometry and Edman sequencing (Patton 2000). It also allows detection of glycoproteins, lipoproteins, low MW proteins, and metalloproteins that are not stained well by other stains. This fluorescent stain is easily visualized with simple UV or blue-light transilluminators, as well as by the Molecular Imager FX™ Pro Plus multiimager and VersaDoc™ imaging systems.

2.4.4 Silver staining

Two popular methods for silver staining are recommended for 2-D analysis. They are based on slightly different chemistries but have similar sensitivities for protein. Bio-Rad's silver stain kit, based on the method of Merril *et al*. (1981), can be as much as 100 times more sensitive than Coomassie Blue R-250 dye staining and allows visualization of heavily glycosylated proteins in gels. Protein spots containing 10–100 ng of protein can be easily seen. Proteins in gels are fixed with alcohol and acetic acid, then oxidized in a solution of potassium dichromate in dilute nitric acid, washed with water, and treated with silver nitrate solution. Silver ions bind to the oxidized proteins and are subsequently reduced to metallic silver by treatment with alkaline formaldehyde. Color development is stopped with acetic acid when the desired staining intensity has been achieved. This method is not compatible with mass spectroscopic analysis since the oxidative step changes protein mass. The Silver Stain Plus stain from Bio-Rad requires only one simultaneous staining and development step and is based on the method developed by Gottlieb and Chavko (1987). Proteins are fixed with a solution containing methanol, acetic acid, and glycerol, and washed extensively with water. The gels are then soaked in a solution containing a silverammine complex bound to colloidal tungstosilicic acid. Silver ions transfer from the tungstosilicic acid to the proteins in the gel by means of an ion exchange or electrophilic process. Formaldehyde in the alkaline solution reduces the silver ions to metallic silver to produce the images of protein spots. The reaction is stopped with acetic acid when the desired intensity has been achieved. Because silver ions do not accumulate in the bodies of gels, background staining is light. Since this method lacks an oxidizing step, visualization of heavily glycosylated proteins and lipoproteins can be less sensitive than with the Merril stain. This method is better for use in proteomics when the end goal is identification by mass spectrometric analysis.

2.5 Image acquisition and analysis

2.5.1 Image acquisition instruments

Before 2-D gels can be analyzed with an image evaluation system, they must be digitized. The most commonly used devices are camera systems, densitometers, phosphor imagers, and fluorescence scanners. All of Bio-Rad's imaging systems are seamlessly integrated with PDQuest™ software, and they can export and import images to and from other software via TIFF files.

2.5.2 Densitometry

Densitometers compare the intensity of a light beam before and after attenuation by a sample. The GS-800™ calibrated imaging densitometer has been customized for analysis of gels, autoradiograms, and blots. The transmittance and true reflectance capabilities allow accurate scans of samples that are either transparent (gels and film) or opaque (blots). It provides high-quality imaging to resolve close spots and a variable resolution feature to preview and crop images. Wet 2-D gels may be scanned with red, green, and blue color CCD technology on the watertight platen.

2.5.3 Storage phosphor and fluorescence scanners

Digitization of 2-D gels stained with fluorescent dyes or radioactive compounds requires specific imaging systems (Patton 2000). The Molecular Imager FX™ Pro Plus system is flexible and expandable. 2-D gels of radiolabeled proteins can be imaged using a Kodak phosphor screen more rapidly and accurately than with film. Popular proteomic fluorescent stains, including SYPRO Ruby protein gel and blot stains and SYPRO Orange protein gel stain can be imaged with single-color and multicolor fluorescence via direct laser excitation. This system permits detection of almost any fluorophore that is excited in the visible spectrum. The internal laser and external laser options allow optimal excitation of single-color or multicolor fluorescent samples. Computer-controlled, user-accessible filter wheels have eight filter slots, allowing detection of many multicolor combinations of dyes (Gingrich *et al.* 2000).

2.5.4 Computer-assisted image analysis of 2-D electrophoretic gels

Computer-assisted image analysis software is an indispensable tool for the evaluation of complex 2-D gels. It allows:

a.   Storage and structuring of large amounts of collected experimental image data
b.   Rapid and sophisticated analysis of experimental information
c.   Supplementation and distribution of data among labs
d.   Establishment of 2-D-protein data banks

Image analysis systems deliver error-fee comprehensive qualitative and quantitative data from a large number of 2-D gels (Miller 1989). PDQuest software from Bio-Rad is a popular analysis tool. Gel analysis of digitized gel images includes spot detection, spot quantitation, gel comparison, and statistical analysis. PDQuest software has the further advantage of seamless integration with any of Bio-Rad's image acquisition instruments, as well as the ability to control the ProteomeWorks™ spot cutter. The advanced annotation feature can be used to label spots with text, URL links, document links, or mass spectrometry data.

2.5.5 Spot detection and spot quantitation

Before the software automatically detects the protein spots of a 2-D gel, the raw image data are corrected and the gel background is subtracted. The process is executed with simple menus and "wizards." PDQuest software models protein spots mathematically as 3-D Gaussian distributions and uses the models to determine absorption maxima. This enables automatic detection a resolution of merged spots. Following this procedure, spot intensities are obtained by integration of the Gaussian function. The mathematical description of the spots is used both for data reduction and for increasing evaluation speed, since reevaluation of data after an image change takes only fractions of a second. The hit rate of automatic spot detection is highly dependent on the quality of the 2-D gels. Correction capabilities of PDQuest software can be used to add undetected spots to the list of spots or to delete spots that arise from gel artifacts.

2.5.6 Gel comparison

The next step in 2-D gel evaluation is the identification of proteins that are present in all gels of a series. This task is made difficult primarily because of inherent irreproducibility in gels, which affects the positions of spots within a gel series. Gel analysis software must detect minor shifts in individual spot position within the gel series. Many software packages for automatic gel comparison are created with the assumption that the relative positions of spots are altered only slightly relative to each other, and allocate the spots on this basis. Prior to automatic gel comparison, PDQuest software selects the best 2-D gel of a gel series as a reference or standard gel and compares all other 2-D gels to this gel. Proteins in a gel series that are not present in the reference gel are added manually so that the reference gel will include all proteins of a gel series.

Before the software can detect and document matching of different spots, a number of landmarks, or identical spots in the gel series, must be manually identified. The landmarking tool speeds the process by making "best guess" assignments of landmark spots to images in the gel series. With PDQuest software, it is possible to simultaneously display up to 100 enlarged details of 2-D gels on the screen. This simultaneous display of all 2-D gels of a test series enables rapid and error-free determination of the fixed points. Using the landmarks, the image analysis software first attempts to compare all spots lying very near these fixed points and then uses the matched spots as starting points for further comparisons. Thus, the entire gel surface is systematically investigated for the presence or absence of matching spots in a gel series. The results of the automatic gel comparison require verification, as does automatic spot detection.

Two tools assist this verification process in PDQuest: Either identical protein spots are labeled with matching letters and allocated section by section, or the deviations in the spot positions of a particular 2-D gel can be displayed as lines that show spot shifts in comparison to the reference gel.

2.5.7 Data analysis

With PDQuest software, all gels of an experiment are viewed as a unit. To compare gels from different experiments, the reference images are compared. In such comparisons, each spot is automatically assigned a number so that identical spots have identical numbers. Experimental data can also be analyzed statistically both parametric and nonparametric tests are available.

2.6 Identification and characterization of 2-D protein spots

2.6.1 Sequence data from 2-D gels

2-DE has the virtually unique capability of simultaneously displaying several hundred gene products. 2-D gels are an ideal starting point for protein chemical identification and characterization. Peptide mass fingerprint or sequence data can be derived following 2-DE with mass spectrometry or amino acid sequence analysis (Eckerskorn *et al.* 1988, Ducret *et al.* 1996). The sensitivity of currently available instruments makes 2-DE an efficient "preparative" analytical method. Most current protein identification depends on mass spectrometry of proteins excised from gels or blots.

2.6.2 Integration of image analysis with automated spot cutting

Image analysis software obtains quantitative and qualitative information about the proteins in a sample, and stores the information in files, which may also contain additional annotations. The ProteomeWorks™ spot cutter expands the capabilities of proteome labs by integrating PDQuest™ image analysis software. The image analysis files acquired by PDQuest direct automated spot cutting. Excised protein spots are deposited into microtiter plates ready for further automated processing. PDQuest software tracks the protein spots through spot cutting and protein identification. Downstream protein spot identifications are generally obtained from peptide mass fingerprint analysis using mass spectrometry. The ProteomeWorks spot cutter is a precision instrument with a small benchtop footprint. It is fully automatic to increase throughput and minimize the amount of hands-on time spent excising protein spots. The spot cutter individually excises even overlapping spots for unique identification.

2.6.3 Automated protein digestion

The ProteomeWorks spot cutter eliminates the first of two bottlenecks for excision and enzymatic digestion of protein spots. Driven by PDQuest software, it enables automated spot excision and deposition of cut gel spots into microtiter plate wells. Isolated proteins from the gel pieces are then digested to release peptides for detailed sequence analysis by mass spectrometry, leading to protein identification. Excised gel spots can be robotically destained, chemically modified (reduced and alkylated), and digested in preparation for either MALDI-TOF-MS or ESI-MS with the Micromass MassPREP station. Each process is executed under fully automated software control with a range of standard protocols enabling high throughput and flexibility. Manual protein digestion is a tedious, time-consuming process that is subject to variability and keratin contamination. Automation of this process with the MassPREP station eliminates a significant bottleneck for high-throughput protein identification.

Operational features of the MassPREP station include a variable temperature control for optimized reduction, alkylation, and digestion of proteins, and onboard cooling capabilities for reagents and peptide digests to ensure reproducible digestion results. The station employs a variety of sample cleaning technologies (MassPREP targets and Millipore ZipTip pipet tips) to prepare peptide digests prior to automated deposition of samples onto a M@LDI or MassPREP target plate. Contamination of peptide samples is also minimized with the MassPREP clean air enclosure.

2.6.4 Rapid, high-throughput protein identification by MALDI-TOF-MS

Peptide mass fingerprinting of protein digest products using matrix assisted laser desorption ionization time of flight mass spectrometry (MALDI-TOF-MS) provides an ideal method for protein identification when samples have been separated by 2-D PAGE. The M@LDI HT is one of a new generation of networked "2-D gel-MS" analyzers for high-throughput protein identification. M@LDI HT is the primary MS data acquisition device of the ProteomeWorks system, and features a fully automated target plate auto-changer for increased throughput. Networking enables distribution of data capture, protein assignment, and result presentation functions of ProteinLynx Global *SERVER* software within a secure clientserver architecture, maximizing computing power to quickly identify proteins. The M@LDI HT enables automated acquisition of optimized mass spectra and the derivation of monoisotopic peptide mass fingerprint information. Interrogation of multiple FASTA databases using Global *SERVER* software following capture of MS results provides rapid identification of proteins that fit the samples' peptide mass fingerprint, along with a confidence score indicating the validity of the identifications. Following MS identification, peptide mass fingerprint spectra and all of the identification results are available through electronic reports. In addition, protein identification results are seamlessly integrated with the gel image in PDQuest software. Using this system, the working time to process data from spot cutting to protein digestion to MS analysis and image annotation is reduced by over 50% compared to manual processing of gel samples, with a corresponding reduction in error. All of the instrumentation and software in this process is part of the integrated ProteomeWorks system, a set of powerful tools for proteomic analysis.

2.6.5 Advanced protein characterization with ESI-LC-MS and MALDI-TOF

MALDI-TOF MS provides an ideal high-throughput solution for protein identification; however, where protein identity is ambiguous, known databases must be searched with a higher degree of sequence information. The Micromass Q-Tof family of MS-MS instruments incorporates quadrupole/ orthogonal acceleration time-of-flight (Q/oa-TOF) technology, enabling exact mass measurement, and acquisition of the highest-level peptide sequence information for de novo sequencing and BLAST analyses. Protein digest samples in microtiter plates, prepared with the MassPREP station, can be transferred directly to the Micromass CapLC (capillary HPLC) system for automated injection into the Q-Tof *micro* for integrated LC-MS-MS under MassLynx software control. The capability for MS to MS-MS switching "on the fly" with the Q-Tof family of instruments maximizes the amount of amino acid sequence information that can be generated with these instruments. MassSeq software also provides the capacity for automated de novo amino acid sequencing based on the MS results.

2.6.6 2DE in identification of bladder carcinoma protein marker (Calreticulin)

Susumu Kageyama et al., 2004 screened proteins as tumor markers for bladder cancer by proteomic analysis (2DE) of cancerous and healthy tissues and investigated the diagnostic accuracy of one such marker, *Calreticulin* (CRT) in urine.

They have produced two important findings in their experiments. The first is that increased production of CRT in bladder cancer tissue which was confirmed by proteome profiling by 2DE. Furthermore, we detected two isoforms of CRT, and full-length CRT which was more useful than cleaved CRT for distinguishing bladder cancer from healthy tissue.

In their study, although visual comparison of 2DE gels of TCC (transitional cell carcinoma) and noncancerous urothelium showed similar expression profiles, 15 protein spots (U-1 to U-15) were more intense in TCC samples (Fig. 1). They identified 10 of the proteins by use of a peptide mass fingerprinting method. One spot among them, with an apparent mass of 55 kDa and pI of 4.3, was identified as CRT (spot U-2 in Fig. 1 A). From NH2-terminal amino acid sequencing, 10 amino acids were sequenced (EPAVYFKEQF), and they were identical to residues 1–10 of mature human CRT according to the sequence homology search.

Further, to validate the 2DE finding of increased production of full-length CRT in bladder cancer tissue, they performed quantitative Western blot analysis in cancerous and healthy tissue using anti-COOH-terminus antibody. They compared CRT band intensities for 22 cancerous with 10 noncancerous tissues. For band quantification, they defined the CRT band derived from a total of 1 µg of heat-shocked HeLa cell extract as 1.0 unit/µg of protein. The mean (SD) concentrations in cancerous and healthy tissue were 1.0 (0.4) and 0.4 (0.3) units/µg of protein, respectively (Mann–Whitney U-test, $P = 0.0003$; Fig. 2). Among these tissue samples, six pairs of cancerous and noncancerous specimens were obtained from the bladders of patients who had undergone radical cystectomy. CRT concentrations were higher in all cancer tissues compared with the corresponding healthy urothelium.

To confirm the presence of isoforms of CRT, they performed two-dimensional Western blotting with two different antibodies: monoclonal antibody FMC75, which was produced against recombinant human CRT; and a polyclonal antibody that was produced using synthesized peptides of the human CRT COOH terminus (amino acids 388–400) as an immunogen. On Western blots with anti- COOH-terminus antibody, only one of the two spots was visualized, whereas both spots became visible on blots incubated with FMC75 (Fig. 3). One was the same as the 55-kDa (pI 4.3) spot, and the other had an apparent molecular mass of 40 kDa and pI of 4.5. This lower molecular- mass spot had the same NH2-terminal amino acid sequence as amino acids 1–10 of mature human CRT as shown by amino acid sequencing. Therefore, they suggested that the higher-molecular-mass spot was the full-length form and the other spot was a cleaved form that is truncated elsewhere in the COOH domain. Production of the full-length CRT in cancer tissue was increased compared with in healthy tissue, but the spots for cleaved CRT in cancerous and healthy urothelium had intensities that were similar and were reproducible on all silverstained 2DE gels.

Subsequently they tried to confirm whether anti-COOH terminus antibody binds to molecules other than full lengt1h CRT and performed immuno-precipitations (Fig. 4). The Western blot of the immuno-precipitate extracted from cancer tissue revealed only one band, and they concluded that anti-COOH-terminus antibody binds specifically to full-length CRT of ~55 kDa. They therefore judged that full-length CRT recognized by anti-COOH terminus antibody is appropriate as a tumor marker.

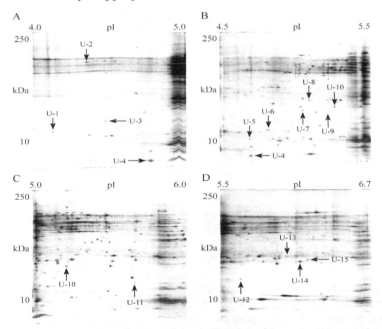

Fig. 1. Silver-stained images of analytical narrow-pH-range 2DE gels of proteins from bladder cancer. (A), pH 4.0–5.0; (B), pH 4.5–5.5; (C), pH 5.0–6.0; (D), pH 5.5–6.7. Arrows indicate spots (U-1 to U-15) containing higher amounts of protein. Spot U-2, ~55 kDa and pI 4.3, was confirmed to be CRT by a peptide mass fingerprinting method and NH2-terminal amino acid sequencing. *(Source: Susumu Kageyama et al., 2004 Clinical Chemistry, 50: 857–866)*

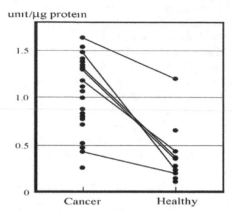

Fig. 2. Quantitative Western blot analysis of cancerous and healthy tissues using anti-COOH-terminus antibody. *Lines* show six pairs of cancerous and healthy specimens obtained from the bladders of patients who had undergone radical cystectomy.*(Source:* Susumu Kageyama *et al., 2004 Clinical Chemistry, 50: 857–866)*

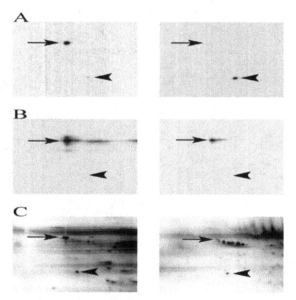

Fig. 3. Close-up sections of two-dimensional Western blot images obtained with two different antibodies, FMC75 (*A*) and anti-COOH terminus antibody (*B*), and proteins in silver-stained pH 4–7 gels (*C*). *Left panels* are bladder cancer and *right panels* are healthy urothelium. *Arrows* indicate full-length CRT (55 kDa; pI 4.3), and *arrowheads* indicate cleaved CRT (40 kDa; pI 4.5). These two CRT forms have the same NH2-terminal amino acid sequence (EPAVYFKEQF), but cleaved CRT is considered to lack the COOH terminus because of no immunoreactivity for anti-COOH-terminus antibody. *(Source: Susumu Kageyama et al., 2004 Clinical Chemistry, 50: 857–866)*

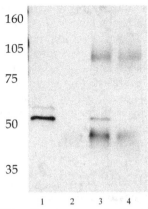

Fig. 4. Western blot with FMC75 antibody. *Arrow* indicates 55-kDa full-length CRT. *Lane 1,* total cell lysate; *lane 2,* extraction from protein A Sepharose beads that did not bind antibody; *lane 3,* immunoprecipitant eluted from beads binding anti-COOH-terminus antibody; *lane 4,* immunoprecipitant extracted from beads binding normal rabbit IgG indicates approximate molecular masses. *(Source: Susumu Kageyama et al., 2004 Clinical Chemistry, 50: 857–866)*

2.6.7 2-DE in identification of pancreatic carcinoma protein marker

Leucine-rich alpha-2-glycoprotein is characterized by its unusually high content of leucine, about 17% by weight. The primary structure of LRG suggests that it may be a membrane associated or membrane-derived protein. Aberrant regulation of LRG has been observed in patients with malignant disease and with virus infection.

Tatsuhiko Kakisaka et al., 2007 examined the plasma LRG expression levels of cancer patients who were not in the acute inflammatory phase. First, they selected cancer patients with normal C-reactive protein (CRP) concentration (Table 1, P6–10) to explore whether the increase in LRG levels paralleled the dynamics of the commonacute phase proteins. Second, they tested plasma from cancer patients with a normal level of CA19-9 a tumor marker commonly used for the diagnosis of pancreatic cancer, to examine the possibility of plasma LRG levels being using in a way complementary to existing tumor markers.

SDS-PAGE/Western blotting using an anti-LRG antibody showed consistent up-regulation of LRG in these patients, who were negative for CRP and/or CA19-9, compared with the noncancer bearing healthy donors (Fig. 5). Therefore, increased amounts of LRG may be independent of the regulation of other acute phase proteins and tumor markers. They also examined plasma samples from chronic pancreatitis patients and found that they tended to express lower LRG levels compared with the samples from pancreatic cancer patients. By correlating the expression levels of LRG with clinical information from a large sample set, they hope to validate the utility of LRG as a biomarker to monitor the status of patients. Some plasma samples from pancreatic cancer patients did not express high LRG levels, leading us to suggest that the examination of plasma LRG levels in combination with the existing biomarkers would increase the specificity and sensitivity of the diagnosis.

Some protein spots on 2D-PAGE gels overlapped across fractions in the anion-exchange chromatography even when they used the step-wise gradient method with system wash between intervals. They considered these overlapping spots to correspond to different isoforms of the same protein, and have therefore counted all protein spots on the 2D-PAGE gels. However, not every differentially expressed protein was considered to be a suitable tumor marker; for example, spots 8 and 11 (transthyretin) were differentially expressed between cancer patients and healthy donors, but they were very minute amounts of the total abundant transthyretin, and it was difficult to extract theseportion of the protein. On the contrary, LRG, which was also differentially expressed between cancer patients and healthy donors, was only expressed in one fraction and was therefore selected as a candidate for a tumor marker of pancreatic cancer.

The use of high-resolution 2D-PAGE with narrow-range IPG gels and large-format second dimension gels could solve this problem to some extent.

Patient informations of validation set 1

Case[a]	Age	Sex	Tumor location	Stage[b]	CA19-9 (U/ml)	CRP (mg/dl)
P6	56	Male	Head	IV	1	0.1
P7	45	Female	Body~tail	IV	3698	0.1
P8	55	Female	Body	III	<1	<0.1
P9	58	Male	Body	IV	25600	0.1
P10	65	Female	Body	III	804	<0.1

Case[a]	Age	Sex
N6	53	Male
N7	51	Female
N8	60	Female
N9	59	Male
N10	64	Female

[a] P: pancreatic cancer patients, N: non-cancer bearing healthy donors.
[b] The Union Internationale Contre le Cancer (UICC) classification [41].

Table 1. Patient informations of validation set 1

2.6.8 2-DE in identification of human gastric carcinoma protein marker

Gastric cancer is the second most common cause of cancer deaths worldwide and due to its poor prognosis, it is important that specific biomarkers are identified to enable its early detection. Through 2-D gel electrophoresis and MALDI-TOF-TOF-based proteomics approaches, Chien-Wei Tseng et al., 2011 found that 14-3-3β, which was one of the proteins that were differentially expressed by 5-fluorouracil-treated gastric cancer SC-M1 cells, was up regulated in gastric cancer cells. 14-3-3β levels in tissues and serum were further validated in gastric cancer patients and controls. The results showed that 14-3-3β levels were elevated in tumor tissues in comparison to normal tissues, and serum levels in cancer patients were also significantly higher than those in controls (Fig. 6). Elevated serum 14-3-3β levels highly correlated with the number of lymph node metastases, tumor size, and a reduced survival rate. Moreover, over-expression of 14-3-3β enhanced the growth, invasiveness, and migratory activities of tumor cells. Twenty-eight proteins involved in anti-apoptosis and tumor progression were also found to be differentially expressed in 14-3-3β-overexpressing gastric cancer cells. Overall, these results highlight the significance of 14-3-3β in gastric cancer cell progression and suggest that it has the potential to be used as a diagnostic and prognostic biomarker in gastric cancer.

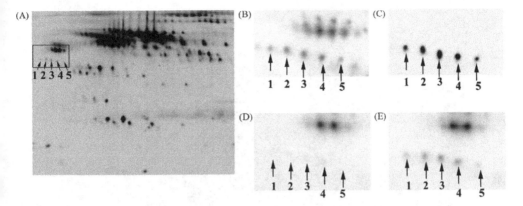

Fig. 5. Elevated level of plasma LRG in pancreatic cancer. The localizations of the five LRG spots are indicated by arrows 1–5 on the 2D image of the 150mM NaCl sample (A). The boxed area was transferred to a nitrocellulose membrane and scanned with a laser scanner to obtain the LRG spots on the membrane (B). The scanned membrane was reacted with an anti-LRG antibody and the antibody–antigen complexes were detected with an ECL system (C). The fluorescent signals of the LRG spots on the 2D-PAGE gels were compared between non-cancer bearing healthy donors (D) and pancreatic cancer patients (E). (*Source: Tatsuhiko Kakisaka et al., 2007 Journal of Chromatography B, 852: 257–267*)

2.6.9 2DE in identification of squamous cervical carcinoma protein markers

Recently, proteomic and genomic approaches to identify tumor markers are undergoing. Hellman and coworkers reported the protein expression patterns in primary carcinoma of the vagina. In relation to HPV, C33A cell line transfected with HPV *E7* gene and proteomic and genomic analyses were performed. But, until now, there was no report of SCC in cervix tissues. Prof. W.S. Ahn and his colleagues at Cancer Research Center of The Catholic University of Korea, South Korea contributed much more to understand the significance of 2DGE in diagnosis of cervical cancer. In general, screening in cervical cancer is progressing to find out candidate genes and proteins, which may work as biological markers and play a role in tumor progression. They examined the protein expression patterns of squamous cell carcinoma (SCC) tissues from Korean women using two-dimensional polyacrylamide gel electrophoresis (2-DE) and matrix-assisted laser desorption/ionization-time of fight (MALDI-TOF) mass spectrometer. A total of 35 proteins are detected in SCC. 17 proteins are up regulated and 18 proteins are down-regulated. Among the proteins identified, 12 proteins (pigment epithelium derived factor, annexin A2 and A5, keratin 19 and 20, heat shock protein 27, smooth muscle protein 22 alpha, alpha-enolase, squamous cell carcinoma antigen 1 and 2, glutathione S-transferase, apolipoprotein a1) are previously known proteins involved in tumor and 21 proteins were newly identified in this study. They concluded that the 2-DE offers total protein expression profiles of SCC tissues and further characterization of proteins that are differentially expressed will give a chance to identify tumor-specific diagnostic markers for SCC (Fig. 7 & 8).

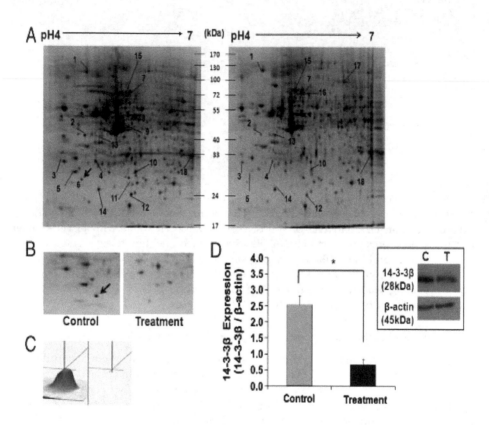

Fig. 6. 14-3-3β is differentially expressed after 5-FU treatment of SC-M1 cells.
(A) Proteins from the 5-FU-treated (right) SC-M1 cells and the untreated control (left)
were compared using 2-DE. Enlarged images and 3-D profiles of 14-3-3b on the gels are
shown in (B) and (C). (D) 14-3-3β expression was significantly reduced after 5-FU
treatment as confirmed by Western blot. (*Source: Chien-Wei Tseng et al., 2011 Proteomics,
11:2423–2439*)

In addition to cervical cancer, the 2-dimensional polyacrylamide gel electrophoresis (2-DE)
has also been used to examine heterogeneity of protein expression in tissues from different
tumors such as bladder, breast, colon–rectum, lung, and ovary. The advantage of 2-DE is
that the complex protein expression is analyzed qualitatively and quantitatively. 2-DE
combined with MALDI-TOF-MS has been applied to identify cancer-specific protein
markers. These markers can provide a basis for developing new methods for early diagnosis
and treatment.

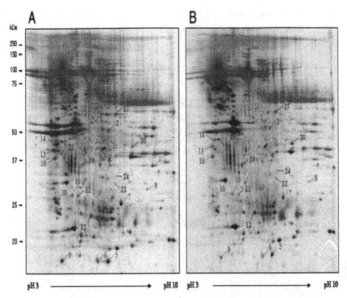

Fig. 7. Comparison of proteome by two-dimensional gel electrophoresis on normal tissues and cervical SCC tissues. Representative examples of 2-DE gels derived from a normal cervix tissue and cervical SCC tissue. Normal cervix (A) and cervical SCC (B) total proteins were separated by 2-DE using IPG strips pH 3–10 in the first and 12% SDS-PAGE in the second dimension. Identified protein spots are indicated by numbers. Proteins down- (A) or up-regulated (B) in cervical SCC are indicated. (*Source: Bae et al., 2005 Gynecologic Oncol., 99:* 26-35)

2.6.10 Breast cancer protein profiling

Franzen *et al.,* 1997 well documented that the two-dimensional electrophoresis (2-DE) analyses of human breast carcinoma reveals the following observations: (i) Analysis of samples from different areas of the same tumor showed a high degree of similarity in the pattern of polypeptide expression. Similarly, analysis of two tumors and their metastases revealed similar 2-DE profiles. (ii) In contrast, large variations have been observed between different lesions with comparable histological characteristics. Larger differences in polypeptide expression are pointed out in between potentially highly malignant carcinomas and comparisons of less malignant lesions. These differences are in the same order of magnitude as those observed comparing a breast carcinoma to a lung carcinoma. (iii) The levels of all cytokeratin forms resolved (CK7, CK8, CK15, and CK18) were significantly lower in carcinomas compared to fibro adenomas. (iv) The levels of high molecular weight tropo-myosins (1–3) were lower in carcinomas compared to fibro adenomas. The expression of tropomyosin-1 is 1.7-fold higher in primary tumors with metastatic spread to axillar lymph nodes compared to primary tumors with no evidence of metastasis ($p < 0.05$). (v) The expression of proliferating cell nuclear antigen (PCNA) and some members of the stress protein family (pHSP60, HSP90, and calreticulin) are higher in carcinomas. We conclude that malignant progression of breast carcinomas results in large heterogeneity in polypeptide expression between different tumors, but that some common themes such as decreased expression of cytokeratin and tropomyosin polypeptides can be discerned (Fig. 9 & 10). (Franzen, *et al.* 1997 *Electrophoresis*, 18 : 582–587.).

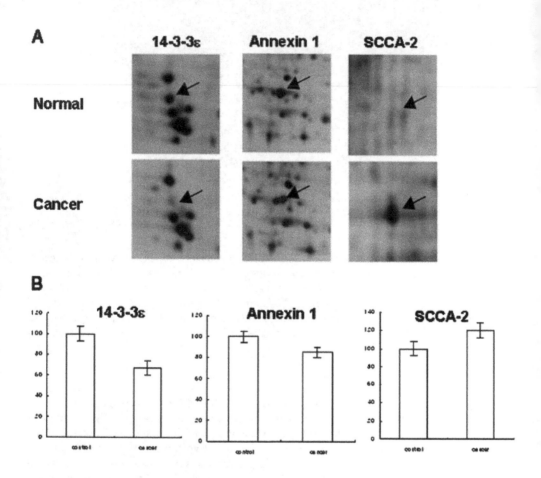

Fig. 8. Protein expression comparisons of normal samples and SCC samples. Up- and down-regulated proteins (14-3-3ε, annexin A1, and SCCA-2) were selected and magnified gel images were presented (A). From the PDQuest 2-D software quantification, the expression difference was statistically meaningful (P value < 0.05) (B). (*Source: Bae et al., 2005 Gynecologic Oncol., 99: 26-35*)

The same gel was post-stained with SyproRuby dye (*right panel*). 100 μg of each lysate from serum-starved cells were analyzed on a 9–16% gradient gel. *Circles* represent differentially expressed proteins detectable by both methods. *Arrows* represent spots detected by SyproRuby but not Cy dye labeling. B, the shift in molecular weight between the modified and unmodified proteins was visualized by image overlaying. The DIGE image (*Blue*) was overlaid with the SYPRO image (*Red*).

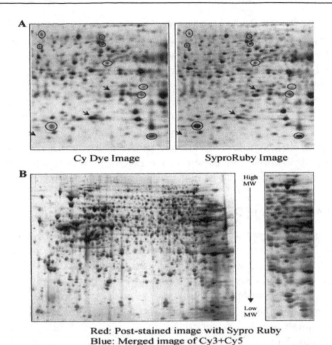

Red: Post-stained image with Sypro Ruby
Blue: Merged image of Cy3+Cy5

Fig. 9. Sensitivity of 2D-DIGE and compatibility with SYPRO gel staining. *A*, comparison of 2D-DIGE imaging and SyproRuby poststaining. Merged Cy dye image of HB4a lysate labeled with Cy3 (*red*) and HBc3.6 lysate labeled with Cy5 (*blue*) (*left panel*).

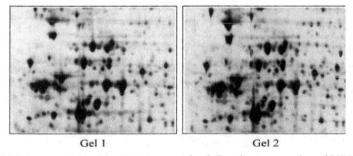

Fig. 10. 2D-DIGE is a reproducible detection method. Duplicate samples of HB4a and HBc3.6 were labeled separately with Cy3 (*red*) and Cy5 (*blue*), respectively.

2.6.11 2-DE protein pattern in classification of carcinoma cells

High-resolution two-dimensional polyacrylamide gel electrophoresis (2-D PAGE) is a powerful research tool for the analytical separation of cellular proteins. The qualitative and quantitative pattern of polypeptides synthesized by a cell represents its phenotype and thus defines characteristics such as the morphology and the biological behavior of the cell. By analyzing and comparing the protein patterns of different cells, it is possible to recognize the cell type and to identify the most typical features of these cells. In applied pathology it is often difficult to identify the tissue of origin and the stage or grade of a neoplasia by cellular morphology analyzed by classical or immunostaining procedures. The protein pattern itself is the most characteristic feature of a cell and should therefore contribute to the identification of the cell type. For this reason, we separated protein fractions originating from different lung tumor cell lines using 2-D PAGE and we compared the resulting patterns on a multivariate statistical level using correspondence analysis (CA) and ascendant hierarchical clustering (AHC). The results indicate that (i) protein patterns are highly typical for cells and that (ii) the comparison of the protein patterns of a set of interesting cell types allows the identification of potentially new marker proteins. 2-D PAGE is thus a unique and powerful tool for molecular cytology or histopathology, unveiling the protein expression level of tissues or cells.

2.6.12 2-DE in understanding Ovarian intratumoral heterogeneity

The process of tumor progression leads to the emergence of multiple clones, and to the development of tumor heterogeneity. One approach to the study of the extent of such heterogeneity is to examine the expression of marker proteins in different tumor areas. Two-dimensional gel electrophoresis (2-DE) is a powerful tool for such studies, since the expression of a large number of polypeptide markers can be evaluated. The tumor cells have been prepared from human ovarian tumors and analyzed by 2-DE and PDQUEST. As judged from the analysis of two different areas in each of nine ovarian tumors, the intratumoral variation in protein expression is low. In contrast, large differences have been observed when the protein profiles of different tumors are compared. The differences in gene expression between pairs of malignant carcinomas are slightly larger than the differences observed between pairs of benign tumors. Hence, the 2-DE analysis of intratumoral heterogeneity in ovarian cancer tissue indicates a low degree of heterogeneity.

2.6.13 Strengths and weaknesses of 2D-PAGE

Electrophoresis is an established technique that has undergone several advances that have enhanced resolution, detection, quantization, and reproducibility. The 2-D SDS-PAGE and 2DDIGE approaches to protein profiling are accessible and economical methods that possess high resolving power and enable the detection of hundreds of proteins on a single gel plate. Although reproducibility has been an issue with 2D-PAGE, especially when profiling two protein mixtures, it has been greatly improved with the use of 2D-DIGE. Resolution has been enhanced by the introduction of IPGs, which enable the analyst to tailor the pH gradient for maximum resolution using ultrazoom gels with a narrow pH gradient range. With modern 2D-PAGE, it is not unusual to resolve two proteins that differ in pI by 0.001 U. Although 2D-PAGE has been limited by its inability to resolve proteins that are too basic or too acidic, too large or too small, this limitation is continuously diminishing. For example, the separation of basic proteins can be analyzed using IPGs in the pH range of 4–12. Separation science is always evolving, and it will not be long before the remaining issues of gel electrophoresis are adequately resolved. The introduction of 2D-DIGE contributed immensely to solving problems of reproducibility and quantitation. The use of imagers and computers allows not only fast data mining, acquisition, and analysis but also spot detection, normalization, protein profiling, background correction, and reporting and exporting of data. As a separation, detection, and quantitation technique, 2D-DIGE is an important tool, especially for clinical laboratories involved in the determination of protein expression levels and disease biomarker discovery. When absolute biological variation between samples is the main objective, as in biomarker discovery, 2D-DIGE is the method of choice. While there has been significant progress in nongel (or solution-based) methods for coupling fractionation methods directly online with MS analysis, 2DPAGE has remained a popular technique for conducting proteomic studies. Though 2D-PAGE, like any fractionation scheme, has its advantages and disadvantages, there is no doubt that it will remain an essential technique for the characterization of proteomes for many years to come.

2.6.14 Two-dimensional electrophoresis for cancer proteomics

Proteome analysis is a direct measurement of proteins in terms of their presence and relative abundance (Wilkins *et al.*, 1996). The overall aim of a proteomic study is characterization of the complex network of cell regulation. Neither the genomic DNA code of an organism nor the amount of mRNA that is expressed for each gene product (protein) yields an accurate picture of the state of a living cell (Lubec *et al.*, 1999), which can be altered by many conditions. Proteome analysis is required to determine which proteins have been conditionally expressed, how strongly, and whether any posttranslational modifications are affected. Two or more different states of a cell or an organism (e.g., healthy and diseased tissue) can be compared and an attempt made to identify specific qualitative and quantitative protein changes. One of the greatest challenges of proteome analysis is the reproducible fractionation of these complex protein mixtures while retaining the qualitative and quantitative relationships. Currently, two-dimensional polyacrylamide gel electrophoresis (2-D PAGE) is the only method that can handle this task (Cutler et al., 1999, Fegatella et al., 1999, Görg et al., 2000), and hence has gained special importance. Since 2-D PAGE is capable of resolving over 1,800 proteins in a single gel (Choe & Lee, 2000), it is important as the primary tool of proteomics research where multiple proteins must be separated for parallel analysis. It allows hundreds to thousands of gene products to be analyzed simultaneously. In combination with computer assisted image evaluation systems for comprehensive qualitative and quantitative examination of proteomes, this electrophoresis technique allows cataloging and comparison of data among groups of researchers.

3. Conclusion

Two-dimensional gel electrophoresis and MALDI-MS are an effective strategy for determining the protein domains present in those gel spots that are observed at significantly lower MW values than are given in the database. While average sequence coverage is only 30%, the peptides detected are confined to a specific region of the protein, such as the protein N- or C-terminal. This information could easily be incorporated into protein identification tables. Regional coverage information is not readily available from either LC-MS/MS analysis of digests of cellular lysates or from epitope-specific antibodies. Some of the protein fragments correspond to chains produced by known cellular processing and activation pathways. Others have been detected as functional and structural domains during in vitro experiments or noted in other in vivo studies, indicating they function intra- or extra-cellularly. By using tools that allow both protein identification and measurement of MW, we can assess the abundance and distribution of protein fragments. Correlation of these results with targeted functional studies on specific proteins will elucidate the biological function of protein fragments.

4. References

Bae, SM., Lee, CH., Cho,YL., Nam, KH., Kim, YW., Kim, CK., Han, BD., Lee, YJ., Chun, HJ. & Ahn, WS. (2005). Two-dimensional gel analysis of protein expression profile in squamous cervical cancer patients. *Gynecologic Oncology* 99: 26-35.

Bjellqvist, B., Ek, K., Righetti, PG., Gianazza, E., Gorg, A., Westermeier, R. & Postel, W. (1982). Isoelectric focusing in immobilized pH gradients: principle, methodology and some applications, *J Biochem Biophys Methods* 6: 317–339.

Bjellqvist, B., Hughes, GJ., Pasquali, C., Paquet, N., Ravier, F., Sanchez, JC., Frutiger, S. & Hochstrasser, D. (1993b). The focusing positions of polypeptides in immobilized pH gradients can be predicted from their amino acid sequences, *Electrophoresis* 14: 1023–1031.

Bjellqvist, B., Sanchez, JC., Pasquali, C., Ravier, F., Paquet, N., Frutiger, S., Hughes, GJ. & Hochstrasser, D. (1993a). Micropreparative two-dimensional electrophoresis allowing the separation of samples containing milligram amounts of proteins, *Electrophoresis* 14: 1375–1378.

Carroll, K., Ray, K., Helm, B. & Carey, E. (2000). Two-dimensional electrophoresis reveals differential protein expression in high- and low-secreting variants of the rat basophilic leukaemia cell line, *Electrophoresis* 21: 2476–2486.

Chevallet, M., Santoni, V., Poinas, A., Rouquie, D., Fuchs, A., Kieffer, S., Rossignol, M., Lunardi, J., Garin, J. & Rabilloud, T. (1998). New zwitterionic detergents improve the analysis of membrane proteins by two-dimensional electrophoresis, *Electrophoresis* 19: 1901–1909.

Choe, LH. & Lee, KH. (2000). A comparison of three commercially available isoelectric focusing units for proteome analysis: The multiphor, the IPGphor and the protean IEF cell, *Electrophoresis* 21: 993–1000.

Cordwell, SJ., Basseal, DJ., Bjellqvist, B., Shaw, DC. & Humphery-Smith, I. (1997). Characterisation of basic proteins from *Spiroplasma melliferum* using novel immobilised pH gradients, Electrophoresis 18, 1393–1398.

Corthals,, GL., Wasinger, VC., Hochstrasser, DF. & Sanchez, JC. (2000). The dynamic range of protein expression: a challenge for proteomic research, *Electrophoresis* 21: 1104–1115.

Cutler, P., Bell, DJ., Birrell, HC., Connelly, JC., Connor, SC., Holmes, E., Mitchell, BC., Monte, SY., Neville, BA., Pickford, R., Polley, S., Schneider, K. & Skehel, JM. (1999). An integrated proteomic approach to studying glomerular nephrotoxicity, *Electrophoresis* 20: 3647–3658.

Ducret, A., Bruun, CF., Bures, EJ., Marhaug, G., Husby, G. & Aebersold, R. (1996). Characterization of human serum amyloid A protein isoforms separated by two-dimensional electrophoresis by liquid chromatography/electrospray ionization tandem mass spectrometry, *Electrophoresis* 17: 866–876.

Eckerskorn, C., Jungblut, P., Mewes, W., Klose, J. & Lottspeich, F. (1988). Identification of mouse brain proteins after two-dimensional electrophoresis and electroblotting by microsequence analysis and amino acid composition analysis, *Electrophoresis* 9: 830–838.

Fegatella, F., Ostrowski, M. & Cavicchioli, R. (1999). An assessment of protein profiles from the marine oligotrophic ultramicrobacterium, *Sphingomonas sp.* Strain RB2256, *Electrophoresis* 20: 2094–2098.

Fountoulakis, M., Takacs, MF., Berndt, P., Langen, H. & Takacs, B. (1999). Enrichment of low abundance proteins of *Escherichia coli* by hydroxyapatite chromatography, *Electrophoresis* 20: 2181–2195.

Franzén, B., Linder, S., Alaiya, AA., Eriksson, E., Fujioka, K., Bergman, AC., Jörnvall, H. & Auer, G. (1997). Analysis of polypeptide expression in benign and malignant human breast lesions. *Electrophoresis* 18: 582–587.

Garfin, DE. (1995). Electrophoretic methods, pp 53–109 in Glasel JA and Duetscher MP (eds) *Introduction to Biophysical Methods for Protein and Nucleic Acid Research*, Academic Press, San Diego.

Garfin, DE. (2000). Isoelectric focusing, pp 263–298 in Ahuja S (ed) *Separation Science and Technology*, Vol 2, Academic Press, San Diego.

Gingrich, JC. Davis, DR. & Nguyen, Q. (2000). Multiplex detection of quantification of proteins on western blots using fluorescent probes, *Biotechniques* 29: 636–642.

Görg, A. (1991). Two-dimensional electrophoresis, *Nature* 349: 545–546.

Görg, A., Obermaier, C., Boguth, G. & Weiss, W. (1999). Recent developments in twodimensional gel electrophoresis with immobilized pH gradients: wide pH gradients up to pH 12, longer separation distances and simplified procedures, *Electrophoresis* 20: 712–717.

Görg, A., Obermaier, C., Boguth, G., Harder, A., Scheibe, B., Wildgruber, R. & Weiss, W. (2000). The current state of two-dimensional electrophoresis with immobilized pH gradients, *Electrophoresis* 21: 1037–1053.

Görg, A., Postel, W., Domscheit, A. & Günther, S. (1989) Methodology of two dimensional electrophoresis with immobilized pH gradients for the analysis of cell lysates and tissue proteins, in Endler AT and Hanash S (eds) Two- Dimensional Electrophoresis. *Proceedings of the International Two-Dimensional Electrophoresis Conference*, Vienna, Nov. 1988, VCH, Weinheim FRG.

Gottlieb, M. & Chavko, M. (1987). Silver staining of native and denatured eukaryotic DNA in agarose gels, *Anal Biochem* 165: 33–37.

Herbert, BR., Molloy, MP., Gooley, AA., Walsh, BJ., Bryson, WG. & Williams, KL. (1998). Improved protein solubility in two-dimensional electrophoresis using tributyl phosphine as reducing agent, *Electrophoresis* 19: 845–851.

Hermann, T., Finkemeier, M., Pfefferle, W., Wersch, G., Kramer, R. & Burkovski, A. (2000). Two-dimensional electrophoretic analysis of *Corynebacterium glutamicu* membrane fraction and surface proteins, *Electrophoresis* 21: 654–659.

Klose, J. (1975). Protein mapping by combined isoelectric focusing and electrophoresis of mouse tissues. A novel approach to testing for induced point mutation in mammals. *Humangenetik* 26, 231–243.

Lubec, G., Nonaka, M., Krapfenbauer, K., Gratzer, M., Cairns, N. & Fountoulakis, M. (1999). Expression of the dihydropyrimidinase related protein 2 (DRP-2) in Down syndrome and Alzheimer's disease brain is downregulated at the mRNA and dysregulated at the protein level, *J Neural Transm Suppl* 57: 161–177.

Macri, J., McGee, B., Thomas, JN., Du, P., Stevenson, TI., Kilby, GW. & Rapundalo, ST. (2000). Cardiac sarcoplasmic reticulum and sarcolemmal proteins separated by two-dimensional electrophoresis: surfactant effects on membrane solubilization, *Electrophoresis* 21: 1685–1693.

Masuoka, J., Glee, PM. & Hazen, KC. (1998). Preparative isoelectric focusing and preparative electrophoresis of hydrophobic *Candida albicans* cell wall proteins with in-line transfer to polyvinylidene difluoride membranes for sequencing, *Electrophoresis* 19: 675–678.

Merril, CR., Goldman, D., Sedman, SA. & Ebert, MH. (1981). Ultrasensitive stain for proteins in polyacrylamide gels shows regional variation in cerebrospinal fluid proteins, *Science* 211: 1437–1438.

Miller, MJ. (1989). Computer-assisted analysis of two-dimensional gel electropherograms, pp 182–217 in Chrambach A, Dunn MJ and Radola BJ (eds) *Advances in Electrophoresis*, Vol. 3, Verlag Chemie, Weinheim.

Molloy, MP. (2000). Two-dimensional electrophoresis of membrane proteins using immobilized pH gradients, *Anal Biochem* 280: 1–10.

Molloy, MP., Herbert, BR., Walsh, BJ., Tyler, MI., Traini, M., Sanchez, JC., Hochstrasser, DF., Williams, KL. & Gooley, AA. (1998). Extraction of membrane proteins by differential solubilization for separation using two-dimensional gel electrophoresis, *Electrophoresis* 19: 837–844.

Morel, V., Poschet, R., Traverso, V. & Deretic, D. (2000). Towards the proteome of the rhodopsin-bearing post-Golgi compartment of retinal photoreceptor cells, *Electrophoresis* 21: 3460–3469.

Nilsson, CL., Larsson, T., Gustafsson, E., Karlsson, KA. & Davidsson, P. (2000). Identification of protein vaccine candidates from *Helicobacter pylori* using a preparative two-dimensional electrophoretic procedure and mass spectrometry, *Anal Chem* 72: 2148–2153.

O'Farrell, PH. (1975). High resolution two-dimensional electrophoresis of proteins. *J Biol Chem* 250, 4007–4021.

Patton, WF. (2000). A thousand points of light: the application of fluorescence detection technologies to two-dimensional gel electrophoresis and proteomics, *Electrophoresis* 21: 1123–1144.

Pennington, SR. & Dunn, MJ. (2001). *Proteomics. From Protein Sequence to Function*, Springer/Bios, New York.

Rabilloud, T. (1998). Use of thiourea to increase the solubility of membrane proteins in two-dimensional electrophoresis, *Electrophoresis* 19: 758–760.

Rabilloud, T. (1999). Solubilization of proteins in 2-DE. An outline, *Methods Mol Biol* 112: 9–19.

Rabilloud, T. (2000). *Proteome Research: Two Dimensional Gel Electrophoresis and Identification Tools*, Springer, Berlin.

Righetti, PG. (1983). *Isoelectric Focusing: Theory, Methodology and Applications*, Elsevier, Amsterdam.

Righetti, PG. (1990). Recent developments in electrophoretic methods, *J Chromatogr* 516: 3–22.

Sanchez, JC., Rouge, V., Pisteur, M., Ravier, F., Tonella, L., Moosmayer, M., Wilkins, MR. & Hochstrasser, DF. (1997). Improved and simplified in-gel sample application using reswelling of dry immobilized pH gradients, *Electrophoresis* 18: 324–327.

Smithies, O. & Poulik, MD. (1956). Two-Dimensional Electrophoresis of Serum Proteins. *Nature* 177, 1033.

Taylor, RS., Wu, CC., Hays, LG., Eng, JK., Yates, JR. & Howell, KE. (2000). Proteomics of rat liver Golgi complex: minor proteins are identified through sequential fractionation, *Electrophoresis* 21: 3441–3459.

Wilkins, MR., Gasteiger, E., Sanchez, JC., Bairoch, A. & Hochstrasser, DF. (1998). Two-dimensional gel electrophoresis for proteome projects: the effects of protein hydrophobicity and copy number, *Electrophoresis* 19: 1501–1505.

Wilkins, MR., Sanchez, JC., Gooley, AA., Appel, RD., Humphery-Smith, I., Hochstrasser, DF. & Williams, KL. (1996). Progress with proteome projects: Why all proteins expressed by a genome should be identified and how to do it. *Biotechnol Genet Eng Rev* 13: 19–50.

Wilkins, MR., Williams, KL; Appel, RD. & Hochstrasser, DF. (1998). Proteome Research: *New Frontiers in Functional Genomics*, Springer, Berlin.

6

Two-Dimensional Gel Electrophoresis and Mass Spectrometry in Studies of Nanoparticle-Protein Interactions

Helen Karlsson, Stefan Ljunggren, Maria Ahrén, Bijar Ghafouri,
Kajsa Uvdal, Mats Lindahl and Anders Ljungman
Linköping University; County Council of Östergötland
Sweden

1. Introduction

1.1 Nanoparticles

Adverse health effects have been associated with the exposure to particulate matter (PM) ever since the London smog in the winter of 1952. Recent estimates attribute about 12,000 excess deaths to have occurred because of acute and persisting effects of the London smog (Bell & Davis, 2001). Over the years a number of epidemiological studies have shown that PM from combustion sources such as motor vehicles contributes to respiratory and cardiovascular morbidity and mortality (Kreyling et al., 2002, 2006, Wick et al., 2010). Especially so do the ultra-fine particles (UFPs) with a diameter less than 0.1 micrometer. UFPs from combustion engines are capable to translocate over the alveolar–capillary barrier (Rothen-Rutishauser et al., 2007). When nano-sized PM (nanoparticles, NP), which are small enough to enter the blood stream, do so they are likely to interact with plasma proteins and this protein-NP interaction will probably affect the fate of and the effects caused by the NPs in the human body. Herein we present results showing that several proteins indeed are associated to NPs that have *in vitro* been introduced to human blood plasma.

NPs are atoms and molecules defined as particles less than 100 nanometers in at least one dimension (Elsaesser & Howard, 2011). Due to the plethora of NPs being produced in various forms (e.g. spherical, fibers, rods, clusters) or by different processes (e.g. flame-spray synthesis, chemical vapor deposition), defining the characteristics of a NP is not an easy task even when it comes to manufactured NPs and when considering those formed unintentionally during processes such as combustion in motor vehicles it becomes an even harder task. This variation in properties according to the respective composition of NPs is also the basis of the wide range of potential applications, from medicine to consumer products. Due to the unique physicochemical properties of nanomaterials, there are plenty of possibilities for NPs to enter the human body, either deliberately as medicines or unintentionally as environmental contaminants and thus potentially cause adverse human health effects (Elsaesser & Howard, 2011; Stern & McNeil, 2008). Although many characteristics have been highlighted as driving the potential adverse health effects associated with NP exposure, it has been specifically the size and increased surface area of NPs that has been concluded as elucidating any such adverse effects observed (Elsaesser & Howard, 2011; Stern & McNeil, 2008).

UFPs from combustion sources, such as motor vehicles, are capable to promote atherosclerosis, thrombogenesis and other cardiovascular events mainly via the ability to induce inflammatory and protrombotic responses (McAuliffe & Perry 2007). Thus, NP induced effects within the lung has been studied over the past twenty years (Mühlfeld et al., 2008; Rothen-Rutishauser et al., 2007). Indeed, exposure to most forms of NPs will initially be via inhalation, especially when considering occupational exposure, and thus will affect the respiratory system. The respiratory system is by far the main port of entry even though the gut and skin are also possible ways of entry. However, NP localization and fate is not only restricted to their portal of entry. NPs can be distributed to organs distal to their site of exposure, so that potential NP toxicity can occur in any secondary site. Research into the possible secondary toxicity of NPs is quite limited. Even so studies investigating the effects of NP translocation to secondary organs have shown that NPs can elicit negative effects to the liver, brain, GI tract (following inhalation), spleen, reproductive systems and the placenta (Kreyling et al., 2002, 2006; McAuliffe & Perry 2007; Wick et al., 2010). NPs toxic mechanisms at the cellular level includes protein misfolding and protein fibrillation causing major problems in the brain and chronic inflammation as a result of nanoparticle exposure, for example in the lung and other organs, via frustrated phagocytosis or production of reactive oxygen species. A vulnerable target for possible toxicological effects of nanoparticles is the fetus. Gold nanoparticles have been shown to cross the maternal-fetal barrier and fullerenes were found to have a fatal effect on mouse embryos (Elsaesser & Howard, 2011; Stern & McNeil, 2008).

When nanoparticles enter the blood vessels, or any biological fluid, e.g. saliva or mucus they are most likely surrounded by a layer of proteins. This dynamic protein "corona" depends on the concentration in the biological fluid and hence the composition of the layer varies in different parts of the body (Cedervall et al., 2007; Lundqvist et al., 2008; Lynch et al., 2007; Walczyk et al., 2010,). Thus, the reactions in the body to such a NP-protein complex is most likely different from that induced by the bare NP and possibly affecting bio-distribution and thereby causing unwanted side effects (Adiseshaiah et al., 2010; Leszczynski 2010,). The function of protein coating is not fully known but since most nanoparticles show strong affinity for proteins, it is of importance to investigate this interaction in different fluids. In blood, plasma proteins constituting the NP corona is possibly affecting a wide range of effects such as phagocytosis via immunoglobulins and complement (Dobrovolskaia & McNeil 2007), coagulation via prothrombin (Dobrovolskaia 2009) and the distribution of lipoproteins (Benderly et al., 2009; Hellstrand et al., 2009; Zensi et al., 2010).

The availability and toxicity of any substance to a biological organism is determined by both the concentration/dose that the organism is exposed to, as well as the "toxicokinetics" of the substance. These include the uptake, transport, metabolism and sequestration to different compartments by the organism, as well as the elimination of the substance from the biological organism. These parameters are essential since the potential toxicity of substances is dependent upon the specific organs or cell types exposed, which form the substance is in (e.g. bound to serum protein, aggregated, dissolved, oxidized), as well as the period of time the substance interacts/remains at the site of primary and secondary exposure. These parameters are influenced by the physical–chemical characteristics of the substance, therefore a detailed characterization of the substance is pivotal in order to allow generalizable conclusions and should therefore be given ample attention. Furthermore, most of these toxicological parameters involve proteins and/or actions carried out by proteins. Thus it is pivotal to the understanding of NP toxicology, and thereby the possibility to predict health effects caused by NPs, to understand NPs interactions with proteins. One of the best, if not the best, technique to separate proteins is two-dimensional gel electrophoresis (2-DE). Preferably, this is combined with peptide mass fingerprinting and MALDI-TOF MS analyses for fast identification of the separated proteins, which then may be followed by tandem MS analyses for sequence information. Here we present our results regarding serum protein interactions with metal-nanoparticles (Al_2O_3, $ZnO/Al6\%$, SiO_2) and Carbon Nanotubes, obtained using 2-DE and MS. Furthermore, we elaborate on and exemplify different aspects of the 2-DE/peptide mass fingerprinting-technique to further improve the approach.

1.2 General introduction to the 2-DE technique

Two-dimensional gel electrophoresis (2-DE) is an excellent method for separation of proteins from most kinds of tissues and complex mixtures of proteins (O`Farrel-1975). Both qualitative characterization of the protein expression, including post-translational modifications and quantitative characterization comparing the protein expression in different individuals or groups, are possible by this technique. Two steps are included, the isoelectric focusing (IEF) step, where the proteins are separated according to their isoelectric point (pI) in a pH-gradient, and the sodium dodecyl sulphate polyacrylamide gel electrophoresis (SDS-PAGE) step, where the proteins are separated according to their molecular weight. Since it is less common that two proteins have the same isoelectric point and molecular weight, this will result in each protein migrating to its own unique position. The 2-DE technique allows, depending on the nature of the sample, the separation of 500-3000 protein spots and the resolution can be improved, e.g. by removal of abundant proteins or by composite gels from overlapping pH-gradients. Proteins separated by gel electrophoresis can be visualized by a number of methods using different types of stains. Various stains interact differently with the proteins and some of the stains used are not even specific for proteins. The degree of sensitivity is also different. Processing data from stained protein gels by computers includes the gel images being digitized by an imaging system and then analyzed using computer software allowing a number of different measurements such as number, size, and intensity of the stained protein spots. Separated proteins are then identified by mass spectrometry (MS). The proteins are in-gel digested and extracted peptides analyzed by peptide mass fingerprinting or peptide sequencing. Two widely used MS instruments used for these respective analyses are matrix assisted laser desorption/ionization-time of flight-mass spectrometry (MALDI-TOF MS) where peptides are transferred from solid phase to gas phase, and electrospray ionization tandem mass spectrometry (ESI MS/MS) where peptides are transferred from liquid phase to gas phase. The key advantage with 2-DE is the ability to separate protein isoforms. On the other hand, very large and hydrophobic proteins are underrepresented in 2-DE and the need of peptide extraction from the in-gel digests may influence to amounts of analytes available for MS protein identification. Nevertheless, combining the separation and analytical ability of the 2-DE technique with the identification power of MS provides a powerful tool in human toxicology.

2. Methods

2.1 Characterization of nanoparticles

Commercial SiO_2 size 0.007μ (S-3051), ZnO/6% Al doped <50nm(677450), and Al_2O_3 <50nm (544833) were purchased from Sigma Aldrich. As comparison, single walled Carbon Nanotubes (CNO, 704121) from Sigma Aldrich was used. Nanoparticles, characterized by manufacturer, were dispersed in Milli-Q water and/or PBS (137 mM NaCl, 2.7 mM KCl, 8.45 mM Na_2HPO_4, 1.47 mM KH_2PO_4 pH 7.3) and sonicated on ice for 10 min. The hydrodynamic sizes of the particles were analyzed by Dynamic Light Scattering (DLS).

DLS measurements were performed using an ALV/DLS/SLS-5022F system (ALV-GmbH, Langen Germany) and a HeNe laser at 632.8 nm with 22 mW output power. The scattering angle was 90° and the temperature 22°C. For temperature stabilizing purposes, samples were placed in a thermostat bath (22°C) for at least 10 minutes prior to the measurements. Samples were diluted in Milli-Q water or PBS. Ultrasonication of two different kinds was performed to decrease the degree of agglomeration; either an ultrasonic bar homogeniser (Sonoplus HD 2200, Bandelin electronics, Germany) was used or samples were placed in an ultrasonic bath (USC300T, VWR, Sweden). The viscosity of PBS was set to 0.9782 mPa's by linear interpolation between tabulated values for 20°C (0.911) and 25 °C (1.023) (Hackley & Clogston, 2007).

Data analysis was performed using a nonlinear fit model via ALV-Regularized Fit (ALV-Correlator Software Version 3.0. using ALV-Regularized in nonlinear fit model http://www.alvgmbh.de/).

2.2 Preparation of human plasma

Human plasma, collected in sodium citrate tubes, was prepared from three healthy volunteers. After cooling, the blood was centrifuged in 800g for 10 min and plasma, free from red blood cells, was drawn from the top of the tube. SiO_2, ZnO/6% Al doped, Al_2O_3 and CNO were then exposed to three plasma samples respectively. In all nanoparticle exposures fresh plasma was used.

2.3 Nanoparticle/plasma incubation

Nanoparticles were dissolved (final concentration 2 mg/ml) in PBS and incubated with 1% plasma at 37°C for 1h. As control, to ensure there was no protein precipitation, one sample was prepared without nanoparticles. Unbound proteins were separated from nanoparticles by centrifugation for 40 min at 50 000g and 4°C. The supernatant was discarded and the particle pellet was washed in Dithiothreitol (DTT, Sigma-Aldrich) 20mM/-acetone buffer followed by a second centrifugation step for 10 min at 50 000g and 4°C. The supernatant was discarded and the pellet was air dried. The pellets containing nanoparticles and attached proteins were then dissolved in denaturing solution containing 9M Urea (Sigma-Aldrich), 65mM DTT and 4% (3-(3-cholamidopropyl)-dimethylamino)-1-propanesulfonate (CHAPS). The solution was incubated at room temperature for 30 min before denatured proteins were separated from the nanoparticles by centrifugation for 30 min at 50 000g and 4°C. The supernatant was collected and the samples were analyzed in triplicates. 20 μl (40μg protein) was applied on the IEF strip in each 2-DE analysis.

2.4 Isolation of lipoproteins

2.4.1 HDL isolation

Preparation of high density lipoprotein (HDL) was performed by a method described by Sattler et al.1994, with slight modifications (Karlsson et al., 2005). Blood samples in EDTA-containing tubes were obtained from healthy volunteers after an overnight fast. After centrifugation (10 min, $700g$) at room temperature, plasma was collected. EDTA (1 mg/mL) and sucrose (final concentration 0.5%) were added to prevent HDL oxidation and aggregation, respectively. Five milliliters of EDTA-plasma adjusted to a density of 1.24 g/mL with solid KBr (0.3816 g/mL) was layered in the bottom of a centrifuge tube (Beckman, Ultraclear tube). The EDTA plasma fraction was gently overlayered with KBr/PBS solution (0.0834 g KBr/mL, total density 1.063 g/mL). In one centrifuge tube, proteins were stained with Coomassie Brilliant Blue to be used as a reference while collecting the HDL fraction. Ultracentrifugation was performed in a Beckman XL-90 equipped with a Ti 70 rotor (fixed angle; Beckman Instruments, Fullerton, CA, USA) for 4 h at 290 000g and 15°C. By this procedure the lipoprotein fractions with a density lower than 1.063 g/mL (low density and very low density lipoprotein) are located at the top of the tube and HDL is located in the middle of the tube. HDL was collected by penetrating the tube with a syringe. To avoid contamination by serum proteins, HDL were then further purified by a second centrifugation step. KBr/PBS solution (0.3816 g KBr/mL) was added to the HDL (total density 1.24 g/mL) and the centrifugation was performed under the same conditions as described above, but for 2 h. HDL was collected from the top of the tube and desalted using desalting buffer (NH_4HCO_3, 12mM, pH 7.1) and PD 10 columns (Sephadex™ G-25 M, GE Healthcare, Buckinghamshire, United Kingdom). Protein concentration in the HDL solution was determined with Bio-Rad protein assay (Bio-Rad, Richmond, CA, USA). Sample (3.5 mL) was lyophilized and dissolved in 0.25 mL sample solution (9 M Urea, 4% CHAPS, 2% Pharmalyte , 65 mM DTT, 1% bromophenol blue) according to Görg et al.1988.

2.4.2 Immunoaffinity chromatography

Anti-ApoA-I antibodies were attached to a 5 mL HiTrap NHS-activated HP column (GE Healthcare) according to manufacturer's instructions. 2.5 mL plasma were desalted by the use of PD-10 columns and the eluted sample were diluted to 4 mL with 50 mM Tris-HCl, 0.15M NaCl, pH 7.5. The ApoA-I coupled immunoaffinity column were equilibrated by allowing 10 column volumes flow through it. Desalted sample were applied into the column and allowed to recirculate for 40 minutes. Loop were disconnected and washed with 10 column volumes of 50 mM Tris-HCl, 0.15M NaCl, pH 7.5 followed by ten column volumes of 50 mM Tris-HCl, 0.5M NaCl, pH 7.5. ApoA-I adsorbed to the column were eluted with 20 mL of 0.1M Glycin-HCl, pH 2.2. Sample was collected in fractions of 0.4 mL in tubes which each contained 20 μL 1M Tris, pH 9.0 for pH-neutralization of the sample. Fractions containing proteins were pooled and desalted using PD-10 columns.

2.5 Albumin and IgG removal

The removal of the high abundance proteins albumin and IgG from plasma was performed using an Albumin and IgG removal kit (GE Healthcare). Briefly, the column was equilibrated with binding buffer (20mM $Na_2H_2PO_4$, 0.15M NaCl, pH 7.4), 50 μl of plasma was diluted to a volume of 100 μl and applied to the column. After 5 min incubation, the depleted sample was collected in an eppendorf tube by centrifugation at 784g. The total protein concentration before/after depletion was determined with Bio-Rad protein assay. After depletion of high abundant proteins the samples were desalted using PD-10 columns.

2.6 2-DE analysis

2-DE was performed using IPGphor and Multiphor (GE healthcare). Briefly, the proteins were resuspended in 150 μL of a 2-DE sample buffer containing 9 M urea, 65 mM DTT, 2% Pharmalyte (GE Healthcare), 4% CHAPS, and 1% bromophenol blue and then centrifuged at 4°C and 23000g for 30 min to remove debris. The supernatant was then mixed with a rehydration buffer consisting of 8 M urea, 4% CHAPS, 0.5% IPG buffer 3-10 NL (GE Healthcare), 19 mM DTT, and 5.5 mM Orange G to a final volume of 350 μL. The first dimension was performed by in-gel rehydration for 12 h in 30 V on 18 cm pH 4-7 linear or pH 3-10 nonlinear IPG strips (Immobiline DryStrips, GE Healthcare). The proteins were then focused at 53000 Vh at a maximum voltage of 8000 V (Görg et al., 2000). The second dimension (SDS-PAGE) was performed by transferring the pI focused proteins (IPG strips) to homogeneous or gradient home-cast gels on gel bonds. The electrophoresis was performed at 40-800 V, 10°C, 20-40 mA, overnight.

2.7 Staining and image analysis

Sypro Ruby (Bio-Rad) staining were done according to manufacturer's instructions. In short, gels to be stained with Sypro Ruby were directly placed in a fixing solution containing 10% methanol and 7% acetic acid for at least 20 minutes after 2-DE. Gels were then washed 3x10 minutes under agitation with Milli-Q water before approximately 400 mL of Sypro Ruby stain were added and incubated in room temperature over night.

Silver staining of gels were done according to Shevchenko et al. 1996, with some few modifications. Proteins were fixed by incubating the gel in 50% methanol and 5% acetic acid for at least 20 minutes directly after 2-DE and then incubated with 50 % Methanol for 5 minutes, followed by Milli-Q water for 10 minutes. In the sensitizing step the gel was incubated with 0.02 % sodium thiosulphate for 1 minute, followed by 2x1 minutes washing with Milli-Q water. The gel was then immersed in 0.1 % silver nitrate solution for 20 minutes before excess of silver was washed away by 2x1 minute in Milli-Q water. Next, the gel was developed in 0.04 % formaldehyde in 2 % sodium bicarbonate solution for 2x1 minute. The exact developing time was optimized depending of the protein amount in the gel. Finally, the reaction was stopped by incubation 1x5 min in 0.5 % glycine and the gel washed with Milli-Q water for 2x20 min.

The images of the protein patterns were analyzed by a CCD (Charge-Coupled Device) camera digitizing at 1340*1040 pixel resolution in a UV scanning illumination mode for Sypro Ruby stained gels or at 1024*1024 pixel resolution in white light mode for silver stained gels using a Flour-S-Multi Imager in combination with a computerized imaging 12-bit system (PDQuest 2-D gel analysis software, version 7.1.1). The unit of the UV light source is expressed in counts while the unit of the white light source is expressed as optical density (OD). Gel images were evaluated by spot detection, spot intensities and geometric properties.

2.8 Isolation of protein spots

Protein spots were excised from the gels using a syringe and transferred to eppendorf tubes. For silver destaining, 25 µL of 100 mM sodium thiosulphate and 25 µL of potassium ferricyanide were added to the gel pieces (Gharahdaghi et al., 1999). When the pieces were completely destained, the chemicals were removed by washing (6×5 min with Milli-Q water) before addition of 50 µL of 200 mM ammonium bicarbonate and incubation for 20 min at room temperature. The gel pieces were washed (3x5 min with Milli-Q water) and dehydrated with 100% acetonitrile (ACN) for 5 min or until the gel pieces were opaque white. After removal of the ACN, the gel pieces were dried in a SpeedVac vacuum concentration system (Savant, Farmingdale, NY). Protein spots excised from Sypro Ruby stained gels were washed in 50% ACN/25 mM ammonium bicarbonate 2x30 min prior to dehydration with 100% ACN.

2.9 Digestion

2.9.1 Tryptic digestion

The protein spots were excised from the gel with a syringe and transferred to small eppendorf tubes (0.5 mL). Proteins from fluorescently stained gels were visualized and excised on a blue light transluminator (DR-180 B from Clara Chemical Research, Denver, CO, USA) wearing darkened amber glasses. After destaining and dehydration, about 25 µL trypsin (20 mg/mL in 25 mM ammonium bicarbonate, Promega, Madison, WI, USA) was added to each gel piece. To minimize autocatalytic activity, the samples were kept on ice for 30 min, prior to incubation in 37° C over night. The supernatant was transferred to a separate tube and the peptides were further extracted from the gel piece by incubation in 50% ACN/5% trifluoroacetic acid (TFA, Sigma-Aldrich) for 5 h at room temperature. The supernatant from the two steps was then pooled and dried in SpeedVac until complete dryness. If not dissolved in 5 µL 0.1% TFA for further MS preparation, the proteins were stored at -70°C.

2.9.2 Glu-C digestion

Glu-C (Roche, Basel, Switzerland) was diluted with 25 mM NH$_4$HCO$_3$, pH 7.8, to a concentration of 20 µg/mL. 25 µL was added to each gel piece and incubated at room temperature over night. The supernatant was then dried in SpeedVac until complete dryness. If not dissolved in 5 µL 0.1% TFA for further MS preparation the proteins were stored at -70°C.

2.9.3 Cyanobromide digestion

One Cynobromide (CNBr, Sigma-Aldrich) crystal was dissolved in 250 µL 70% TFA (in dH$_2$O). 25 µL CNBr in 70% TFA was added to the dried gel piece and incubated in darkness in room temperature overnight. The supernatant was then dried in SpeedVac until complete dryness. If not dissolved in 5 µL 0.1% TFA for further MS preparation the proteins were stored at -70°C.

2.9.4 Endoproteinase Asp-N digestion

Asp-N (P3303, Sigma-Aldrich) was diluted with 100 mM NH$_4$HCO$_3$ pH 8.5 to an enzyme concentration of 8 µg/mL. 25 µL was added to one tube containing one gel piece. To minimize autocatalytic activity, the samples were kept on ice for 30 min, prior to incubation in 37° C over night. The supernatant was then dried in SpeedVac until complete dryness. If not dissolved in 5 μL 0.1% TFA for further MS preparation the proteins were stored at -70°C.

2.9.5 Enzymatic deglycosylation

HDL (500 μg) was lyophilized and dissolved in 100 μL 2-DE sample buffer (described earlier).The sample was then incubated with 20 U of PNGase F (Sigma-Aldrich, P7367) in 37° C over night. After incubation, sample were stored in -20°C until 2-DE.

HDL was desalted in a PD-10 column and eluted in a 50 mM Na_2HPO_4, pH 6.0 buffer. A volume corresponding to 1200 μg of HDL proteins were then incubated with 10 U of Neuraminidase (Sigma-Aldrich, N3786) in 37° C for 4 hours. After incubation, sample was desalted in a PD-10 column with desalting buffer (NH_4HCO_3, 12mM, pH 7.1) and subsequently frozen in -70°C. Sample was lyophilized before 2-DE.

2.10 ZipTip

After digestion and drying of peptide samples, some were desalted and purified by the use of ZipTip$_{C18}$® pipette tips (Millipore, Billerica, MA, USA). Samples were diluted up to 10 μL with 0.1% TFA. A ZipTip was wet by loading 3x10 μL with 50% ACN and discarding the liquid. Ziptip was equilibrated by loading 3x10 μL of 0.1% TFA and discarding the fluid before peptide sample was carefully loaded into the ZipTip by pipetting. The ZipTip was washed with 5x10 μL of 0.1% TFA before peptides were eluted with 10 μL 50% ACN.

2.11 Mass spectrometry

Peptides obtained after digestion were mixed 1:1 with matrix, α-Cyano-4-hydroxycinnamic acid (CHCA, 0.02 mg/mL) or 2,5-dihydroxybenzoic acid (DHB, 0.02 mg/mL) in 70% ACN/0.3% TFA, and then spotted onto a stainless steel target plate. Analyses of peptide masses were performed using MALDI TOF-MS (Voyager DE PRO; Applied Biosystems) equipped with a 337 nm N2 laser operated in reflector mode with delayed extraction. Positive ionization, a delay time of 200 ns, and an accelerating voltage of 20 kV were used to collect spectra in the mass range of 600–3600 Da. Data processing of the spectra was performed in a Data Explorer TM Version 4.0 (Applied Biosystems). External mass calibration with a standard peptide mixture and internal calibration using known trypsin autolysis peaks (m/z 842.5100, 1045.5642, 2211.1046) were also performed prior to the database search. For tandem MS analysis, the digested peptides were dried and dissolved in 10 μL 0.1% TFA. The peptides were desalted and purified by using ZipTip$_{C18}$® columns. Elution was acidified by the addition of 1% formic acid. About 2 μL of the sample was applied into a silver-coated glass capillary and analyzed by a hybrid (triple quadrupole-TOF) mass spectrometer (API Q-STAR Pulzer; Applied Biosystems) equipped with a nanoelectrospray ion source (MDS-Protana, Odense, Denmark) operated in the nanopositive mode. Data processing was performed with Analyst QS software (Applied Biosystems). Fragmentation spectra were interpreted manually.

2.12 Database search

Peptide masses (major peaks) in the spectra were submitted to database search. NCBI and Swiss-Prot were used with Aldente or MS-Fit as search engines. Restrictions were human species, mass tolerance >75 ppm in most of the searches, maximum one missed cleavage, and cysteine modification by carbamidomethylation. MS-Digest, MS product, and BLAST search was used for protein identification of the derived tags resulting from amino acid sequencing with MS/MS. In peptide mass fingerprinting, protein matches with p-values below 0.05 are used and with LC-MS/MS analyses an FDR $\leq$ 1 % is considered significant.

2.13 Western blot

Plasma proteins were separated on 2-DE. Proteins were then transferred to a 0.2 µm PVDF membrane. After blocking with 5% non-fat dry milk in Tris buffered saline (TBS) overnight, the membrane was washed two times with Tween-Tris buffered saline (TTBS, pH 7.5) and then incubated overnight with primary rabbit anti human C-III antibodies (Abcam, 21032, 1:5000) in 2% non-fat dry milk in TTBS (pH 7.5) at room temperature. After washing four times with TTBS, the membrane was further incubated for 1h with secondary goat anti rabbit antibodies conjugated with horse radish peroxidase (HRP, 170-6515, BioRad,1:40 000) in 2% non-fat dry milk in TTBS (pH 7.5) at room temperature. In order to visualize the proteins the PVDF membrane was treated with ECL Plus Western Blotting Detection System (GE Healthcare) and then exposed to X-ray film (AGFA Medical, Mortsel, Belgium).

3. Results and discussion

3.1 Nanoparticle characterization

DLS, which is also known as Photon Correlation Spectroscopy or Quasi-Electron Light Scattering, is a technique used to study the size and size distribution of particles suspended in a liquid. The technique is based on the scattering of light of particles in diffusive random (brownian) motion. The average displacement for the Brownian motion is defined by the translational diffusion coefficient (D). The particle diffusive motion in liquid is size dependent, and a larger particle has a slower motion as compared to a smaller particle. This brownian motion can be investigated by irradiating the sample with a coherent laser and studying the intensity fluctuations of the scattered light (Finsy, 1994).

Particle sizing can be done in several ways. Typically the information retrieved from different techniques is to some extent diverse, as each technique is sensitive to it's specific properties of the particles. That means that a technique which is based upon the scattering intensity does not deliver the same size or size distribution as a technique that is based upon the projected area or the density of a nanoparticle. For nano sized particles, transmission electron microscopy (TEM) is frequently used in purpose to study the size and the shape of particles. In TEM, the sample preparation together with the measurement is relatively time consuming and furthermore the measurement is limited only to a very small fraction of the sample. This means that a lot of replicates must be studied in order to achieve good statistics. A dynamic light scattering measurement on the other hand, is fast and convenient as it usually takes only a few minutes to perform. Data recording procedure is thus short but the analysis and interpretation requires knowledge and care. DLS measurements must be performed with highly diluted solutions, to avoid multiple scattering phenomenons and misleading artifacts are frequently present in DLS studies.

Particle sizes obtained when measuring with DLS are by default larger than those obtained when analyzing the material with TEM. The size calculated from the translational diffusion coefficient in DLS generally is referred to as the hydrodynamic diameter, e.g the diameter of a sphere having the same diffusion coefficient as the particle, while the size of the nanoparticles obtained from TEM is the core size of the nanoparticle investigated. It should be noted that in many cases when the sample consists of a mixture of nanoparticles with a range of sizes and/or mixture of particle shapes, results should be taken as an estimation only but clearly trends can be observed.

The autocorrelation function of the scattered intensity results in an average value of the product of the intensity at time t and the intensity at a time delay later t+dt. The value obtained from the correlation function is large for short delays, since the intensities are highly correlated. The value will be low for longer delays; i.e. the autocorrelation can be described as a decaying function of time delay. From the autocorrelation function, the diffusion constant D can be determined. Furthermore, by using Stokes-Einstein equation the corresponding size distribution is calculated according to:

$$d = k_B T / 3\pi\eta D \tag{1}$$

where k_B is the Boltzman constant, T the temperature, η the viscosity of the solvent and d the hydrodynamic diameter of the particles. This implies that the temperature must be constant during the measurement and the viscosity of the sample solvent must be known. It should be noted that the formula shown above (Equation 1) is valid only for non-interacting spherically shaped particles, i.e experimental data are fitted to a model assuming spherical particles (Finsy, 1994).

Figure 1 shows five different functions representing the typical information retrieved in DLS in the actual measurement and by using algorithms. The measured data in DLS is the correlation function (Figure 1A). This function holds information about the diffusion of particles in the sample and can be transformed to a graph showing the decay time of light scattering fluctuations (Figure 1B). From the decay time distribution function, the values of the particle radius (Figure 1C) can be calculated using the Stokes-Einstein equation.

The amount of light that is scattered from a particle is dependent on the particle size. According to the Rayleigh theory (Sorensen, 2008), the scattering factor roughly is proportional to the sixth power of the particle size. This means that a small particle scatter light less than a larger particle and thus different weights have to be applied to transform the intensity weighted data to a useful size distribution. These weights can be mass based (Figure 1D) or number based (Figure 1E). Powder based samples that are dispersed in a liquid often are severely aggregated. Ultrasonic baths can be used to decrease the aggregation in the solution. Choice of solvent is also important and it clearly affects the capability of dispersing nanoparticles.

A number of examples of DLS results for commercial particles are shown in Figure 2, 3 and 4. The size distributions of Al_2O_3 nanoparticles , based upon number weighted fits of the data are shown in Figure 2. A set of samples were dispersed both in Milli-Q water and PBS. The size and size distribution were measured as a function of concentration. According to the supplier these particles are < 50 nm in size as measured by TEM, which should be taken as the core size of the nanocrystals in the Al_2O_3 material. A lot larger hydrodynamic diameters are achieved in our measurements, which show that the water based sample is composed of at least two populations with hydrodynamic radius of about 100 nm and 200-300 nm respectively. The smallest radius from this DLS study was achieved for Al_2O_3 dispersed in Milli-Q water and ultrasonicated with a bar homogenizer (radius approx. 30 nm). Also for samples dispersed in PBS, ultrasonication with a bar homogenizer indicates decrease of the aggregation and introduction of populations with smaller radius.

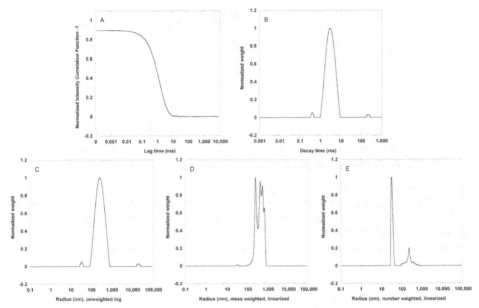

Fig. 1. Example of the typical information retrieved in a DLS measurement.
The measurement is performed on Al doped ZnO nanoparticles dissolved in MilliQ water.
A) shows the correlation function which has a high correlation (close to 1) for really short lag times but decays to zero with a rate dependent on the particle size distribution in the sample. B) shows the normalized distribution function of the Decay time. C) shows the normalized distribution of the unweighted radius and the corresponding normalized distributions of the mass weighted radius and the number weighted radius are shown in D) and E) respectively.

Figure 3 shows the size distribution of Aluminium doped ZnO nanoparticles dispersed in water and PBS. Bar ultrasonication introduces smaller sized populations, in consistence with the results in Figure 2. The hydrodynamic radii of these particles in MilliQ water are around 30 nm as largest according to Figure 3. This could be compared with the information from the supplier that these particles are < 50 nm as measured by TEM. Again hydrodynamic diameter is always larger that the core size of nanocrystals obtained from TEM. In this case the sample is also aggregated in solution, which produces even larger sizes and size distributions.

The SiO_2 particles are 7 nm sized as primary particles according to the supplier based on calculations using the surface area as measured by the nitrogen adsorption method of Brunauer (Brunauer et al., 1938). The supplier also states that these particles commonly form

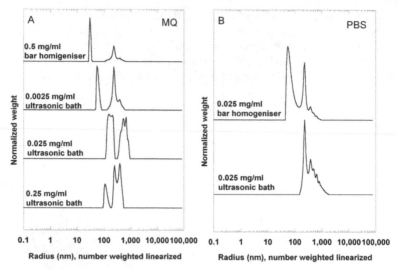

Fig. 2. DLS measurements of Al$_2$O$_3$ nanoparticles.
The normalized distribution of the number weighted radii of particles in Al$_2$O$_3$ nanopowder diluted to different concentrations in A) MilliQ and B) PBS as measured by DLS. As marked in the figure, two different kinds of ultrasonic treatment were used to decrease the aggregation before performing the measurement; either an ultrasonic bath or a bar homogenizer.

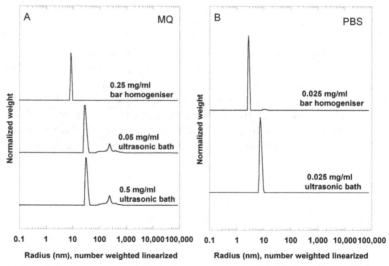

Fig. 3. DLS measurements of ZnO (Aluminium 6% doped) nanoparticles.
The normalized distribution of the number weighted radii of particles in ZnO (Aluminium 6% doped) nanopowder diluted to different concentrations in A) MilliQ and B) PBS as measured by DLS. As marked in the figure, two different kinds of ultrasonic treatment were used to decrease the aggregation before performing the measurement; either an ultrasonic bath or a bar homogenizer.

some hundreds of nanometer long chainlike branched aggregates. The size distributions obtained in our DLS measurements are presented in Figure 4. For all three concentrations in Milli-Q water the particle size is below 100 nm in radius. When SiO_2 particles are dispersed in PBS (Figure 4B), large aggregates are definitely present, which are partly removed when ultrasonicated with a bar homogenizer. Multiple scattering, surface charge on the nanoparticles and water solubility should be considered when further evaluating these data. Furthermore, it is known that these specific samples (SiO_2) are inhomogenous and the sample is thus far from ideal, i.e does not contain spherical shaped particles. It is shown in previous studies that long chainlike branched aggregates are present.

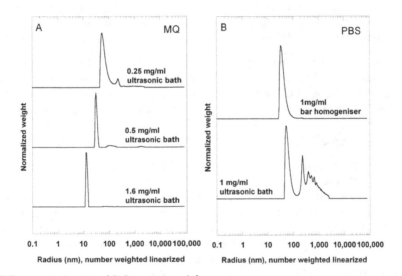

Fig. 4. DLS measurements of SiO_2 nanoparticles.
The normalized distribution of the number weighted radii of particles in SiO_2 nanopowder diluted to different concentrations in A) MilliQ and B) PBS as measured by DLS. As marked in the figure, two different kinds of ultrasonic treatment were used to decrease the aggregation before performing the measurement; either an ultrasonic bath or a bar homogenizer.

In conclusion, it could be said that size and size distribution of nanoparticle samples could be estimated by DLS. Valuable information as the trends in size distribution connected to sample preparation methods and choice of solvent can be obtained. Sample preparation methods are indeed very important as well as choice of solvent. Care should be taken when choosing fitting model and the model-inbuilt parameters. Inhomogeneous samples are less straight forward to analyze. Presence of aggregates is easily detected. In summary information obtained from DLS is important for everybody that is doing research on nanoparticles in liquids. The numbers given as product information i.e. the size and size distribution, are often relevant for the core-size of the nanocrystals within the material. However the nanoparticles are most often not soluble to that extent. Nanoparticles obtained in dry state and then dispersed in liquid usually form aggregates as shown in this study.

3.2 Nanoparticle-plasmaprotein interactions

In a previous study performed by us, the inflammatory response in human monocyte derived macrophages after exposure to wear particles generated from the interface of studded tires and granite containing pavement (Karlsson et al. 2011) was investigated. Particle characterization showed that dominating peaks in the EDX spectra were Silica and Aluminium. Particles of nanosize were also present (SMPS), but it was not possible to characterize them due to low abundance. As a result of their very small diameter (< 0.1 μm), inhaled nanoparticles are believed to be predominantly agglomerated and deposited in the periphery of the lungs, where they interact with cells such as macrophages and epithelial cells (Beck-Speier et al. 2005) but they may also translocate into the circulation, which is a critical step, since their fate *in vivo* in not known. Investigating plasma protein-nanoparticle interactions with a toxico-proteomic approach is a useful tool to improve our knowledge about the effects of nanoparticles of different origin, size and surface properties in biological systems.

In purpose to mimic a potential exposure to airborn nanoparticles translocated into the circulation, commercial SiO_2 and Al_2O_3 were mixed with plasma proteins. As comparison, commercial ZnO (Al-doped 6%) and a non-metal oxide; single walled Carbon Nanotubes was used. All preparations were performed in triplicates with three different subjects exposed to each type of particles. The protein patterns resulting from the three different exposures of commercial SiO_2, Al_2O_3, ZnO and CNO respectively were identical.

Particle characterization and estimation of particle agglomeration prior to exposure is crucial. In a recent study of nanoparticle-plasma protein interactions (Deng et al. 2009), the DLS spectra indicated large agglomerates prior to plasma protein exposure. Most likely, and in line with the authors suggestions (Deng et al. 2009), complexes with hydrodynamic size of 10000-100 000 nm do not result in the same protein patterns as the interactions of smaller particles/agglomerates and plasma proteins. In our optimized protocol, with different particle origin, less gentle bar sonication instead of in water bath and thereby reduced hydro dynamic sizes of the agglomerates (Fig 2-4) - an altered pattern of interacting proteins was indeed found (Figure 5, Table 1).

Interestingly, the interaction of SiO_2 and CNO with plasma proteins resulted in very similar protein patterns despite their different properties (Fig 5). It has to be stated though, that the fate of CNO in the lung may not be translocation into the circulation due to the tube like structure, but CNO is also of interest for medical applications (Wu et al. 2011).

The transport proteins Albumin and Alpha-2-HS-glycoprotein interacted with SiO_2, CNO and ZnO but not Al_2O_3 while Transferrin interacted with SiO_2, Al_2O_3 and notably also CNO but not ZnO. Supporting our results, the binding of albumin to single walled CNO has previously been described to promote uptake by the scavenger receptor in RAW cells (Dutta et al., 2007). In line, intravenous administration of CNO has in a different study resulted in high localization in the liver (Cherukuri et al., 2006). Another possible way for nanoparticles into the cells, are as Transferrin/particle complexes that are able to enter the cells via the Transferrin receptor. The Transferrin receptor is an interesting and relevant target in cancer research since its expression is increased in tumor cells. In a recent study, Transferrin covalently attached to silica nanoparticles carrying a hydrophobic drug caused an increase in mortality of the targeted cancer cells compared to cells exposed to nontargeted particles and free drug (Ferris et al. 2011). What demand further studies though, is the fate of the silica particles in a longer perspective, taken up by tumor cells as well as other cells. A third transport protein, the thyroxin transporting protein Transthyretin, also known to bind toxic components in the blood stream (Hamers et al., 2011), was found to interact with CNO and SiO_2. This finding confirms a previous study that pointed out that silica interaction (inhaled) with plasma Transthyretin is contributing to the stabilization of fibroids in rat lungs (Kim et al., 2005). The possible effects of CNO interaction with Transthyretin remains to be investigated.

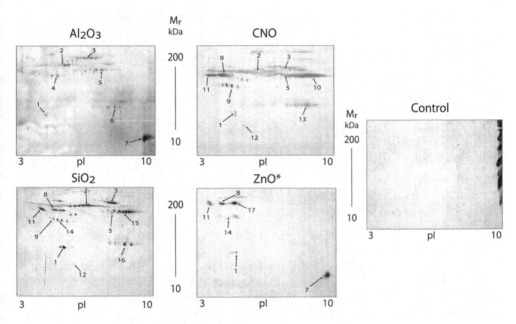

Fig. 5. Plasma protein binding profiles of different nanoparticles.
Four different nanoparticles; Al_2O_3, CNO, ZnO (* Aluminium-doped 6%) and SiO_2, were mixed with human plasma and isolated through ultracentrifugation. The protein contents were then separated by 2-DE and silver stained. Bound proteins were identified by MS as shown in Table 1.

The inflammatory marker Fibrinogen was found to interact with SiO_2, Al_2O_3 and CNO but interestingly not with ZnO (Fig 5). Fibrinogen binds foreign substances in the circulation and promotes macrophage activation - a mechanism that may result in retention of particle/protein complexes in the intima with accompanying cardiovascular complications (Shulz et al., 2005). On the other hand, Lysozyme C, a well known anti-bacterial agent, was only found on Al_2O_3 and ZnO. Under the present condition, it was unfortunately not determined if the binding to ZnO occurred due to the 6% Al doping of the ZnO. Lysozyme has previously been found to interact with nano-TiO_2 particles (Xu et al., 2010). They reported that the coexistence of nano-TiO_2 particles and Lysozyme resulted in the transition of Lysozyme conformation from α-helix into β-sheet secondary structure and a substantial inactivation of Lysozyme. Moreover the β-sheets are able to induce the formation of amyloid fibrils, a process which plays a major role in pathology.

Number in Figure 2	Protein	Found in Nano particle	Uniprot AccessionNumber	pI[a]	Mw[a] (Da)	Matched Peaks[b]	Sequence Coverage (%)	MOWSE Score
1	ApoA-I	Al2O3, CNO, SiO2, ZNO	P02647	5.2	24500	28	74	9.65e+10
2	Albumin	Al2O3, CNO, SiO2	P02768	6.0	68000	30	47	4.20e+14
3	Transferrin	Al2O3, CNO, SiO2	Q53H26	6.7	77000	43	65	1.23e+22
4	Fibrinogen γ	Al2O3	P02679	5.4	50000	15	32	1.39e+6
5	Fibrinogen β	Al2O3, CNO	P02675	7.0	55000	17	37	5.83e+8
6	Ig Light chain	Al2O3	Q0KKI6	7.4	30000	5	32	2779
7	Lysozyme C	Al2O3, ZNO	P61626	9.4	15000	7	33	12888
8	α1-AT	CNO, SiO2, ZNO	P01009	5.0	55000	31	63	5.36e+18
9	Haptoglobin	CNO, SiO2	P00738	5.1	46000	8	20	2115
10	DKFZ	CNO	Q6N096	8.3	55000	14	34	1.10e+7
11	Alpha-2-HS	CNO, SiO2, ZNO	P02765	4.8	55000	6	14	681
12	Transthyretin	CNO, SiO2	P02766	5.4	16000	7	51	69267
13	Ig KC	CNO	Q6PJF2	8.0	30000	10	55	409936
14	ApoA-IV	SiO2, ZNO	P06727	5.1	45000	10	23	72729
15	Igγ	SiO2	P01859	7.7	55000	7	23	79647
16	Igγ-1 chain	SiO2	P01857	8.5	35000	12	40	3.74e+7
17	Amyloid βA4	ZNO	B4DJT9	5.2	60000	20	15	53.1

Table 1. Identification of nanoparticle bound plasma proteins by peptide mass fingerprinting after 2-DE.
The table shows identified proteins with Uniprot accession number, isoelectric point (pI), molecule weight (Mw), number of peptide masses matched, sequence coverage and MOWSE score. [a] Isoelectric point (pI) and molecular weight (Mw) as estimated on gels. [b] Matched peak masses with a mass error tolerance of 75 ppm.

Furthermore, some antigen binding proteins; IgKC, DKFZ and IgLC, were also found to interact with Al_2O_3, CNO and SiO_2 but were not detectable on ZnO (Fig 5). Immunoglobulins are able to activate the complement system but they also often represent unspecific binding during protein purification caused by insufficient washing. Our results compared to previous findings indicates that increased hydrodynamic size of nanoparticle agglomerates seems to correlate to increased amounts of immunoglobulins. Overall ZnO was not binding as many proteins as the other particles and may even bind less without being Al-doped but an interesting finding in the ZnO preparation was a protein only described on transcript level and highly similar to the protein Amyloid β A4 precursor. This family of proteins acts as chelators of metal ions such as iron and zinc. They are also able to induce histidine-bridging between beta-amyloid molecules resulting in beta-amyloid-metal aggregates and it has been reported that extracellular zinc-binding increases binding of heparin to Amyloid β A4 (Uniprot 2011).

The protease inhibitor Alpha-1-antitrypsin, was found in the CNO, SiO_2 and ZnO preparations but not in Al_2O_3 while another antioxidant, Haptoglobin was found only on CNO and SiO_2. To our knowledge, the binding of Alpha-1-antitrypsin to the nanoparticles in this study has not been described previously but an increase of Haptoglobin has been reported in a study investigating acute phase proteins as biomarkers for predicting the exposure and toxicity of nanomaterials (Higashisaka et al., 2011).

At last, the HDL associated Apo A-I, with well known anti-endotoxin activity (Henning et al., 2006) and receptor interaction properties was found in all preparations but was most abundant after SiO_2 exposure. Apo A-I may be acting as a scavenger clearing the particles from the blood stream via the scavenger class B-I receptor (SRBI). The SRBI receptor is mainly located on the liver and plays an important role in cholesterol efflux (Verger et al., 2011) but is also present on other cells (Mooberry et al. 2010). Apolipoproteins in general are of interest for the pharmaceutical industry as carriers of nanoparticle bound drugs for brain uptake (Kreuter et al., 2005) since apo E and apo B-100 are taken up by the cells via receptor mediated endocytosis. The hypothesis is that the nano-particle/apolipoprotein complex mimics the natural lipoprotein particle. The identity of Apo A-I was, as the other proteins, confirmed by peptide mass fingerprinting using MALDI TOF MS (Table 1) and one dominating Apo A-I peptide in the MS spectra (Figure 6A) was in addition sequenced by MS/MS (Figure 6B) to further confirm the identity.

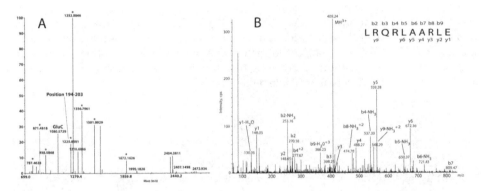

Fig. 6. Identification of nanoparticle bound apoliprotein A-I with mass spectrometry after 2-DE.
A: Peptide mass fingerprint spectrum obtained by MALDI-TOF mass spectrometry after endoproteinase GluC digestion. Asterisks represent peaks corresponding to peptide masses of apo A-I. B: Sequencing of a triply charged peptide (m/z 409.9) corresponding to position 194-203 of apo A-I by collision induced disassociation (CID) in a tandem mass spectrometer. The amino acid sequence with ions corresponding to the different fragments is shown in the upper right corner.

Overall the binding of plasma proteins to nanoparticles, based on previous and our findings, seems to vary with origin, surface properties, size and thereby also diameter of agglomerates. Particle characterization prior to exposure for plasma proteins or cells is therefore extremely important to receive reliable results that are possible to interpret. Interacting proteins under the present conditions are dominated by proteins involved in the immune defense and reverse transport to the liver but notably also proteins mediating brain uptake.

4. Methodological considerations and improvements of 2-DE and MALDI-TOF MS

4.1 Plasma sample preparations

Sample preparation is an important step that influences the separation of proteins with 2-DE. A common problem in most biological samples is the presence of salt ions. In 2-DE, salt concentrations >10 mM affects the isoelectric focusing step and markedly reduces the effectiveness of the charge separation. There are several easy ways to remove salts, e.g. by precipitation of the proteins or by gelfiltration in small desalting columns. Plasma samples are possible to analyze directly with 2-DE since the high protein concentration allows a simple dilution of the sample to lower the salt concentration. However, as illustrated in figure 7, a desalting step besides the dilution still improves the protein pattern. Another well-known problem with biological fluids is the presence of a few highly abundant proteins that may prevail over the low abundant proteins. In plasma, albumin and immunoglobulin G (IgG) constitutes a very large proportion (about 75%) of the total protein loaded on the gels. This may lead to the proportion of low abundant proteins being below the detection limit. Also, the staining of the abundant proteins may interfere with proteins with similar molecular mass and pI. It is therefore advisable to remove albumin and IgG and there are several commercial removal kits available, usually based on antibodies directed towards albumin and IgG. As shown in figure 7, such sample preparation step removes a large fraction of these proteins and increases the proportion of the other proteins in the sample. However, it is important to realize that this step also introduces unspecific removal of proteins and it is our experience that this unwanted loss of proteins varies considerable between the different available removal kits.

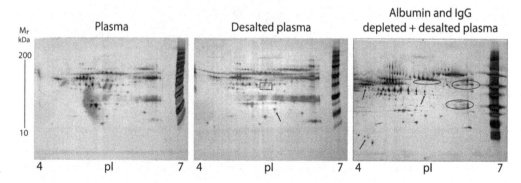

Fig. 7. Preparation of plasma for 2-DE
Untreated plasma (left), desalted plasma (middle) and desalted plasma after removal of albumin and immunoglobulin G (IgG), (right), were separated with 2-DE and silver stained. Arrows indicate protein or protein clusters with increased abundance after treatment compared to untreated plasma. The rectangle indicate an area with improved resolution after desalting. Positions for albumin and IgG chains are indicated with rings.

Plasma contains a wide variety of proteins, many of which are not detectable with 2-DE without further fractionation. In view of the lipid metabolism, important sub-fractions of plasma to study are the lipoproteins. As shown in table 1, one protein that interacts with nanoparticles is apo A-I, the major constituent of HDL. This implicates the need of more investigations of HDL as a possible target of nanoparticles that may influence the cholesterol metabolism and increase the risk of cardiovascular disease. HDL can be isolated based on density, size or protein content (e.g. apo A-I) using ultracentrifugation, size-exclusion chromatography or immune-affinity chromatography, respectively, each technique with its own merits and drawbacks. Thus, the rather harsh conditions during ultracentrifugation in high salt gradients may remove weakly associated proteins while the rather mild conditions during chromatography may favor unspecific co-purification of proteins with the cholesterol particle. We have previously mapped the protein content of HDL isolated by two-step density gradient ultracentrifugation (Karlsson et al., 2005). In this study we have compared ultracentrifugation and anti-apo A-I affinity chromatography to isolate HDL from the same plasma sample. As shown in figure 8, more proteins were obviously identified in immune-affinity purified HDL. However, some of the proteins must be considered as possible plasma contaminants as they were not, as e.g. the apolipoproteins, enriched in the HDL fraction.

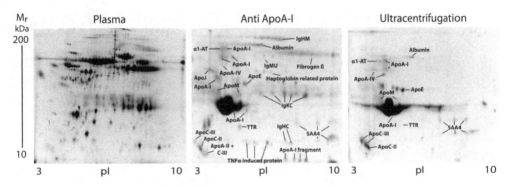

Fig. 8. Subfractionation of plasma with regard to HDL
Plasma (left), HDL purified according to apo A-I with immunoaffinity chromatography (middle) and HDL isolated according to density with ultracentrifugation (right) were separated with 2-DE and silver stained. Proteins were identified with mass spectrometry.

4.2 Protein detection

Gel-separated proteins can be visualized by several commonly used staining methods, including dyes (e.g. Coomassie Brilliant Blue and colloidal Coomassie), metals (e.g. silver staining) and fluorescent probes (e.g. Sypro staining and Cy-dyes) (Rabilloud, 2000). The stains interact differently with the proteins and have different limitations with regard to sensitivity, linear range, compatibility with mass spectrometry and type of proteins that stain best. In general, for staining of complex protein samples, silver staining can be considered the most sensitive technique (1-5 ng protein) and Coomassie Brilliant blue the least (50-100 ng) while the sensitivity of colloidal coomassie and the fluorescent dyes are in between. However, it is important to bear in mind that the different stains interact differently with the proteins and therefore one protein may stain very well with one staining method but not with the other (Fig. 9). For example, silver ions react with negatively charged groups and therefore stain glycoproteins containing negative sialic acid very well.

One the contrary, Sypro ruby that binds to proteins through hydrophobic interactions stains hydrophilic glycoproteins quite poorly. Being most sensitive, silver staining is obviously very useful for proteomic approaches. However, the advantage with silver is hampered by its rather low linear range, making it less suitable for quantification than the other staining techniques. For plasma analyses we have therefore adopted a double staining strategy. As illustrated in figure 9, the 2-DE gel is first stained with Sypro Ruby and the proteins are quantified within a high dynamic range. The gel then can be destained and restained by silver to detect additional proteins. As these additional proteins are less abundant their intensities usually are within the limited linear range of the silver staining technique. The proteins may then be selected for MS analyses. As the sample preparation protocol is more time-consuming for silver stained gels than Sypro stained gels it is convenient to pick as many proteins as possible after the first staining step. In this step it is also important to check the optimal destaining time before MS analyses, with plasma samples at least 90 minutes (fig 9). If the samples are not fully destained the signal to noise in the spectra are reduced.

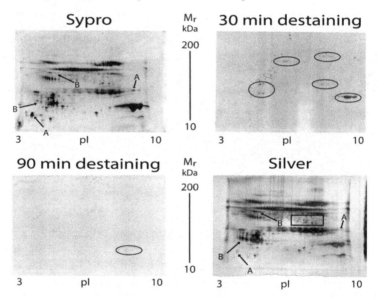

Fig. 9. Double staining of proteins
Plasma proteins were separated with 2-DE and first stained with Sypro Ruby. The gel was then destained with 25 mM ammonium bicarbonate/50 % acetonitrile buffer and the efficiency of the process checked after different time intervals. Finally, the proteins were re-stained with silver. A: Proteins more stained by Sypro compared to silver. B: Glycoproteins more stained by silver. The rectangle shows an area containing additional proteins detected by silver. Residual Sypro staining of proteins after destaining are indicated with rings.

4.3 Protein identification with peptide mass fingerprinting using MALDI-TOF MS

Peptide mass fingerprinting with MALDI-TOF MS is an excellent and robust technique for fast identification of plasma proteins after 2-DE (Lahm & Langen, 2000). However, several peptide peaks needs to be detected in the spectra with a high mass accuracy to avoid false positive results. There are several approaches to consider for improving the data from the analyses, such as digestion protocol, purification of peptides and choice of matrix.

4.3.1 Alternative digestion of samples

One of the most widely used ways to digest proteins before MS analyses is by trypsin, which cleaves C-terminally of lysine and arginine (not followed by proline). Although lysine and arginine often are distributed in the protein sequences in a way that provides sufficient number of peptides for identification after trypsin cleavage, this is not always so. Furthermore, it is sometimes necessary to use alternative digestion protocols in order to find specific peptides to e.g. characterize differences between protein isoforms. In these cases alternative enzymes, e.g. Asp-N (cleaves N-terminally of aspartic acid and cystein), Glu-C (C-terminally of glutamic acid) or chemical induced cleavage by CNBr that hydrolyzes C-terminally of methionine, is needed. In this study we used Asp-N as a complement to trypsin when identifying the plasma protein serum amyloid A4 (SAA4). This combined digestion approach generated almost 95 % sequence coverage of the protein (figure 10). SAA4 is a constitutively expressed protein which can be found in HDL as differently charged isoforms (fig 14, Karlsson et al., 2005). One explanation to these isoforms could be a small truncation of SAA4 in which one lysine and one tyrosine is removed C-terminally and thereby making the protein more acidic (Farwig et al., 2005). However, by using Asp-N we were able to detect both the intact C-terminal and the intact N-terminal peptide that were not possible after trypsin digestion, ruling out the presence of truncated SAA4 in our sample (figure 10). Besides the use of Asp-N to study SAA4, we have also used Glu-C to analyze apolipoprotein A-I (figure 6) and CNBr to study serum amyloid A-1/2 isoforms (figure 13).

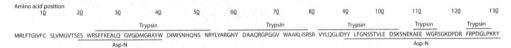

Fig. 10. Alternative digestion to improve sequence coverage of Serum amyloid A4 (SAA4).

Sequence coverage obtained through peptide mass fingerprinting with MALDI-TOF mass spectrometry by the use of trypsin and endoproteinase Asp-N as indicated by the lines. Sequence coverage (without the signal peptide in position 1-18) was 79.5 % with trypsin and 57.1 % with Asp-N. The combined sequence coverage was 93.8%.

4.3.2 Peptide sample cleaning

Peptide samples after in-gel digestion can be cleaned by adsorption to C-18 containing pipette tips (ZipTip®). This clean-up procedure done manually is rather time-consuming but is absolutely necessary before electrospray-quadrupole MS. On the other hand, with a MALDI-TOF instrument, being more insensitive to salts and other contaminants, it is not that obvious. To investigate the possible advantage with ziptip cleaning before MALDI-TOF MS we picked 17 sypro stained proteins after 2-DE. The proteins were digested with trypsin and an aliquot of the obtained peptide solution was purified by ZipTip® (50% ACN elution solution according to the protocol recommended by the supplier), mixed with the matrix CHCA and spotted on the MALDI-target plate and another aliquot of the peptide solution was mixed directly with the matrix and spotted on the same plate. All samples were then analyzed with MALDI-TOF MS with the same settings and with the laser induced collection of spectra in an automatic mode. The spectra were then used for NCBI database search with MS-fit using the same settings for all samples. All 17 proteins were identified with peptide mass fingerprinting in both ziptip cleaned and untreated samples. As shown in Table 2, ziptip cleaning significantly improved the peak intensities, signal to noise ratio and the mass accuracy. This illustrates that removal of salts and other contaminants that will compete with the peptides in the spectra increases the intensities of the peptide peaks and thereby increases the accuracy of the mass determinations and, as a consequence, also increases the reliability of the identifications. On the other hand, the number of peptides and sequence coverage found in ziptip cleaned samples were about the same as in the untreated samples (Table 2). In general, there was a clear tendency that in ziptip cleaned samples more peptides were detected in the lower mass region (<1000 m/z) while fewer peptides were detected in the higher mass region (>2000 m/z). This suggests that the removal of salt ions and low molecular chemicals with subsequent improved signal to noise increases the possibility to detect low molecular mass peptides but that this beneficial effect is counteracted by adsorption of larger peptides to the solid phase of the ziptip. To test this, 10 protein samples were sequential eluted from the ziptip with increasing ACN concentration, up to 90 %. Indeed, this procedure increased the number of peptides found and the sequence coverage increased from 52 +/- 16 % in the untreated samples to 58 +/- 16 in the ziptip cleaned samples (p<0.05). The effect varied among the different proteins but was in some samples quite profound, almost 2 times higher sequence coverage. It can be concluded from these experiments that purification of peptide samples with ziptip improves the results with MALDI-TOF MS. However, when it comes to the identification of proteins with peptide mass fingerprinting the beneficial effect of the cleaning procedure is quite limited as the number of peptides found, using the standard protocol, is not increased. Therefore, considering the work-load needed for the ziptip procedure, it is doubtful if it is practical to routinely clean samples with ziptip before MALDI-TOF analyses. However, for selected, low abundant, samples it can most likely make a significant difference for the identification. In these cases, sequential elution of the peptides from the ziptip with increasing acetonitrile concentrations is recommended.

	Average error (ppm)	Sequence coverage (%)	Number of peptides	Signal/noise ratio	Peak Intensity
Without Ziptip (n=17)	11.2 +/- 3.6	24.9 +/- 12.0	8 +/- 4	100 +/- 130	3500 +/- 2000
Ziptip (n=17)	7.9 +/- 4.4	24.1 +/- 11.4	8 +/- 3	200 +/- 480	6300 +/- 4000
Statistical significance	0.01	No	No	<0.001	<0.001

Table 2. Influence of sample cleanup of in-gel digested proteins on peptide mass fingerprinting data obtained with MALDI-TOF MS.

Proteins were separated by 2-DE and in-gel digested by trypsin. The same peptide samples were then purified by ZipTip, mixed with the matrix and spotted on the MALDI plate or directly mixed with the matrix and spotted on the plate. Statistical interpretations were done by Wilcoxons signed rank sum test.

4.3.3 Choice of matrix

To enhance the quality of the MALDI-TOF mass spectra and the number of desorbed peptides there are several matrices that could be considered to use. Different matrix compounds, both acidic and basic, have proved to work in sample preparation for MALDI mass analyzers. The far most commonly used matrix for peptide mass fingerprinting is CHCA, which is recommended for peptides with mass ions below 2500 Da (Beavis et al., 1992). Alternative matrices also used are sinapinic acid, mostly for masses higher than 25 kDa (Lewis et al., 2000) and DHB, originally suggested for glyco- or phosphopepides that are difficult to ionize (Strupat et al., 1991), but later also proven useful for silver stained proteins (Ghafouri et al., 2007). As illustrated in figure 11, CHCA and DHB have very different crystal structures on the target plate. Whereas CHCA usually has a homogeneously distributed spot appearance, making it ideal for automatic laser induced peptide desorption, the DHB crystals are needle shaped and often aggregated into fan-like structures directed from the outside towards the centre of the sample spot. Interestingly, we have found that peptides appear to be enriched in the base of the DHB structures (figure 11), significantly increasing the signal to noise ratio in spectra obtained from these areas. This is illustrated by the identification of transthyretin, one of the proteins that interact with silica nanoparticles and carbon nanotubes (figure 5). With DHB, the improved signal to noise in the spectrum displayed twice the number of peptides than with CHCA (figure 12). This dramatically increased the sequence coverage from 39 % obtained with CHCA to 72 % obtained with DHB. Thus, the use of DHB for low-abundant silver stained proteins from 2-DE is recommended.

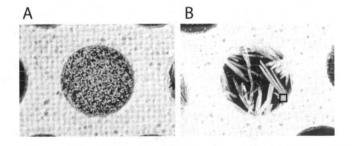

Fig. 11. Crystals of α-cyano-4-hydroxycinnamic acid (CHCA) and 2,5-dihydroxybenzoic acid (DHB).
A; CHCA and B; DHB as matrix on a MALDI plate. The marked area indicates the position of the laser where the best signal to noise was obtained with DHB.

4.4 Separation and characterization of isoforms

One of the main challenges for human proteomics is to identify and characterize co- and post-translational modifications to be able to study their relevance and place in systems biology. Most human proteins are expressed as different isoforms often depending on post-translational modifications. Two common modifications in plasma are truncations and glycosylations and here we have used the ability of 2-DE to separate such isoforms based on differences in isoelectric point and molecular mass.

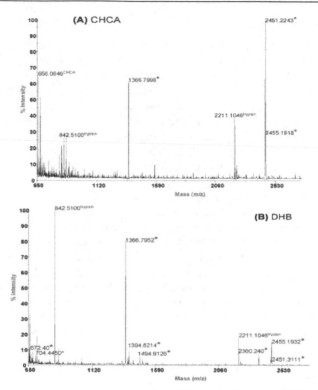

Fig. 12. Comparison of α-cyano-4-hydroxycinnamic acid (CHCA) and 2,5-dihydroxybenzoic acid (DHB) as matrices in MALDI-TOF MS of a silver stained protein.
Transthyretin was identified with sequence coverage of 39 % with CHCA (A) versus 72 % with DHB (B). Peptide peaks marked with an asterisk were matched to the theoretical masses with an accuracy <50 ppm.

Serum amyloid A1 and A2 (SAA1 and SAA2, respectively) are two acute phase proteins, which share more than 95 % sequence identity. Both SAA1 and SAA2 can also be expressed as an alpha- and a beta-form, which are discriminated from the others only in one amino acid position (Strachan et al., 1989). SAA1 and SAA2 are associated to HDL (Karlsson et al., 2005) and are heavily induced by endotoxins (Levels et al., 2011), which is highly relevant in particle toxicology to discriminate between different environmental agents (Karlsson et al. 2011). Based on differences in isoelectric points we were able to separate four isoforms of SAA1/2 in HDL (figure 13A). By peptide mass fingerprinting after trypsin digestion we identified two of the isoforms as SAA1α with pI 5.5 and 6 and two as SAA2α with pI 7 and 8 (figures 13A and 13B). The theoretical pI of SAA1α and SAA2α is 5.9 and 8.3, respectively. Thus, the pI of one of the isoforms of SAA1 and of SAA2 corresponded to the theoretical values while the other two had an acidic shift (pI 6→5.5 in SAA1α and pI 8→7 in SAA2α). N-terminal truncations of SAA1 and SAA2 that would produce such acidic shifts have previously been described (Ducret et al., 1996) and we therefore focused the MS analyses on the N-terminal peptide. As SAA1 and SAA2 contain arginine at the N-terminus we used CNBr, which cleaves before methionines, as an alternative digestion agent to detect the full length N-terminal peptide. These analyses showed that the more acidic isoforms of SAA1 and SAA2 comprised a mixture of truncations with the loss of one, two or four amino acids N-terminally; des-Arg, des-Arg-Ser and des-Arg-Ser-Phe-Phe, respectively, with the loss of arginine being the main explanation to the acidic shifts (figure 13D). On the other hand, the native peptide was only found in the more basic isoforms of SAA1 and SAA2. In total, four forms of SAA1α and four forms of SAA2α was identified. Interestingly, studies of these small molecular mass variants of SAA1 and SAA2 with SELDI-TOF MS indicates population cluster differences in HDL related to the truncations in response to endotoxin (Levels et al., 2011).

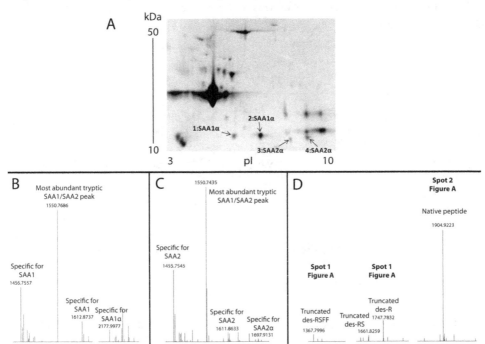

Fig. 13. Identification of serum amyloid A isoforms by 2-DE and MALDI-TOF MS.
A: HDL proteins were separated by 2-DE and stained by Sypro Ruby. Arrows indicate the two isoforms of serum amyloid A1α (SAA1α) and the two isoforms of serum amyloid A2α (SAA2α) identified by peptide mass fingerprinting. B and C: MS spectra after trypsin digestion of SAA1α and SAA2α, respectively, with specific masses indicated. D: MS spectra after CNBr digestion of SAA1α. Masses corresponding peptides from N-terminal truncated variants of the protein (protein spot 1) and the mass corresponding to the N-terminal peptide of the native protein (protein spot 2) are indicated.

2-DE makes it possible to separate proteins based on the degree of glycosylation. Hydrophilic sugars affect the binding of SDS and usually render the proteins an apparent higher molecular mass in the second dimension and the presence of negative sialyl-groups makes the proteins more acidic in the first dimension. We have therefore adapted two simple 2-DE mobility shift assays to demonstrate glycosylation of proteins and applied these to study glycosylated isoforms of plasma proteins in HDL. In the first we use endoglycosidase PNGase to cleave N-linked oligosaccharides from the protein backbone and the second is based on enzymatic removal of sialic acid with neuraminidase. As shown in figure 14A, SAA4 is usually expressed in HDL as 6 isoforms, three with molecular masses about 18k and three with molecular masses about 11k. After PNGase treatment it was clearly shown that the 18k isoforms are depending on N-linked glycosylation (Fig 14A). Another glycosylated protein in HDL is apo C-III that can be found as three isoforms; one di-sialylated, one mono-sialylated and one minor non-sialylated form (Bruneel et al., 2008). This was demonstrated by treatment with neuraminidase, which induced a mobility shift with the loss of the two sialylated isoforms and a substantial increase of the non-sialylated apo C-III form (figure 14B).

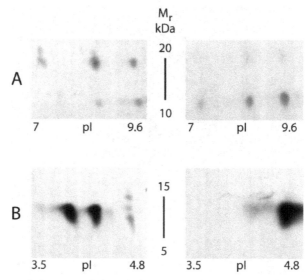

Fig. 14. 2-DE mass and charge mobility shift assays to demonstrate glycosylated protein isoforms.
A: SAA4 analyzed by 2-DE and silver stained. N-linked glycosylated serum amyloid A4 (SAA4) isoforms shown by deglycosylation with PNGase. B: Apo C-III analyzed by 2-DE and Western blots. Sialylated apo C-III isoforms shown by desialylation with neuraminidase.

5. Conclusions and future research

Overall the binding of plasma proteins to nanoparticles, based on our findings, seems to vary with origin, surface properties and size of the particles. A large portion of the interacting proteins we identified by 2-DE/MS are proteins involved in the immune defense and reverse cholesterol transport to the liver, but we also identified proteins mediating brain uptake. Most likely these protein patterns of the nanoparticles represent a mixture of particle-protein and protein-protein interactions. Extensive research in this field is therefore needed before conclusions could be drawn regarding potential health effects of nanoparticles and their associated protein "corona". One major difficulty to overcome is how to characterize the particles used in different studies in such a way that comparisons and generalized conclusions are allowed. Most types of nanoparticles seem to form aggregates, especially so in water suspensions and the "corona" seen might be heavily influenced by the size/diameter of these aggregates rather than by other particle characteristics. Thus, characterization of particles and the aggregates they form prior to exposure of plasma proteins, cells or other biological systems is therefore extremely important. One way of doing that, as we have showed herein, is by DLS. These analyses gives valuable information about the trends in size distribution connected to sample preparation methods and choice of solvent. Sample preparation methods are indeed very important as well as choice of solvent and care should be taken when choosing fitting model and the model-inbuilt parameters. Thus, information obtained from DLS is important for everybody that is doing research on nanoparticles in liquids. The numbers given as product information i.e. the size and size distribution are often relevant for the core-size of the nanocrystals within the material. However the nanoparticles are most often not soluble to that extent. Consequently, nanoparticles obtained in dry state and then dispersed in liquid usually form aggregates as shown in this study.

Given ample attention to the characterization of the NPs used, future studies of the NP-protein complex behavior in different biological systems are needed. Questions that need to be addressed are which properties of the NPs that govern the protein "corona" formed around the NPs in biological fluids and how these complexes interact with endothelial cells, platelets, cells of the immune system etc. One interesting finding in this study is Amyloid β A4, not previously identified in plasma, which was only associated to ZnO particles. This protein may act as a chelator forming metal-amyloid aggregates and needs further attention in toxicological studies.

In summary, the improved 2-DE/MS protocols shown herein underline this proteomic approach as a powerful tool in human nano-particle toxicology. Furthermore, thorough characterisation of the particles studied, e.g. with DLS, is crucial to evaluate the results.

6. References

Adiseshaiah, P.; Hall, J. & McNeil, S. E. (2010). Nanomaterial Standards for Efficacy and Toxicity Assessment. *Wiley Interdisciplinary Reviews – Nanomedicine and Nanobiotechnology*, Vol.2, No.1, pp. 99–112, ISSN 1939-0041

Beavis R.; Chaudhary T. & Chait B. (1992). α-Cyano-4-hydroxycinnamic acid as a matrix for matrixassisted laser desorption mass spectrometry. *Organic Mass Spectrometry*, Vol.27, No.2, pp. 156-158

Beck-Speier, I.; Dayal, N.; Karg, E.; Maier, K.; Schumann, G.; Schulz, H.; Semmler, M.; Takenaka, S.; Stettmaier, K.; Bors, W.; Ghio, A.; Samet, J. & Heyder, J. (2005). Oxidative stress and lipid mediators induced in alveolar macrophages by ultrafine particles. *Free Radical Biology and Medicine*, Vol.38, No.8, pp. 1080-1092, ISSN 0891-5849

Bell M. & Davis D. (2001). Reassessment of the lethal London Fog of 1952: Novel Indicators of Acute and Chronic Consequences of Acute Exposure to Air Pollution. *Environmental Health Perspectives*, Vol.109, pp. 389-394, ISSN 0091-6765

Benderly, M.; Boyko, V. & Goldbourt, U. (2009). Apolipoproteins and Long-Term Prognosis in Coronary Heart Disease Patients. *American Heart Journal*, Vol.157, No.1, pp. 103–110, ISSN 1097-6744

Brunauer, S.; Emmett, P. & Teller, J. (1938). Adsorption of Gases in Multimolecular Layers. *Journal of American Chemical Society*, Vol.60, pp.309-319.

Cedervall, T.; Lynch, I.; Lindman, S.; Berggard, T.; Thulin, E.; Nilsson, H.; Dawson, K. & Linse, S. (2007). Understanding the Nanoparticle-Protein Corona Using Methods to Quantify Exchange Rates and Affinities of Proteins for Nanoparticles. *Proceedings of the National Academy of Sciences of the U.S.A.* Vol.104, No.7, pp. 2050–2055, ISSN 0027-8424

Cherukuri, P.; Gannon, C.; Leeuw, T.; Schmidt, H.; Smalley, R.; Curley, S. & Weisman R. (2006). Mammalian pharmacokinetics of carbon nanotubes using intrinsic near-infrared fluorescence. *Proceedings of the National Academy of Sciences of the U.S.A*, Vol.103, No. 50, pp. 18882-18886, ISSN 0027-8424

Da Silva, E.; Tsushida, T. & Terao, J. (1998). Inhibition of mammalian 15-lipoxygenase-dependent lipid peroxidation in low-density lipoprotein by quercetin and quercetin monoglucosides. *Archives of Biochemistry and Biophysics*, Vol.349, No.2, pp. 313-320, ISSN 0003-9861

Deng, Z.; Mortimer, G.; Schiller, T.; Musumeci, A.; Martin, D. & Minchin R. (2009). Differential plasma protein binding to metal oxide nanoparticles, *Nanotechnology*, Vol.20, No.45, ISSN 1361-6528

Dobrovolskaia, M. & McNeil, S. (2007). Immunological Properties of Engineered Nanomaterials. *Nature Nanotechnolology*, Vol.2, No.8, pp. 469–478, ISSN 1748-3395

Dobrovolskaia, M.; Patri, A.; Zheng, J.; Clogston, J.; Ayub, N.; Aggarwal, P.; Neun, B.; Hall, J. & McNeil, S. (2009). Interaction of Colloidal Gold Nanoparticles with Human Blood: Effects on Particle Size and Analysis of Plasma Protein Binding Profiles. *Nanomedicine*, Vol.5, No.2, pp. 106–117, ISSN 1549-9642

Ducret A.; Bruun C.; Bures E.; Marhaug G.; Husby G. & Aebersold R. (1996). Characterization of human serum amyloid A protein isoforms separated by two-dimensional electrophoresis by liquid chromatography/electrospray ionization tandem mass spectrometry. *Electrophoresis*, Vol.17, No.5, pp. 866-876, ISSN 0173-0835

Dutta, D.; Sundaram, S.; Teeguarden, J.; Riley, B.; Fifield, L.; Jacobs, J.; Addleman, S.; Kaysen, G.; Moudgil, B. & Weber, T. (2007). Adsorbed proteins influence the biological activity and molecular targeting of nanomaterials. *Toxicological Sciences*, Vol.100, No.1, pp. 303-315, ISSN 1096-6080

Elsaesser, A. & Howard, C. (2011). Toxicology of nanoparticles. *Advanced Drug Delivery Reviews*, ISSN 1872-8294

Farwig Z.; McNeal C.; Little D.; Baisden C. & Macfarlane R. (2005). Novel truncated isoforms of constitutive serum amyloid A detected by MALDI mass spectrometry. *Biochemical and Biophysical Research Communications* Vol.332, No.2, pp. 352–356

Ferris, D.; Lu, J.; Gothard, C.; Yanes, R.; Thomas, C.; Olsen, J.; Stoddart, J.; Tamanoi, F. & Zink, J. (2011). Synthesis of biomolecule-modified mesoporous silica nanoparticles for targeted hydrophobic drug delivery to cancer cells. *Small*, Vol.7, No.13, pp. 1816-1826, ISSN 1613-6829

Finsy R. (1994). Particle Sizing by Quasi-elastic light scattering, *Advances in colloid and interface science*, Vol.52, pp. 79-143

Ghafouri, B.; Karlsson, H.; Mortstedt, H.; Lewander, A.; Tagesson, C. & Lindahl, M. (2007). 2,5-Dihydroxybenzoic acid instead of alpha-cyano-4-hydroxycinnamic acid as matrix in matrix-assisted laser desorption/ionization time-of-flight mass spectrometry for analyses of in-gel digests of silver-stained proteins. *Analytical Biochemistry*, Vol.371, No.1, pp. 121–123, ISSN 0003-2697

Gharahdaghi F.; Weinberg C.; Meagher D.; Imai B. & Mische S. (1999). Mass spectrometric identification of proteins from silver-stained polyacrylamide gel: a method for the removal of silver ions to enhance sensitivity. *Electrophoresis*, Vol.20, No.3, pp. 601-605, ISSN 0173-0835

Görg, A.; Postel, W. & Gunther, S. (1988). The current state of two-dimensional electrophoresis with immobilized pH gradients. *Electrophoresis*, Vol.9, No.9, pp. 531-546, ISSN 0173-0835

Görg ,A.; Obermaier, C.; Boguth, G.; Harder, A.; Scheibe, B.; Wildgruber, R.; and Weiss, W. (2000). The current state of 2-dimensional electrophoresis with immobilized pH gradients. *Electrophoresis*. Vol. 21, No. 12 pp.1037-53

Hackley V. & Clogston J. (2007). Measuring the Size of Nanoparticles in Aqueous Media Using Batch-Mode Dynamic Light Scattering, *NIST-NCL Joint Assay Protocol PCC-1 Version 1.0*, Available from http://ncl.cancer.gov/NCL_Method_NIST-NCL_PCC-1.pdf

Hamers, T.; Kamstra, J.; Cenijn, P.; Pencikova, K.; Palkova, L.; Simeckova, P.; Vondracek, J.; Andersson, P.; Stenberg, M. & Machala, M. (2011) In vitro toxicity profiling of ultrapure non-dioxin-like polychlorinated biphenyl congeners and their relative toxic contribution to PCB mixtures in humans. *Toxicological Sciences*, Vol121, No.1, pp. 88-100, ISSN 1096-0929

Hellstrand, E.; Lynch, I.; Andersson, A.; Drakenberg, T.; Dahlback, B.; Dawson, K.; Linse, S. & Cedervall, T. (2009). Complete High-Density Lipoproteins in Nanoparticle Corona. *FEBS Journal*, Vol.276, No.12, pp. 3372–3381, ISSN 1742-4658

Henning, M.; Garda, H. & Bakas, L. (2006). Biophysical characterization of interaction between apolipoprotein A-I and bacterial lipopolysaccharide. *Cell Biochemistry and Biophysics*, Vol.44, No.3, pp. 490-496, ISSN 1085-9195

Higashisaka, K..; Yoshioka, Y.; Yamashita, K.; Morishita, Y.; Fujimura, M.; Nabeshi, H.; Nagano, K.; Abe, Y.; Kamada, H.; Tsunoda, S.; Yoshikawa, T.; Itoh, N. & Tsutsumi, Y. (2011). Acute phase proteins as biomarkers for predicting the exposure and toxicity of nanomaterials. *Biomaterials*, Vol.32, No.1, pp. 3-9, ISSN 1878-5905

Karlsson H.; Leanderson P.; Tagesson C, & Lindahl M. (2005). Lipoproteomics II: mapping of proteins in high-Density lipoprotein using two-dimensional gel electrophoresis and mass spectrometry. *Proteomics*, Vol.5, No.5, pp. 1431–1445, ISSN 1615-9853

Karlsson, H.; Lindbom, J.; Ghafouri, B.; Lindahl, M.; Tagesson, C.; Gustafsson, M. & Ljungman, A. (2011). Wear particles from studded tires and granite pavement induce pro-inflammatory alterations in human monocyte-derived macrophages: a proteomic study. *Chemical Research in Toxicology*, Vol.24, No.1, pp. 45-53, ISSN 1520-5010

Kim, Y.; Chung, S. & Lee, S. (2005). Roles of plasma proteins in the formation of silicotic nodules in rats. *Toxicology Letters*, Vol.158, No1, pp. 1-9, ISSN 0378-4274

Kreuter, J.; Michaelis, K.; Dries, S. & Langer, K. (2005). The role of apolipoproteins on brain uptake of nanoparticle-bound drugs, *Proceedings of 15th International symposium on Microencapsulation*, Parma, Italy, September 2005

Kreyling, W.; Semmler, M.; Erbe, F.; Mayer, P.; Takenaka, S. & Schulz, H. (2002). Translocation of ultrafine insoluble iridium particles from lung epithelium to extra pulmonary organs is size dependent but very low. *Journal of Toxicology and Environmental Health.* Vol.65, No.20, pp. 1513-1530, ISSN 1528-7394

Kreyling, W.; Semmler-Behnke, M. & Moller, W. (2006). Ultrafine particle-lung interactions: does size matter? *Journal of Aerosol Medicine,* Vol.19, No.1, pp. 74–83, ISSN 0894-2684

Lahm H. & Langen H. (2000). Mass spectrometry: a tool for the identification of proteins separated by gels. *Electrophoresis,* Vol.21, No.11, pp. 2105–2114, ISSN 0173-0835

Leszczynski, J. (2010). Bionanoscience: Nano Meets Bio at the Interface. *Nature Nanotechnology,* Vol.5, No.9, pp. 633–634, ISSN 1748-3395

Levels, J.; Geurts, P.; Karlsson, H.; Maree, R.; Ljunggren, S.; Fornander, L.; Wehenkel, L.; Lindahl, M.; Stroes, E.; Kuivenhoven, J. & Meijers, J. (2011). High-density lipoprotein proteome dynamics in human endotoxemia. *Proteome Science,* Vol.9, No.1, ISSN 1477-5956

Lewis, J.; Wei, J. & Siuzdak G. (2000). Matrix-assisted Laser Desorption/Ionization Mass Spectrometry in Peptide and Protein Analysis. In: *Encyclopedia of Analytical Chemistry,* Meyers, R., pp. 5880–5894, John Wiley & Sons Ltd, ISBN 978-0-471-97670-7, Chichester

Lundqvist, M.; Stigler, J.; Elia, G.; Lynch, I.; Cedervall, T. & Dawson, K. (2008). Nanoparticle Size and Surface Properties Determine the Protein Corona with Possible Implications for Biological Impacts. *Proceedings of the National Academy of Sciences of the U.S.A,* Vol.105, No.38, pp. 14265–14270, ISSN 1091-6490

Lynch, I.; Cedervall, T.; Lundqvist, M.; Cabaleiro-Lago, C.; Linse, S. & Dawson, K. (2007). The Nanoparticle-Protein Complexas a Biological Entity; a Complex Fluids and Surface Science Challenge for the 21st Century. *Advances in Colloid and Interface Science,* Vol.134-135, ISSN 0001-8686

McAuliffe, M. & Perry, M. (2007). Are nanoparticles potential male reproductive toxicants? A literature review. *Nanotoxicology.* Vol.1, No.3, pp. 204-210

Mooberry, L.; Nair, M.; Paranjape, S.; Mc Conathy, W. & Lacko, A. (2010). Receptor mediated uptake of pacilitaxel from synthetic high density lipoprotein nancarrier. *Journal of drug targeting,* Vol.18, No.1, pp. 53-58, ISSN 1029-2330

Mühlfeld, C.; Rothen-Rutishauser, B.; Blank, F.; Vanhecke, D.; Ochs, M. & Gehr, P. (2008). Interactions of nanoparticles with pulmonary structures and cellular responses. *American Journal of Physiology - Lung Cellular and Molecular Physiology.*Vol. 294, No.5, pp. L817-L829, ISSN 1040-0605

O'Farrell, P.H. (1975). High-resolution two-dimensional electrophoresis of proteins. *J Biol Chem.* 250, pp. 4007–4021

Rabilloud, T. (2000. Detecting proteins separated by 2-D gel electrophoresis. *Analytical Chemistry.* Vol.72, No.1, pp. 48A-55A, ISSN 0003-2700

Rothen-Rutishauser, B.; Muhlfeld, C.; Blank, F.; Musso, C. & Gehr, P. (2007). Translocation of particles and inflammatory responses after exposure to fine particles and nanoparticles in an epithelial airway model. *Particle and Fibre Toxicology,* Vol.4, No.9, ISSN 1743-8977

Sattler, W.; Mohr, D. & Stocker, R. (1994). Rapid isolation of lipoproteins and assessment of their peroxidation by high-performance liquid chromatography postcolumn chemiluminescence. *Methods in Enzymology*, Vol.233, pp. 469-489, ISSN 0076-6879

Schulz, H.; Harder, V.; Ibald-Mulli, A.; Khandoga, A.; Koenig, W.; Krombach, F.; Radykewicz, R.; Stampfl, A.; Thorand, B. & Peters, A. (2005). Cardiovascular effects of fine and ultrafine particles. *Journal of Aerosol Medicine*, Vol.18, No.1, pp. 1-22, ISSN 0894-2684

Shevchenko, A.; Wilm, M.; Vorm, O. & Mann, M. (1996). Mass spectrometric sequencing of proteins silver-stained polyacrylamide gels. *Analytical Chemistry*, Vol.68, No.5, pp. 850-858, ISSN 0003-2700

Shevchenko, A.; Chernushevich, I.; Shevchenko, A.; Wilm, M. & Mann, M. (2002). De novo sequencing of peptides recovered from in-gel digested proteins by nanoelectrospray tandem mass spectrometry. *Molecular Biotechnology*, Vol.20, **No.1**, pp. 107-118, ISSN 1073-6085

Sorensen, C. (2008). Scattering and Absorption of Light by Particles and Aggregates, *Handbook of Surface and Colloid Chemistry Third Edition*, Birdi K., CRC Press, ISBN 978-0-8493-7327-5, Boca Raton, Florida, US

Stern, S. & McNeil, S. (2008). Nanotechnology safety concerns revisited. *Toxicological Sciences*, Vol.101, No.1, pp. 4-21, ISSN 1096-6080

Strachan A.; Brandt W.; Woo P.; van der Westhuyzen D.; Coetzee G. & de Beer M. (1989). Human serum amyloid A protein. The assignment of the six major isoforms to three published gene sequences and evidence for two genetic loci. *Journal of Biological Chemistry*, Vol.264, No.31, pp. 18368-18373, ISSN 0021-9258

Strupat K.;,Karas M. & Hillenkamp F. (1991). 2,5-Dihydroxybenzoic acid: a new matrix for laser desorption−ionization mass spectrometry. *International Journal of Mass Spectrometry and Ion Processes*, Vol.111, pp. 86-102, ISSN 0168-1176

Uniprot. (2011). Amyloid beta A4 protein, 20th October 2011, Available from http://www.uniprot.org/uniprot/P05067

Vergeer, M.; Korporaal, S.; Franssen, R.; Meurs, I.; Out, R.; Hovingh, G.; Hoekstra, M.; Sierts, J.; Dallinga-Thie, G.; Motazacker, M.; Holleboom, A.; Van Berkel, T.; Kastelein, J.; Van Eck, M. & Kuivenhoven, J. (2011). Genetic variant of the scavenger receptor class B-I in humans. *The New England Journal of Medicine*, Vol.364, No.2, pp. 136-45, ISSN 1533-4406

Walczyk, D.; Bombelli, F.; Monopoli, M.; Lynch, I. & Dawson, K. (2010). What the Cell "Sees" in Bionanoscience. *Journal of the American Chemical Society*, Vol.132, No.16, pp. 5761–5768, ISSN 1520-5126

Wick, P.; Malek, A.; Manser, P.; Meili, D.; Maeder-Althaus, X.; Diener, L.; Diener, P-A.; Zisch, A.; Krug, H.F. & von Mandach, U. (2010). Barrier Capacity of Human Placenta for Nanosized Materials. *Environmental Health Perspectives*. Vol.118, No.3, pp. 432-436, ISSN 1552-9924

Wu, W. & Jiang, X. (2011). A practical strategy for constructing nanodrugs using carbon nanotubes as carriers. *Methods in Molecular Biology*, Vol.751, pp. 565-582, ISSN 1940-6029

Xu, Z.; Liu, X.; Ma, Y. & Gao, H. (2010). Interaction of nano-TiO2 with lysozyme: insights into the enzyme toxicity of nanosized particles. *Environmental Science and Pollutional Research International*, Vol.17, No.3, pp. 798-806, ISSN 1614-7499

Zensi, A.; Begley, D.; Pontikis, C.; Legros, C.; Mihoreanu, L.; Buchel, C. & Kreuter, J. (2010). Human Serum Albumin Nanoparticles Modified with Apolipoprotein a-I Cross the Blood-Brain Barrier and Enter the Rodent Brain. *Journal of Drug Targeting*, Vol.18, No.10, pp. 842–848, ISSN 1029-2330

Part 4

Other Applications of
Gel Electrophoresis Technique

Enzymatic Staining for Detection of Phenol-Oxidizing Isozymes Involved in Lignin-Degradation by *Lentinula edodes* on Native-PAGE

Eiji Tanesaka, Naomi Saeki, Akinori Kochi and Motonobu Yoshida

Kinki University

Japan

1. Introduction

Lignocellulose is the most abundant organic compound in the terrestrial environment. Nonetheless, with the exception of basidiomycetous fungi, most organisms are either unable to degrade lignocellulose, or if they can, they do so with difficulty (Kirk & Fenn, 1982). Wood-decomposing basidiomycetes can be grouped into two categories: white-rot and brown-rot fungi. White-rot fungi have cellulases and lignin-degrading enzymes that decompose most cell wall components, whereas brown-rot fungi have enzymatic systems that selectively degrade cellulose and hemicelluloses, leaving brown shrunken lumps of tissue composed mainly of a loose lignin matrix (Enoki et al., 1988; Highley et al., 1985; Highley & Murmanis, 1987; Kirk & Highley, 1973). The name 'white-rot' is derived from the bleaching effect that this fungus has when degrading wood; the lignin-degrading enzymes that they secrete have the effect of promoting lignin loss and exposing the white cellulose fibrils. White-rot fungi are known to produce polyphenol oxidases (phenoloxidases), which, when the fungi are plated on agar media containing gallic or tannic acids, change the color of the agar to a dark reddish-brown in what is referred to as Bavendamm's polyphenol oxidase test or Bavendamm reaction (Bavendamm, 1928, as cited in Jørgensen & Vejlby, 1953). Based on this reaction, phenoloxidases are considered to be one of putative lignin-degrading enzymes (Higuchi 1990). Laccase (Lcc, EC 1.10.3.2), catechol oxidase (EC 1.10.3.1) and tyrosinase (monophenol monooxygenase, EC 1.14.18.1) are phenoloxidases with considerable overlap in their substrate affinities (Burke & Cairney, 2002). Lcc catalyze the reduction of O_2 to H_2O using a range of phenolics, aromatic amines, and other electron-rich substances as hydrogen donors (Thurston, 1994). Similar phenol-oxidizing activities are also observed in peroxidases (EC 1.11.1.x), which use H_2O_2 as an electron donor. Lignin peroxidase (ligninase, LiP, EC 1.11.1.14) was first discovered in *Phanerochaete chrysosporium* in which the H_2O_2-dependent C_α-C_β cleavage of non-phenolic lignin model compounds was first described (Tien & Kirk, 1983, 1984). Manganese peroxidase (MnP, EC 1.11.1.13) also strongly degrades lignin model compounds and the reaction is mediated by H_2O_2 and Mn^{2+} (Glenn et al., 1983; Glenn & Gold, 1985; Kuwahara et al., 1984). Whereas lignin can effectively be oxidized by LiP directly, as reviewed previously (Cullen & Kersten, 2004;

Gold & Alic, 1993), Mn^{2+} is considered to be an important physiological substrate for MnP. Further, while LiP expression has been observed in certain white-rot fungi (e.g. *Phanerochaete chrysosporium* and *Phlebia radiata*) under specific culture conditions (e.g., temperature, agitation, and nutritional constraints), MnP expression has been observed in a wide range of white-rot fungi (Gold & Alic, 1993), including cultivated edible fungi, such as *Agaricus bisporus* (Bonnen et al., 1994), *Ganoderma lucidum, Lentinula edodes*, and *Pleurotus* spp. (Orth et al., 1993).

The shiitake mushroom, *Lentinula edodes* (Berk.) Pegler, a white-rot basidiomycete, is one of the most valuable, cultured, edible mushrooms in the world (Chang & Miles 1989). Shiitake mushrooms were traditionally cultivated on Fagaceae logs, but they are now grown on sawdust-based media. The ability of white-rot basidiomycetes to degrade wood components, especially lignin, therefore affects both culture-time to harvesting and yields (Kinugawa & Tanesaka, 1990; Ohga & Kitamoto, 1997; Smith et al., 1988; Tanesaka et al., 1993). Although *L. edodes* secretes the lignin-degrading enzymes laccase (Lcc) and MnP when cultivated on sawdust-based media (Buswell et al., 1995; Leatham, 1985; Makker et al., 2001), it does not usually secrete these enzymes in liquid media. It was previously reported that the main isozyme produced by *L. edodes* cultured on sawdust was the manganese peroxidase, LeMnP2 (Sakamoto et al., 2009). In addition, we previously reported that a β-O-4 lignin model compound, 4-ethoxy-3-methoxyphenylglycerol-β-guaiacyl ether (Umezawa & Higuchi, 1985) was effectively degraded by *L. edodes* under MnP-induced conditions, but not under Lcc-induced conditions (Kochi et al., 2009). These observations supported the hypothesis that these enzymes, particularly MnP, play an important role in degrading sawdust during cultivation, and corroborating reports that the expression and properties of these enzymes is likely to influence mycelial growth and fruit body development (Smith et al., 1988; Wood et al., 1988). Several reports have been published on the purification and characterization of the lignin-degrading enzymes secreted by *L. edodes* using sophisticated biochemical procedures (Forrester et al., 1990; Nagai et al., 2002, 2003, 2007; Sakamoto et al., 2008, 2009). However, these methods are impracticable for routine isozyme analysis during breeding trials. Methods for isozyme detection by electrophoresis using enzyme catalytic properties - referred to as "protein activity staining" or "enzymatic staining" - are well established in histochemical studies and genetics (Pasteur et al., 1988). It was expected that Lcc, peroxidases (Per, EC 1.11.1.7), and MnP bands could be distinguished on the same gel by subtraction of newly appeared bands produced by sequential enzymatic staining. In practice, however, unexpected bands frequently appeared on gels exposed to conventional Lcc staining solutions. Indeed, in samples exhibiting strong MnP activity without Lcc activity, no additional bands appeared in subsequent staining procedures for either Per or MnP. We recently reported improved methods for enzymatic staining using native-PAGE to distinguish between Lcc and MnP isozymes induced in liquid cultures of *L. edodes* (Saeki et al., 2011).

In this chapter, we describe an assay system for the induction and identification of phenol-oxidizing enzymes produced by *L. edodes* grown under liquid culture conditions. In addition, the assay system was used to compare the glycosylation characteristics of these extra- and intracellular isozymes, as well as their modes of inheritance within monokaryotic progenies and β-O-4 lignin model compound degradation characteristics under Lcc- and MnP-induced conditions. Based these findings, the potential application of this assay system to elucidate the ligninolytic mechanisms employed by this fungus is also discussed.

Enzymatic Staining for Detection of Phenol-Oxidizing Isozymes Involved in Lignin-Degradation by
Lentinula edodes on Native-PAGE

151

2. Experimental procedures

2.1 Terminology

We use the term "phenol-oxidizing enzymes" to describe all phenoloxidases and peroxidases. We do so because of the ability of these enzymes to utilize the same substrates and produce the same catalytic products as described in the Introduction.

2.2 Fungi and culture conditions

The Hokken 600 variety of *L. edodes* (Hokken Co., Ltd., Tochigi, Japan; hereafter referred to as H600) and monokaryotic progenies derived from basidiospores were used in this study. To induce the phenol-oxidizing enzymes, mycelia were cultured in MYPG liquid medium (2.5 g malt extract; 1.0 g yeast extract; 1.0 g peptone; 5.0 g glucose in 1,000 ml of distilled water) supplemented with sawdust extract (MYPG-S). This sawdust extract was produced by adding 1 g of *Castanopsis cuspidata* (Thunb. ex Murray) Schottky sawdust to 30 ml of distilled water and then autoclaving the mixture for 15 min before filtering through filter paper (No. 1, Advantec, Tokyo) and collecting the extract. MYPG-S liquid media samples were then prepared from MYPG liquid medium by adding half a volume of sawdust extract instead of distilled water, which gave an MYPG-S extract that contained 500 mg sawdust in 30 ml media. Mycelia were sub-cultured at 25°C on MYPG 1% agar plates. After 14 days, three mycelial disks measuring 3 mm in diameter were harvested from the plates and used to inoculate 30 ml MYPG-S liquid in a 100 ml flask which was then statically cultured at 25°C. MnP activity was induced during culture on MYPG-S. Lcc activity was induced by adding 2 mM $CuSO_4 \cdot 5H_2O$ to the same media seven days after initial inoculation.

2.3 Enzyme assay

A schematic representation of the strategies employed to distinguish between individual phenol-oxidizing enzymes by subtractive activity assays (Szklarz et al., 1989) and sequential enzymatic staining of gels using native-PAGE is shown in Fig. 1.

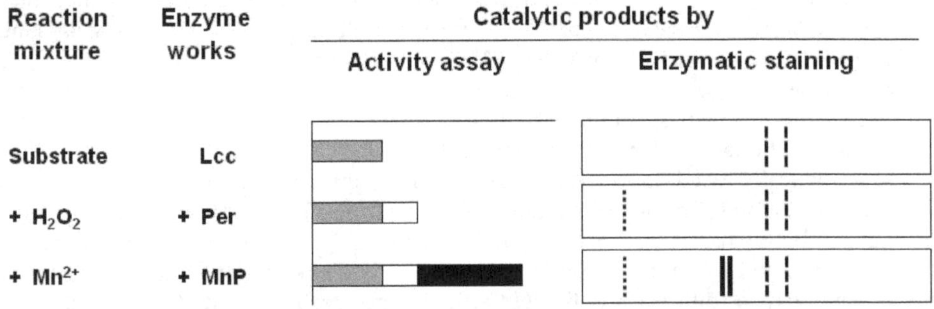

Reaction mixture	Enzyme works	Catalytic products by	
		Activity assay	Enzymatic staining
Substrate	Lcc		
+ H_2O_2	+ Per		
+ Mn^{2+}	+ MnP		

Fig. 1. Strategy underlying the subtractive activity assay and sequential enzymatic staining of a gel to distinguish between individual phenol-oxidizing enzymes. Grey, open, and solid squares represent the activities of Lcc, Per and MnP, respectively. Broken, dotted, and solid lines represent the sequential enzymatic staining of Lcc, Per and MnP in a gel, respectively.

To assay the activities of extracellular enzymes, 100 µl of culture liquid was sampled every two to three days during culture and centrifuged at 13,000 rpm for 10 min; this supernatant was used as a crude enzyme solution. The crude enzyme solution was assayed for Lcc, Per and MnP in identical 5 ml test tubes containing the following reaction mixtures: Lcc assay mixture consisted of 0.1 mM o-dianisidine in 0.1 M sodium tartrate buffer (pH 5.0), with additional H_2O_2 (final concentration 0.1 mM) added to the Lcc assay mixture to assess Per activity. Additional $MnSO_4 \cdot 5H_2O$ (final concentration 0.1 mM) was added to the Per assay mixture to assess MnP activity. Aliquots (20 µl) of crude enzyme solution were added to test tubes containing 980 µl of each reaction mixture, which were then incubated at 37°C for 10 min. The reactions were stopped by the addition of 50 µl of 40 mM NaN_3. To inactivate the enzymes in the control tubes, sodium azide was added to the control tubes containing the Lcc assay mixture before incubation. Catalytic products of the reaction were spectrophotometrically assayed using o-dianisidine as a substrate, and the activity of enzyme products was estimated by subtracting the respective absorbance values at 460 nm: i.e., Lcc activity = Lcc assay minus the control assay; Per activity = Per assay minus the Lcc assay; and MnP activity = MnP assay minus the Per assay, respectively. One unit (U) of enzyme activity was defined as the amount of enzyme required to catalyze 1 µmol of o-dianisidine in 1 min (ε_{460} = 29,400 $M^{-1}cm^{-1}$: Paszczynski et al., 1988).

2.4 Native PAGE and enzymatic staining

Each of the phenol-oxidizing isozymes was detected by native PAGE as described previously (Saeki et al., 2011). Briefly, whole cultures were filtered through a nylon stocking to separate the mycelia from the culture liquid. The collected mycelia were then ground with a ceramic mortar and pestle in two volumes (v/w) of crushing buffer (0.05 M Tris-HCl, pH 7.2, 0.1% β-mercaptoethanol) before being centrifuged at 12,000 rpm for 10 min. The resulting supernatant was considered to represent the intracellular enzyme sample. To prepare the extracellular enzyme sample, the culture liquid was centrifuged at 13,000 rpm for 10 min, and the supernatant was filtered (No.2 filter paper, Advantec) and then concentrated 15-fold by ultrafiltration using the centrifugal filter unit, Centriprep YM-10 (10-kDa cut-off membrane, Millipore, MA). Aliquots containing 15 µl enzyme sample, 1.5 µl glycerol and 1.5 mg bromophenol blue (BPB) as a dye marker were then loaded into the wells of 12.5% (for intracellular) or 17.5% (for extracellular) polyacrylamide gels. Native-PAGE gels were run at 15 mA for 15 min followed by 25 mA for 3–4 h. After electrophoresis, the gel was sequentially incubated at 37°C for 30 min in three different staining solutions. The first staining solution was an improved enzymatic staining solution containing additional ethylenediaminetetraacetic acid (EDTA) (for Lcc and Per) to remove the Mn^{2+} typically used in conventional staining solutions. Staining for enzymes was performed as follows. To the Lcc staining solution (LccS+EDTA), 1.8 mM o-dianisidine, 0.1 mM acetate buffer (pH 4.0) containing 130 mM EDTA (LccS+EDTA), an additional H_2O_2 (final concentration 1.0 mM) was added to produce the Per staining solution (PerS+EDTA). In the same way, additional $MnSO_4 \cdot 5H_2O$ (final concentration 0.1 mM) was added to the PerS without EDTA to produce the MnP staining solution (MnPS). The gels were rinsed with distilled water between each staining procedure to remove the previous staining solutions, particularly the EDTA from PerS+EDTA used for MnP staining. Isozyme nomenclature employed an (-e) or (-i) in Lcc-e, MnP-e or MnP-i to indicate whether the Lcc and MnP enzymes were extra- or intracellular. Numerals in parenthesis, e.g., MnP-e (52),

Enzymatic Staining for Detection of Phenol-Oxidizing Isozymes Involved in Lignin-Degradation by
Lentinula edodes on Native-PAGE

153

MnP-e (57) etc. indicated the relative mobility of each isozyme relative to the mobility of the
bromophenol blue used as a dye.

2.5 Glycosidase treatment

To purify the enzymes in the crude enzyme solutions prior to electrophoresis, 1 ml acetone
was added to a 100 μl aliquot of the enzyme solution and kept at −20°C for 3 h to precipitate
the proteins. The proteins were then resuspended in 100 μl of 10 mM phosphate buffer (pH
6.0). To determine whether the enzymes were glycosylated, the protein suspension in
phosphate buffer was incubated with glycosidase (Glycosidases 'Mixed', Seikagaku
Biobusiness Corp., Japan) at final concentrations of 0.25–2.0% (w/v) with a protease
inhibitor (Complete, Mini, EDTA-free; Roche Diagnostics, Germany) at 37°C overnight.
Effects of the glycosidase treatment on activities of each of the isozymes were then examined
by enzymatic staining after native-PAGE as described in section 2.4 above.

2.6 Identification of isozymes by mass spectrometry

Distinguishing between isozymes was performed as described previously (Saeki et al. 2011).
Briefly, after native-PAGE had been conducted on the same sample solution in adjacent
lanes, each gel was then subjected to enzymatic staining and Coomassie brilliant blue (CBB)
staining. Bands of interest, such as those exhibiting the same mobility as bands in the
enzymatic staining experiments, were then excised from the CBB-stained gel using a sterile
surgical blade and placed in 1.5 ml microcentrifuge tubes. To remove the CBB dye, each
polyacrylamide gel section was then repeatedly washed with 50, 30 and 50% v/v
acetonitrile containing 25 mM NH$_4$HCO$_3$ under sonication for 20 min with a micromixer
(Taitec, Tokyo), before finally being washed with 100% acetonitrile without NH$_4$HCO$_3$ for 5
min. The sections of polyacrylamide gel were then vacuum-dried for 5 min and recovered in
100 μl of 50 mM NH$_4$HCO$_3$ (pH 7.8) containing 10 ng/μl trypsin (Trypsin Gold, Mass
Spectrometry Grade, Promega, WI) on ice for 30 min. Any extra trypsin solution was then
removed and the sections of gel were incubated at 37°C for 16 h. The tryptic fragments in a
gel were then extracted by immersing the gel sections in 50 μl of extraction buffer consisting
50% acetonitrile and 5% trifluoroacetic acid under sonication (Ultrasonic cleaner, SU-3T,
Shibata, Japan) for 20 min. The extraction buffer was placed into new tube and replaced
with 25 μl of fresh extraction buffer. The extraction process was repeated a further three
times and the collected buffer containing the tryptic fragments was finally concentrated to
approximately 5 μl by drying under vacuum. Analysis of the tryptic peptides by tandem
mass spectrometry was performed on a nanoelectrospray ionization quadrupole time-of-
flight (Q-TOF) hybrid mass spectrometer (Q-TOF Premier, Waters Micromass, MA) coupled
with a nano-HPLC (Cap-LC; Waters Micromass). The peptides were separated on a BEH
130-C18 column (1.7 μm, 100 μm × 100 mm, Nano Ease, Waters, MA) at flow rate of 0.2
μl/min according to the manufacturer's instructions. The peptide sequences thus obtained
were then either matched automatically to proteins in a non-redundant database (National
Center for Biotechnology Information, NCBI, www.ncbi.nlm.nih.gov) using the Mascot
MS/MS ions search algorithm (Mascot Server version 2.2, Matrix Science), or BLAST
searches were manually performed against the DNA Data Bank of Japan database (DDBJ,
www.ddbj.nig.ac.jp). Mascot search was also performed to calculate the false discovery rate
(FDR) on acquired MS/MS data against decoy database.

2.7 RNA isolation and Northern blot analysis

Total RNA was extracted from mycelia after varying incubation periods under MnP- and Lcc- induced conditions using TRIzol Reagent (Invitrogen, CA). cDNA was synthesized from total RNA using an RNA PCR Kit Ver.3.0 (Takara Bio, Japan), and amplified using the primer set LeMnP2En5f (5'-TCCGACAGTGTCAATGACCTCGCTC) and LeMnP2En13r (5'-GTCAGTGGTGAGATTTGGGAAGGGC), which were designed based on the highly conserved *lemnp2* region (DDBJ Acc. No. AB306944; Sakamoto et al., 2009). A fragment measuring approximately 700 bp was then extracted from the 1% agarose–formaldehyde gel, purified, and sequenced using an ABI Prism 3100-*Avant* Genetic Analyzer (Applied Biosystems, CA) according to the manufacturer's instructions. Fragments with sequences matching *lemnp2* were then labeled using a PCR-based digoxigenin (DIG)-dUTP labeling kit (PCR DIG Probe Synthesis Kit, Roche Diagnostics) according to the manufacturer's instructions. The resulting DIG-labeled probe, *lemnp2N*, was then used for Northern hybridization following blotting of 10 µg of total RNA onto a Hybond-N$^+$ membrane (GE Healthcare, Switzerland) using an established protocol (Sambrook et al., 1989).

2.8 DNA isolation and Southern blot analysis

Genomic DNA was extracted from mycelia using cetyltrimethylammonium bromide (CTAB) isolation buffer (J. J. Doyle & J. L. Doyle, 1987, as cited in Milligan, 1998). To prepare the DIG-labeled probe, genomic DNA was used as a template for PCR with the primer set, LeMnP2En5f2 (5'-TCAGGAAAATTCCCGACTAT) and LeMnP2En12r (5'-GAACCTCGATG CCATCAA); this primer set was designed to amplify the region from exon 5 to exon 12 of *lemnp2*, including introns (DDBJ Acc. No. AB306944; Sakamoto et al., 2009), and the resulting probe was named *lemnp2S*. In addition, to examine cross-hybridization between *lemnp2* and a relative of the manganese peroxidases LeMnP1 coded by *lemnp1*, a probe for *lemnp1* was also prepared as described above using the primer set LeMnP1f1 (5'-GATTCCTGAGCCTTTCG) and LeMnP1r (5'-TTCGGGACGGGAATAAC); this primer set was designed to amplify the regions from exon 7 to exon 15 of *lenmp1* including introns (DDBJ Acc. No. AB241061; Nagai et al., 2007) and the resulting probe was named *lemnp1S*. These two probes were then used for Southern blot analysis (Sambrook et al., 1989).

2.9 Degradation assay of β-*O*-4 lignin model compound

2.9.1 Culture experiment

To assay the abilities of MnP and Lcc to degrade the β-*O*-4 lignin model compound, 4-ethoxy-3-methoxyphenylglycerol- β-guaiacyl ether (Fig. 2; β-*O*-4 compound, hereafter) was synthesized according to the method previously described (Umezawa & Higuchi 1985).

To prepare media containing the β-*O*-4 compound to media, 300 µg of the β-*O*-4 compound was diluted in 50 µl acetone and added to 30 ml MYPG-S. Whole liquid culture media was collected 14, 21 and 42 days after inoculation and assayed for phenol-oxidizing enzyme activity. The β-*O*-4 compound was then recovered from the liquid culture media by the addition of two volumes of ethyl acetate to separate the aqueous phase, before evaporating the ethyl acetate off and then precipitating the compound. To improve subsequent chromatographic analysis, the recovered β-*O*-4 compound was silylated using N-*O*-bis-(trimethylsilyl) trifluoroacetamide (BSTFA) to form a trimethylsilyl (TMS) derivative. This

Enzymatic Staining for Detection of Phenol-Oxidizing Isozymes Involved in Lignin-Degradation by
Lentinula edodes on Native-PAGE

155

TMS derivative was then subjected to gas chromatography-mass spectrometry (GC-MS) analysis (6890N, Agilent Technologies, CA), which was fitted with an capillary column (HP-5 MS, 30 m × 0.25 mm i.d., 0.25-μm; J&W Scientific, CA) coupled to an MS (JMS-K9, JEOL, Japan) according to the manufacturer's instructions. Helium was used as the carrier gas at 1.5 ml/min. GC oven conditions consisted of 150°C for 1 min, initially ramped at 10°C/min to 200°C and then at 5°C/min to 250°C. The electron impact mass spectra were obtained at an acceleration energy of 70 eV. The degradation rate (%) of the β-O-4 compound was calculated using the rate of quantities of the TMS derivative before and after culture, compensating for the recovery of the β-O-4 compound with 4,4'-dimethoxybenzoin (anisoin), which was used as an internal standard.

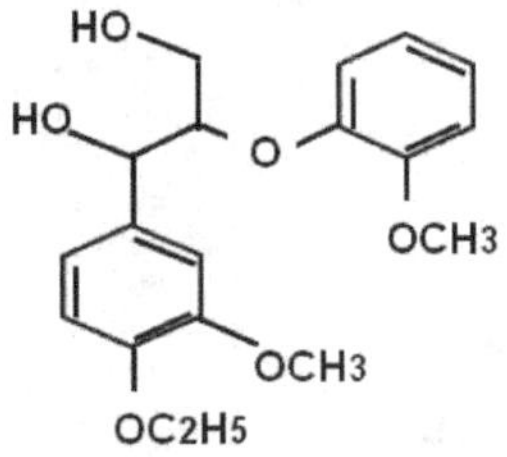

Fig. 2. β-O-4 lignin model compound used in this study; 4-ethoxy-3-methoxyphenylglycerol-β-guaiacyl ether (from Umezawa & Higuchi, 1985).

2.9.2 Incubation experiment

Degradation of the β-O-4 compound was also examined by incubation with the extracellular enzyme solution, which was prepared using the same procedures used for electrophoresis (section 2.4) with the following slight modifications. The extracellular enzyme solutions were diluted with 0.2 M sodium tartrate buffer (pH 5.0, final concentration of 0.1 M) and distilled water to bring the volume to 10 ml and keep the activity of MnP, Lcc and the mixed solution (MnP+Lcc) at 17 U/ml. For the MnP and MnP+Lcc reactions, additional H_2O_2 (final concentration 0.1 mM) and $MnSO_4 \cdot 5H_2O$ (final concentration 0.1 mM) were added to the reaction for Lcc. Then, 500 μg of the β-O-4 compound in 50 μl of acetone was added to the 10 ml enzyme solution and incubated at 37°C with agitation at 100 rpm for up to 10 days. The rate of degradation of the β-O-4 compound was evaluated using GC-MS as above.

3. Results and discussion

3.1 Selective induction of phenol-oxidizing enzyme

Phenol-oxidizing enzyme activities under different culture conditions are shown in Fig. 3. Neither MnP nor Lcc was induced when mycelia were cultured on MYPG liquid medium without sawdust extract (data not shown). Under MnP-induced conditions (i.e. when mycelia were cultured on MYPG+S), MnP activity increased suddenly on day 21, before reaching a maximum activity (95 U/ml) on day 35 and then decreasing thereafter (Fig. 3a).

We previously found that supplementing the MYPG liquid medium with wood chips or sawdust from members of the Fagacae, C. cuspidata or Fagus crenata Blume, induced MnP activity (Yoshikawa et al., 2004). The results of the present study show that sawdust extracts produced by autoclaving sawdust in hot water (section 2.2) induce MnP activity (Fig. 3a). Compared with mycelial growth on the MYPG (without extract) medium, the sawdust extract also had a marked effect on the promotion of mycelial growth (MYPG-S). Although less marked than that observed on MYPG-S (100 mg/30 ml), extracts produced using 100 mg sawdust in 30 ml media also promoted mycelial growth; however, MnP activity was not induced in cultures grown in MYPG-S media with lower extract concentrations for up to 35 days (data not shown). These observations suggested that MnP was induced by specific functional compounds in the sawdust extract, and not only due to mycelial growth.

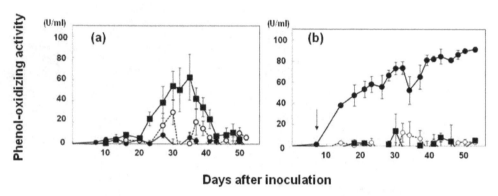

Fig. 3. Changes of phenol-oxidizing activities of (•) Lcc, (o) Per and (■) MnP in a liquid culture medium of *L. edodes* under (a) MnP-induced (MYPG-S) or (b) Lcc-induced conditions (MYPG-S with 2mM CuSO₄ 5H₂O) (reprinted from Saeki et al., 2011). The arrow in panel (b) indicates the day of 2 mM CuSO₄ 5H₂O addition. Values are means with standard errors (vertical bars) for three replicate cultures.

Fourteen days after inoculation in MYPG-S containing Cu^{2+}, Lcc activity was detected (7 days after the addition of 2 mM CuSO₄ 5H₂O) (Fig. 3b). This Lcc activity increased gradually after day 52 while MnP activity was completely suppressed. Lcc has been shown to be induced by aromatic compounds and metallic ions such as copper (Collins & Dobson, 1997; Saparrat et al., 2002; Scheel et al., 2000; Shutova et al., 2008; Soden & Dobson, 2001). Indeed, copper has been reported to be a strong laccase inducer in the white-rot fungi *Pleurotus ostreatus* (Palmier et al., 2000), *Trametes pubescens* (Galhaup & Haltrich, 2001; Galhaup et al., 2002), and *T. versicolor* (Collins & Dobson, 1997). In *Trametes pubescens*, the transcription of the laccase gene is induced within 10 h after the addition of 2 mM CuSO₄ (Galhaup et al., 2002). Under the two culture conditions employed in this study, either Lcc or MnP activity were detected, but not both (Fig. 3). This finding suggests that the induction of MnP and Lcc are controlled by a negative feedback system, i.e., Lcc-induction suppresses MnP production, or more specifically, the addition of CuSO₄ 5H₂O suppresses MnP production. Although the addition of several of the aromatic compounds that were tested did not induce Lcc - 2-methoxyphenol (guaiacol), 2,6-dimethoxyphenol (DMP), 4-anisidine, hydroquinone, or 1,2-benzenediol (catechol) - these substances except for DMP were observed to suppress MnP activity (data not shown).

Enzymatic Staining for Detection of Phenol-Oxidizing Isozymes Involved in Lignin-Degradation by
Lentinula edodes on Native-PAGE

157

3.2 Isozyme detection and identification

Of CBB staining bands subjected to protein identification, proteins with FDR (q≤0.05) were described below. Number of entry (MS/MS data) was ranged from 68 to 89. In the extracellular enzyme sample (culture liquid) prepared under MnP-induced conditions, two MnP-e isozyme bands, MnP-e (52) and MnP-e (57) in Fig. 4a, were detected. These two MnP isozymes were identified as the manganese peroxidase, LeMnP2, a major MnP isozyme that is secreted into sawdust medium by *L. edodes* (Sakamoto et al., 2009). Other enzyme, exo-β-1,3-glucanase, was also detected under MnP-induced conditions (Fig. 4b).   In the extracellular enzyme samples prepared under Lcc-induced conditions, two major Lcc isozyme bands, Lcc-e (61) and Lcc-e (67), were detected together with broad tailing smears (Fig. 4c). These two isozymes were identified as being laccases (Lcc1; Fig. 4d), and are known to be an extracellular laccase produced by *L. edodes* (Nagai et al., 2002; Sakamoto et al., 2008). These results, combined with enzyme assay data and results of isozyme detection using PAGE, indicate that MnP and Lcc isozyme detection using the improved LccS+EDTA, PerS+EDTA and MnPS enzymatic staining methods can be used to successfully distinguished between each of the phenol-oxidizing enzymes (Saeki et al., 2011).

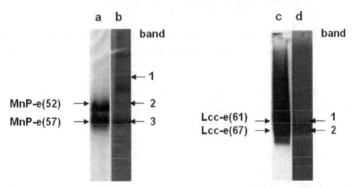

Fig. 4. Protein bands detected by (**a, c**) enzymatic staining and corresponding (**b, d**) CBB staining on native-PAGE (reprinted from Saeki et al., 2011). Lanes (**a**) and (**b**) show bands detected under MnP-induced conditions at 22 days; lanes (**c**) and (**d**) show bands detected under Lcc-induced conditions at 30 days. Protein bands identified by Q-TOF mass spectrometry in (**b**): band 1, exo-β-1,3-glucanase; band 2, manganese peroxidase (LeMnP2); and band 3, manganese peroxidase (LeMnP2) and in (**d**): band 1, laccase (Lcc1); band 2, laccase (Lcc1).

3.3 Comparisons of intracellular and extracellular Lcc isozymes

Expression patterns of extra- and intracellular Lcc isozymes during culture under Lcc-induced conditions are shown in Fig. 5.

Three major extracellular Lcc isozymes were detected: Lcc-e (61), which was expressed from day 12 (5 days after the addition of $CuSO_4$ $5H_2O$), had a constant intensity with a broad smear tails, an Lcc-e (67) from day 17, and another Lcc-e (74) from day 22. All of these enzymes were expressed until the end of culture on day 47. The observed changes in the total band intensity of the three Lcc extracellular isozymes was generally associated with changes in Lcc activity in the culture liquid (refer Fig. 3). Three major intracellular Lcc isozymes, Lcc-i (61), Lcc-i (67) and Lcc-i (74), were also detected, and all exhibited the same mobilities as their respective extracellular Lcc isozyme counterparts. The intracellular Lcc were coincidentally expressed with the extracellular Lcc isozymes.

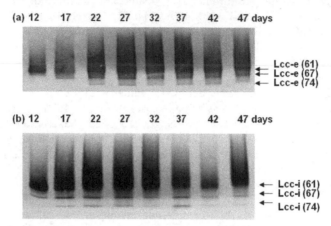

Fig. 5. Laccase isozyme banding patterns detected as (a) extracellular and (b) intracellular isozymes during culture under Lcc-induced conditions. Days after inoculation are shown above each lane.

Although we successfully extracted total RNA from mycelia under Lcc-induced conditions to examine the transcription of Lcc1, extraction of native (undigested) total RNA was unsuccessful. Native total RNA, which was prepared from mycelia under MnP-induced conditions (described in section 3.4 below), was degraded considerably quicker and to a greater extent after the addition of small amounts of cell lysate obtained under Lcc-induced conditions compared to when cell lysate obtained under MnP-induced conditions was added (data not shown). This relatively quicker degradation of native total RNA suggests a relatively high internal RNase activity in the cell lysate of the Lcc-induced condition, which may be attributed to the decrease observed in mycelial growth after the addition of $CuSO_4 \cdot 5H_2O$ and subsequent induction of Lcc, as well as the antagonistic expression of Lcc in the mycelial contact zone of any adjacent and competing basidiomycetes or other fungi (Iakovlev & Stenlid, 2000; Mercer, 1982; White & Boddy, 1992).

3.4 Manganese peroxidase gene transcription

The sequence of the fragment amplified from cDNA, which was prepared from a mixed pool of total RNAs obtained from the mycelia of 10-, 15- and 18-day-old cultures under MnP-induced conditions, was identical to that of *lemnp2a* but slightly different from *lemnp2b* (data not shown), corroborating the results obtained from the protein identification deduced by Q-TOF mass spectrometry (section 3.2). The finding that these sequences were similar also indicated that hot-water sawdust extracts induced the secretion of the same isozyme, LeMnP2, which is a major MnP isozyme that is secreted into sawdust media (Sakamoto et al. 2009). The results of the Northern blotting experiments with *lemnp2* are shown in Fig. 6. Under MnP-induced conditions, a detectable amount of *lemnp2* mRNA was present at

Enzymatic Staining for Detection of Phenol-Oxidizing Isozymes Involved in Lignin-Degradation by
Lentinula edodes on Native-PAGE

159

10–19-days during the initial stage of culture, with transcription increasing from day 22 to day 25 and then decreasing at day 28. These changes in *lemnp2* transcription occurred several days prior to the changes observed in MnP activities in the liquid culture medium.

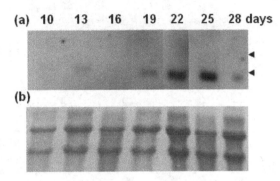

Fig. 6. Northern blot analysis of *lemnp2* gene transcript under (a) MnP-induced conditions and (b) ribosomal RNA used as a loading control. Days after inoculation are shown above each lane. Arrowheads indicate position of 26S and 18S rRNA.

3.5 Comparisons between intracellular and extracellular MnP isozymes

Expression patterns of extra- and intracellular MnP isozymes during culture under MnP-induced conditions are shown in Fig. 7. The extracellular MnP isozymes, MnP-e (52) and MnP-e (57), were strongly expressed during the initial stage of culture on days 11 to 19, before gradually decreasing until day 43. Four major bands were considered to be intracellular MnP isozymes, and of these, two bands, MnP-i (52) and MnP-i (57), exhibited the same mobility as extracellular MnP isozymes, while the other two bands, MnP-i (63) and MnP-i (66), were strictly intracellular. The intracellular MnP isozymes were expressed during the initial stage of culture, either several days before, or coincident with, the expression of the extracellular MnP isozymes. Compared to the intracellular MnP isozymes, the extracellular MnP isozymes maintained relatively high activities for up to 43 days of culture. However, changes in the intensities of bands that were neither extracellular nor intracellular MnP isozymes coincided with changes in MnP activity in the liquid culture medium during culture. Although we have no experimental data to explain why this may have occurred, it is worth noting that the intracellular enzyme solution (cell lysate) did not exhibit any phenol-oxidizing activities when assayed spectrophotometrically. In addition, the addition of intracellular enzyme solution to extracellular enzyme solutions caused marked inactivation of the latter (data not shown). Taken together, these observations either imply that the cell lysate contained a specific inhibitor of phenol-oxidizing enzymes when these enzymes and the inhibitor in cell lysate were not separated on a gel, or that there was an error in the manner in which the different experimental culture lots were processed, including the replicate flasks.

While treatment with glycosidase completely inactivated the two strictly intracellular MnP isozymes, glycosidase treatment had no effect on the activities of the extracellular MnP isozymes (Fig. 8). This finding indicates that the intracellular isozymes were active as glycosylated proteins, and implies that a relationship exists between the secretion of MnP and the simultaneous expression of β-glucanase detected by Q-TOF mass spectrometry.

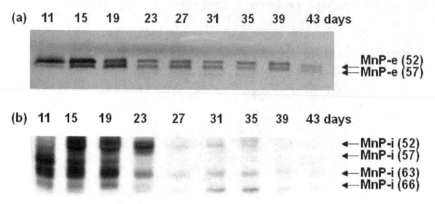

Fig. 7. Manganese peroxidase isozyme banding patterns detected as (a) extracellular and (b) intracellular isozymes during culture under MnP-induced conditions. Days after inoculation are shown above each lane.

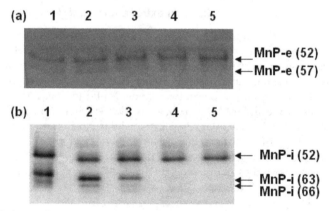

Fig. 8. Effects of glycosidase treatment on enzymatic staining of (a) extracellular and (b) intracellular MnP isozymes expressed under MnP-induced conditions (at 20 days). Lanes differ according to concentration of glycosidase: lane 1, control (0%); lane 2, 0.25%; lane 3, 0.5%; lane 4, 1.0%; and lane 5, 2.0%, respectively. MnP-i (57) was not detected.

3.6 MnP isozymes in monokaryons

Four monokaryotic strains (#317, #208, #305 and #105), each carrying the mating type factor A_1B_1, A_1B_2, A_2B_1 and A_2B_2, respectively, were derived from basidiospores from dikaryon H600. Although MnP activities of the monokaryons were very weak compared to the MnP activities of H600 (refer Fig. 7), both of the extracellular MnP isozymes (MnP-e (52) and MnP-e (57)) that were detected in the parent dikaryon were also detected in monokaryons, irrespective of their mating-type factors (Fig. 9). This finding suggests that, although these

Enzymatic Staining for Detection of Phenol-Oxidizing Isozymes Involved in Lignin-Degradation by
Lentinula edodes on Native-PAGE

161

isozymes may be encoded by different loci, the isozymes are not under allelic control (i.e. they are not allozymes).

Fig. 9. Extracellular manganese peroxidase isozymes expressed by monokaryotic progenies of H600. Lanes represent strains with mating type factor in parentheses: lane 1, #317 (A_1B_1); lane 2, #208 (A_1B_2); lane 3, #305 (A_2B_1) and lane 4, #105 (A_2B_2).

The results of the Southern blotting experiment of *lemnp2* on *Hin*d III-digested monokaryon genomes are shown in Fig. 10. There was no *Hin*d III restriction site in the amplified region (*lemnp2S*) of the H600 genomic DNA, as expected from the database analysis of different *L. edodes* stock SR-1 (DDBJ Acc. No. AB306944, Sakamoto et al. 2009). Two *lemnp2* hybridization signals appeared at positions between 564-2322 bp in all four of the monokaryotic strains (lanes 1-4 in Fig. 10), and all of the strains exhibited the same two hybridization signals observed in the H600 parent dikaryon (lane 5 in Fig. 10). However, single intense and weak hybridization signals of *lemnp1* were observed using another probe, *lemnp1S*, at different positions between 2322-6557 bp in H600 (lane 6 in Fig. 10), indicating that the two probes did not cross-hybridize with each other. Conversely, it is likely that the weak hybridization signals that appeared between 2322-4631 bp (lane 2 in Fig. 10) were cross-hybridization products between the two probes (see lane 6 in Fig. 10). These observations, combined with the observation of two isozymes being expressed by all of the monokaryons assayed in this study, suggest that there are two copies of *lemnp2* in the haploid genome of *L. edodes*. Nevertheless, to confirm whether the lemnp2 gene is indeed duplicated as proposed here, further analysis will need to be undertaken to assign lemnp2 to a genetic linkage map or on chromosomal DNA which separated by contour-clamped homogeneous electric fields (CHEF) gel electrophoresis. Indeed, such attempts at combining assignments of quantitative trait loci (QTL) related to wood and lignin degradation in fungi would facilitate the identification of new genes involved in another ligninolytic system.

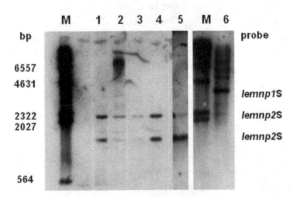

Fig. 10. Southern blot analysis with probe *lemnp2S* on genomic DNAs (digested with *Hin*d III). Lanes represent strains: lane M, size marker (λ/*Hin*d III); lane 1, #317; lane 2, #208; lane 3, #305; lane 4, #105, lane 5, H600; and lane 6, H600 (probed with *lemnp1S*).

3.7 Degradation of β-O-4 lignin model compound

We performed preliminarily examinations of the degradation of a β-O-4 lignin model compound under MnP- and Lcc-induced conditions (culture experiment) and the degradation of the model compound by incubation with enzyme solutions (incubation experiment) (Table 1).

Culture conditions	Days after inoculation in the culture experiment			Days of incubation in the incubation experiment[1]	
	14	21	42	4	10
MnP-induced	1.4 (nd)[3]	13.4 (32.2)[3]	20.0 (46.8)	16.8 (9.4)	23.8 (6.3)
Lcc-induced	1.1 (nd)	1.8 (47.0)	4.2 (58.5)	3.5 (8.0)	6.9 (3.1)
Enzyme mix[2]				16.0 (12.0 for MnP) (5.0 for Lcc)	22.9 (7.0 for MnP) (3.3 for Lcc)

[1]: Extracellular enzyme solution at an initial activity adjusted to 17 U/ml
[2]: Mixture of the extracellular enzyme solutions (MnP+Lcc), each at an initial activity of 17 U/ml
[3]: Numerals in parentheses represent enzyme activities (U/ml) , nd = not detected

Table 1. Degradation rate (%) of β-O-4 lignin model compound under MnP- or Lcc-induced conditions (culture experiment) and after incubation with enzyme solutions prepared from given culture conditions (incubation experiment) (Data from Kochi et al., 2009)

Under MnP-induced conditions, the β-O-4 compound was not degraded at all during the initial stages of the culture experiment. Indeed, effective degradation only occurred after day 21 when MnP activities suddenly increased; by day 42, 20.0% of the β-O-4 compound had been degraded. Conversely, no degradation of the β-O-4 compound was observed under Lcc-induced conditions until day 42 (4.2%). In the incubation experiment with MnP solution, the β-O-4 compound was effectively degraded in the initial 4 days of incubation (16.8%), with degradation increasing very gradually thereafter and then decreasing markedly near the end of the experiment; i.e., 23.8% at day 10 and only 7% of the compound was degraded in the latter 6 days. Conversely, degradation of the β-O-4 compound incubated with Lcc solution was detectable, but weak, until 10 days after inoculation (6.9%). The change in the degradation rates of the β-O-4 compound incubated with a mixture of the MnP and Lcc enzyme solutions (each at an initial activity of 17 U/ml) were similar to the degradation patterns of the MnP solution alone. This similarity indicated that no additive or multiplier effects could be attributed to the interaction of the two enzymes on the degradation of the β-O-4 compound. Compared to the initial period of the incubation experiment, the shallow slope of degradation rate in the latter period of the incubation was partly attributable to decreased enzyme activities over the course of the experiment (Table 1). Unfortunately, because we conducted this experiment without a protease-inhibitor, the decrease in enzyme activities was observed in enzyme solutions containing both MnP and Lcc, as well as the mixed MnP+Lcc solutions. In addition, laccase is also capable of degrading non-phenolic lignin model compounds in systems incorporating naturally occurring or synthetic redox mediators (Johannes & Majcherczyk, 2000; Srebotnik & Hamel 2000; Tanaka et al., 2009). Nevertheless, based on the above results, manganese peroxidase (LeMnP2) appears to be more important than laccase (Lcc1) in lignin degradation by *L. edodes*. Prior to the discovery of Lip and MnP, one of the major catabolites formed by the degradation of the β-*O*-4 dimer by *P. chrysosporium*, 2-guaiacoxyethanol (II), was identified (Enoki et al., 1980). The results described above suggest that the assay system developed in this study is well suited for identifying phenol-oxidizing isozymes involved in the degradation of lignin model compounds. Further, these phenol-oxidizing isozymes have been effective for elucidating the mechanisms involved lignin degradation, and this role is likely to extend into the future (Cullen & Kersten, 2004).

Although we attempted to identify other MnP isozymes using another commercial Japanese Shiitake variety, "Bridge 32" (The General Environmental Technos Co. Ltd., Osaka, Japan), which is also used in sawdust cultivation, the variety exhibited the same extracellular MnP isozyme patterns as H600 (data not shown). The estimated heritability (h^2), which is the ratio of the additive genetic components of variance to the phenotypic components of variance, of the variety's wood-degrading ability was relatively low (32.2%) compared to the heritabilities estimated for other traits in crosses of H600 and Bridge 32 (Tanesaka et al., 2007). This low heritability may be attributable to the low allelic variation that exists between the MnP isozymes of the two varieties. The MnP that is produced by *L. edodes* when it is cultured on sawdust media (Buswell et al., 1995; Leatham 1985; Makker et al., 2001; Sakamoto et al., 2009), and which degrades the β-*O*-4 lignin model compound (Kochi et al., 2009), is likely to be critical for mycelial growth and fruit-body development during sawdust cultivation. The system presented here for assaying phenol-oxidizing enzymes under liquid culture conditions could therefore provide a practical screening method for examining isozymes of value in mushroom cultivation, particularly since the assay system targets the wood-degrading ability and the genomic characteristics of the genes involved in lignin-degradation.

4. Conclusions

When cultivated on sawdust-based media, the white-rot basidiomycete *Lentinula edodes* frequently produces the lignin-degrading enzymes MnP and Lcc. In this study, MnP produced by *L. edodes* was induced in a liquid culture supplemented with a sawdust extract of *Castanopsis cuspidata*. Lcc activity was induced by the addition of 2 mM $CuSO_4$ $5H_2O$ into the same media 7 days after initial inoculation. In addition to employing native-PAGE and sequential enzymatic staining to detect the MnP and Lcc secreted by *L. edodes*, we also compared the expression of intra- and extracellular MnP isozymes. To distinguish between the phenol-oxidizing enzymes after native-PAGE, the gel was sequentially stained using an improved enzymatic staining solution (referred to as LccS+EDTA). In addition to containing 0.1 mM acetate buffer (pH 4.0) for Lcc detection, the staining solution contained 1.8 mM *o*-dianisidine as the substrate and 130 mM EDTA to eliminate Mn^{2+} contamination. Subsequently, 0.1 mM H_2O_2 was added to the LccS+EDTA for Per detection (PerS+EDTA), and 0.1 mM $MnSO_4$ $5H_2O$ was added to the PerS, without EDTA, for MnP detection (MnPS). The two extracellular isozyme bands, MnP-e (52) and MnP-e (57), detected in culture medium under MnP-induced conditions, were both identified as manganese peroxidase (LeMnP2). Similarity, the bands Lcc-e (61) and Lcc-e (67), which were detected under Lcc-induced conditions, were both identified as laccase (Lcc1) by Q-TOF mass spectrometry. Four major, intracellular, MnP isozyme bands were detected in mycelial extracts obtained from *L. edodes* cultured under MnP-induced conditions. Of these isozyme bands, two exhibited the same mobilities as extracellular MnP isozymes, while the other two bands, MnP-i (63) and MnP-i (66), were strictly intracellular. The intracellular MnP isozymes were expressed during the initial stage of culture, either several days before, or coincident with, the expression of the extracellular MnP isozymes. Compared to intracellular MnP isozymes, the extracellular MnP isozymes maintained relatively high activities for up to 40 days of culture. While glycosidase treatment of crude enzyme solutions prior to electrophoresis had no effect on the activities of the extracellular MnP isozymes, such treatment completely inactivated the two strictly intracellular MnP isozymes, implying that the intracellular isozymes were active as glycosylated proteins. Both of the extracellular MnP isozymes detected in the dikaryon were also detected in monokaryotic progeny, suggesting that although these isozymes may be encoded by different loci, they are not under allelic control. Southern blot analysis revealed that the probe *lemnp2* region hybridized with the four of the monokaryotic strains used, all of which exhibited the same two hybridization signals that were observed in the parent dikaryon. These observations suggest that there are two copies of *lemnp2* in the *L. edodes* haploid genome. Moreover, degradation assays involving the addition of the β-*O*-4 lignin model compound in cultures under MnP- and Lcc-induced conditions suggest that, rather than laccase (Lcc1), manganese peroxidase (LeMnP2) is a critical enzyme for lignin degradation in *L. edodes*.

In response to the crucial role played by basidiomycetous fungi in the carbon cycle by degrading lignocelluloses, considerable effort has focused on the functional genomics related to the enzymatic systems and mechanisms involved in lignin degradation, particularly in a few model fungus species. Nevertheless, fungal succession on dead logs and leaf litter in nature show that complete degradation of lignocelluloses is a commensal and competitive process affected by numerous fungi. The assay system presented here would be practical and convenient, not only as a method of screening isozymes of value in mushroom breeding and cultivation, but also for evaluating the lignin-degrading abilities of fungi and assessing the antagonistic interactions of different strains under experimental conditions.

5. Acknowledgments

We are grateful to Professor Toshiaki Umezawa from Kyoto University for his kind support and synthesis of the β-*O*-4 lignin model compound. This work was supported by the "Academic Frontier" Project for Private Universities, with a matching fund subsidy from the Ministry of Education, Culture, Sports, Science and Technology (2004–2008), Japan. The Article Processing Charges for this chapter were provided by a fund from The General Environmental Technos Co. Ltd., Osaka, Japan.

6. References

Bonnen, A.; Anton, L. & Orth, A. (1994). Lignin-degrading enzymes of the commercial button mushroom, *Agaricus bisporus*. *Appl Environ Microbiol*, Vol.60, No.3, (March 1994), pp. 960-965, ISSN 0099-2240

Enzymatic Staining for Detection of Phenol-Oxidizing Isozymes Involved in Lignin-Degradation by
Lentinula edodes on Native-PAGE

165

Burke, R. & Cairney, J. (2002). Laccase and other phenol oxidases in ecto- and ericoid mycorrhizal fungi. *Mycorrhiza*, Vol.12, No.3, (June 2002), pp. 105-116. ISSN 0904-6360

Buswell, J.; Cai, Y. & Chang, S-T. (1995). Effect of nutrient and manganese on manganese peroxidase and laccase production by *Lentinula (Lentinus) edodes*. *FEMS Microbiol Lett*, Vol.128, No.1, (April 1995), pp. 81-88, ISSN 0378-1097

Chang, S-T. & Miles, P. (1989). *Edible mushrooms and their cultivation*, CRC Press, Inc., ISBN 0-8493-6758-X, Boca Raton, Florida, USA

Collins, P. & Dobson, A. (1997). Regulation of laccase gene transcription in *Trametes versicolor*. *Appl Environ Microbiol*, Vol.63, No.3, (September 1997), pp. 3444-3450, ISSN 0099-2240

Cullen, D. & Kersten, P. (2004). Enzymology and molecular biology of lignin degradation. In: *The Mycota III; Biochemistry and molecular biology. 2nd edition,* Brambl, R., & Marzluf, G. (Eds), pp. 249–273, Springer, ISBN 3-540-42630-2, Berlin, Germany

Enoki, A.; Gwendolyn, P., Goldsby, G. & Gold, M. (1980). Metabolism of the lignin model compounds veratryl glycerol-β-guaiacyl ether and 4-ethoxy-3-methoxyphenylglycerol-β-guaiacyl ether. *Arch Microbiol*, Vol.125, No.3, (April 1980), pp. 227-232, ISSN 0302-8933

Enoki, A.; Tanaka, H. & Fuse, G. (1988). Degradation of lignin-related compounds, pure cellulose, and wood components by white-rot and brown-rot fungi. *Holzforschung*, Vol.42, No.2, (January 1988), pp. 85-93, ISSN 0018-3830

Forrester, I.; Grabski, A., Mishra, C., Kelley, B., Strickland, W., Leatham, G. & Burgess, R. (1990). Characteristics and N-terminal amino acid sequence of a manganese peroxidase purified from *Lentinula edodes* cultures grown on commercial wood substrate. *Appl Microbiol Biotechnol*, Vol.33, No.3, (June 1990), pp. 359-365, ISSN 0175-7598

Galhaup, C.; Goller, S., Peterbauer, C., Strauss, J. & Haltrich, D. (2002). Characterization of the major laccase isozyme from *Trametes pubescens* and regulation of synthesis by metal ions. *Microbiology*, Vol.148, No.7, (July 2002), pp. 2159-2169, ISSN 1350-0872

Galhaup, C. & Haltrich, D. (2001). Enhanced formation of laccase activity by the white-rot fungus *Trametes pubescens* in the presence of copper. *Appl Microbiol Biotechnol*, Vol.56, No.2, (July 2001), pp. 225–232, ISSN 0175-7598

Glenn, J.; Morgan, M., Mayfield, M., Kuwahara, M. & Gold, M. (1983). An extracellular H2O2-requiring enzyme preparation involved in lignin biodegradation by the white rot basidiomycete *Phanerochaete chrysosporium*. *Biochem Biophys Res Commun*, Vol.114, No.3, (August 1983), pp. 1077-1083, ISSN 0006-291X

Glenn, J. & Gold, M. (1985). Purification and characterization of an extracellular manganese (II)-dependent peroxidase from lignin-degrading basidiomycete, *Phanerochaete chrysosporium*. *Arch Biochem Biophys*, Vol.242, No.2, (November 1985), pp. 329-341, ISSN 0003-9861

Gold, M. & Alic, M. (1993). Molecular biology of the lignin-degrading basidiomycete *Phanerochaete chrysosporium*. *Microbiological Review*, Vol.57, No.3, (September 1993), pp. 605-622, ISSN 0146-0749

Highley, T.; Murmanis, L. & Palmer, J. (1985). Micromorphology of degradation in western hemlock and sweetgum by the brown-rot fungus *Poria placenta*. *Holzforschung*, Vol. 39, No.2, (January 1985), pp. 73-78, ISSN 0018-3830

Highley, T. & Murmanis, L. (1987). Micromorphology of degradation in western hemlock and sweetgum by the white-rot fungus *Coriolus versicolor*. *Holzforschung*, Vol. 41, No.2, (January 1987), pp. 67-71, ISSN 0018-3830

Higuchi, T. (1990). Lignin biochemistry: biosynthesis and biodegradation. *Wood Sci Technol*, Vol.24, No.1, (March 1990), pp. 23-63, ISSN 0043-7719

Iakovlev, A. & Stenlid, J. (2000). Spatiotemporal patterns of laccase activity in interacting mycelia of wood-decaying basidiomycete fungi. *Microbial Ecology*, Vol.39, No. 3, (April 2000), pp. 236-245, ISSN 0095-3628

Johannes, C. & Majcherczyk, A. (2000). Natural mediators in the oxidation of polycyclic aromatic hydrocarbons by laccase mediator system. *Appl Environ Microbiol*, Vol.66, No.2, (February 2000), pp. 524-528, ISSN 0099-2240

Jørgensen, E. & Vejlby, K. (1953). A new polyphenol oxidase test. *Physiologia Plantarum*, Vol.6, No.3, (July 1953), pp. 533-537, ISSN 0031-9317

Kinugawa, K. & Tanesaka, E. (1990). Changes in the rate of CO_2 release from cultures of three basidiomycetes during cultivation. *Trans Mycol Soc, Japan*, Vol.31, No.4, (December 1990), pp. 489–500, ISSN 0029-0289

Kirk, T. & Fenn, P. (1982). Formation and action of the ligninolytic system in basidiomycetes, In: *Decomposer basidiomycetes: their biology and ecology*, Frankland, J., Hedger, J. & Swift, M. (Eds.), pp. 67-90, Cambridge University Press, ISBN 0-521-24634-2, New York, USA

Kirk, T. & Highley, T. (1973). Quantitative changes in structural components of conifer woods during decay by white- and brown-rot fungi, *Phytopathology*, Vol.63, No. 11, (November 1973), pp. 1338-1342, ISSN 0031-949X

Kochi, A.; Saeki, N., Tanesaka, E. & Yoshida, M. (2009). Degradation of β-O-4 lignin model compound by Shiitake, *Lentinula edodes*. *J Crop Res* Vol.54, (November 2009) pp. 131–136, ISSN 1882-885X

Kuwahara, M.; Glenn, J., Morgan, M. & Gold, H. (1984). Separation and characterization of two extracellular H_2O_2-dependent oxidases from ligninolytic cultures of *Phanerochaete chrysosporium*. *FEBS Lett*, Vol.169, No.2, (April 1984), pp. 247-250, ISSN 0014-5793

Leatham, G. (1985). Extracellular enzymes produced by the cultivated mushroom *Lentinus edodes* during degradation of a lignocellulosic medium. *Appl Environ Microbiol*, Vol.50, No.4, (October 1985), pp. 859-867, ISSN 0099-2240

Makker, R.; Tsuneda, A., Tokuyasu, K. & Mori, Y. (2001). *Lentinula edodes* produces a multicomponent protein complex containing manganese (II)-dependent peroxidase, laccase and β-glucosidase. *FEMS Microbiol Lett*, Vol.200, No.2, (June 2001), pp. 175–179, ISSN 0378-1097

Mercer, P. (1982). Basidiomycete decay of standing trees, In: *Decomposer basidiomycetes: their biology and ecology*, Frankland, J., Hedger, J. & Swift, M. (Eds.), pp. 143-160, Cambridge University Press, ISBN 0-521-24634-2, New York, USA

Milligan, B. (1998). Total DNA isolation. In: *Molecular Genetic analysis of population, 2nd edition*, Hoelzel, A. (Ed.), pp. 29-64, Oxford University Press, ISBN 0-19-963635-4, New York

Nagai, M.; Sato, T., Watanabe, H., Saito, K., Kawata, M. & Enei, H. (2002). Purification and characterization of an extracellular laccase from the edible mushroom *Lentinula edodes*, and decolorization of chemically different dyes. *Appl Microbiol Biotechnol*, Vol.60, No. 3, (November 2002), pp. 327–335, ISSN 0175-7598

Nagai, M.; Kawata, M., Watanabe, H., Ogawa, M., Saito, K., Takesawa, T., Kanda, K. & Sato, T. (2003). Important role of fungal intracellular laccase for melanin synthesis : purification and characterization of an intracellular laccase from the edible mushroom *Lentinula edodes* fruit bodies. *Microbiology*, Vol.149, No.9, (September 2003), pp. 2455-2462, ISSN 1350-0872

Enzymatic Staining for Detection of Phenol-Oxidizing Isozymes Involved in Lignin-Degradation by
Lentinula edodes on Native-PAGE

167

Nagai, M.; Sakamoto, Y., Nakade, K. & Sato, T. (2007). Isolation and characterization of the gene encoding a manganese peroxidase from *Lentinula edodes*. *Mycoscience*, Vol.48, No.2, (April 2007), pp.125-130, ISSN 1340-3540

Ohga, S. & Kitamoto, Y. (1997). Future of mushroom production and biotechnology. *Food Reviews International*, Vol.13, No.3, (October 1997), pp. 461–469, ISSN 8755-9129

Orth, A.; Royse, D. & Tien, M. (1993). Ubiquity of lignin-degrading peroxidases among various wood-degrading fungi. *Appl Environ Microbiol*, Vol.59, No. 12, (December 1993), pp. 4017-4023, ISSN 0099-2240

Palmieri, G.; Giardina, P., Bianko, C., Fontanella, B. & Sannia, G. (2000). Copper induction of laccase isozymes in the ligninolytic fungus *Pleurotus ostreatus*. *Appl Environ Microbiol*, Vol.66, No.3, (March 2000), pp. 920-924, ISSN 0099-2240

Pasteur, N.; Pasteur, G., Bonhomme, F., Catalan, J. & Britton-Davian, J. (1988). *Practical isozyme genetics*, Ellis Horwood Ltd., ISBN 0-7458-0501-9, Chichester, UK

Paszczynski, A.; Crawford, R. & Huynh, V-B. (1988). Manganese peroxidase of *Phanerochaete chrysosporium*: Purification. In: *Methods in enzymology; Vol.161, Biomass, Part B, Lignin, pectin and chitin*, Wood, W. & Kellogg, S. (Eds.), 264-270, Academic Press, ISBN 0-12-182062-9, San Diego, CA, USA

Saeki, N.; Takeda, H., Tanesaka, E. & Yoshida, M. (2011). Induction of manganese peroxidase and laccase by *Lentinula edodes* under liquid culture conditions and their isozyme detection by enzymatic staining on native-PAGE. *Mycoscience*, Vol.52, No.2, (March 2011), pp. 132-136, ISSN 1340-3540

Sakamoto, Y.; Nakade, K., Yano, A., Nakagawa, Y., Hirano, T., Irie, T., Watanabe, H., Nagai, M. & Sato, T. (2008). Heterologous expression of lcc1 from *Lentinula edodes* in tobacco BY-2 cells results in the production an active, secreted form of fungal laccase. *Appl Microbiol Biotechnol*, Vol.79, No.6, (July 2008), pp. 971–980, ISSN 0175-7598

Sakamoto, Y.; Nakade, K., Nagai, M., Uchimiya, H. & Sato, T. (2009). Cloning of *Lentinula edodes* lemnp2, a manganese peroxidase that is secreted abundantly in sawdust medium. *Mycoscience*, Vol.50, No.2, (March 2009), pp. 116-122, ISSN 1340-3540

Sambrook, J.; Fritsch E. & Maniatis, T. (1989). *Molecular cloning: a laboratory manual, Book 2* (2nd edition), Cold Spring Harbor Laboratory Press, ISBN 0-87969-309-6, New York, USA

Saparrat, M.; Guillén, F., Arambarri, A., Martínez, A. & Martínez, M. (2002). Induction, isolation, and characterization of two laccases from the white rot basidiomycete *Coriolopsis rigida*. *Appl Environ Microbiol*, Vol.68, No.4, (April 2002), pp. 1534–1540, ISSN 0099-2240

Scheel, T.; Höfer, M., Ludwig, S. & Hölker, U. (2000). Differential expression of manganese peroxidase and laccase in white-rot fungi in the presence of manganese or aromatic compounds. *Appl Microbiol Biotechnol*, Vol.54, No.5, (November 2000), pp. 686-691, ISSN 0175-7598

Shutova, V.; Revin, V. & Myakushina Y. (2008). The effect of copper ions on the production of laccase by the fungus *Lentinus (Panus) tigrinus*. *Appl Biochem Microbiol*, Vol.44, No.6, (November 2008), pp. 683–687, ISSN 0003-2697

Smith, J.; Fermor, T. & Zadrazil, F. (1988). Pretreatment of lignocellulosics for edible fungi. In: *Treatment of lignocellulosics with white rot fungi*, Zadrazil, F. & Reiniger, P. (Eds.), pp. 3-13, Elsevier Applied Science, ISBN 1-85166-241-3, Essex, England

Soden, D. & Dobson, A. (2001). Differential regulation of laccase gene expression in *Pleurotus sajor-caju*. *Microbiology*, Vol.147, No.7, (July 2001), pp. 1755–1763, ISSN 1350-0872

Srebotnik, E. & Hammel, K. (2000). Degradation of nonphenolic lignin by the laccase/1-hydroxybenzotriazole system. *J Biotechnol*, Vol.81, No.2-3, (August 2000), pp. 179-188, ISSN 0021-9193

Szklarz, G.; Antibus, R., Sinsabaugh, R. & Linkins, A. (1989). Production of phenol oxidase and peroxidases by wood-rotting fungi. *Mycologia*, Vol.81, No.2, (April 1989), pp. 234-240, ISSN 0027-5514

Tanaka, H.; Koike, K., Itakura, S. & Enoki, A. (2009). Degradation of wood and enzyme production by *Ceriporiopsis subvermispora*. *Enzyme Microbial Technol*, Vol.45, No.5, (November 2009), pp. 384-390, ISSN 0141-0229

Tanesaka, E.; Masuda, H. & Kinugawa, K. (1993). Wood degrading ability of basidiomycetes that are wood decomposers, litter decomposers, or mycorrhizal symbionts. *Mycologia*, Vol.81, No.13, (July 1993), pp. 347-354, ISSN 0027-5514

Tanesaka, E.; Nakajima, Y., Uchida, T., Ikeda, Y. & Yoshida, M. (2007). Variation of wood degrading ability among hybrid stocks of Shiitake mushroom, *Lentinula edodes*, and their cultivation traits on sawdust media, *Breeding Research, Proceedings of Japanese Society of Breeding 2007 112th Conference*, Vol.9, No.2, pp. 111, ISSN 1344-7629, Yamagata, Japan, September 22-23, 2007

Thurston, C. (1994). The structure and function of fungal laccases. *Microbiology*, Vol.140, No.1, (January 1994), pp. 19-26, ISSN 1350-0872

Tien, M. & Kirk, T. (1983). Lignin-degrading enzyme from the hymenomycete *Phanerochaete chrysosporium* Burds. *Science*, Vol.221, No. 4611, (August 1983), pp. 661-663, ISSN 0036-8075

Tien, M. & Kirk, T. (1984). Lignin-degrading enzyme from *Phanerochaete chrysosporium*: Purification, characterization and catalytic properties of a unique hydrogen peroxide-requiring oxygenase. *Proc Natl Acad Sci USA*, Vol. 81, No.8, (April 1984), pp.2280-2284, ISSN 0027-8424

Umezawa, T. & Higuchi, T. (1985). A novel C_α-C_β cleavage of lignin model dimer with rearrangement of the β-aryl group by *Phanerochaete chrysosporium*. *FEBS Lett*, Vol.192, No.1 (November 1985), pp. 293-298, ISSN 0014-5793

White, N. & Boddy, L. (1992). Extracellular enzyme localization during interspecific fungal interactions. *FEMS Microbiol Lett*, Vol.98, No.1 (November 1992), pp.75-79, ISSN 0378-1097

Wood, D.; Matcham, S. & Fermor, T. (1988). Production and function of enzymes during lignocellulose degradation, In: *Treatment of lignocellulosics with white rot fungi*, Zadrazil, F. & P. Reiniger, (Eds.), pp. 43-49, Elsevier Applied Science, ISBN 1-85166-241-3, Essex, England

Yoshikawa, T.; Tanesaka E. & Yoshida M (2004) Determination of ligninolytic culture conditions in expression of *Lentinula edodes* manganese peroxidase, *Breeding Research, Proceedings of Japanese Society of Breeding 2004 105th Conference*, Vol.6, No.1, pp. 15, ISSN 1344-7629, Tokyo, Japan, March 30-31, 2004

8

Protection Studies by Antioxidants Using Single Cell Gel Electrophoresis (Comet Assay)

Pınar Erkekoglu
Hacettepe University, Faculty of Pharmacy, Department of Toxicology, Ankara
Turkey

1. Introduction

Oxidation-reduction reactions, simply referred as "redox" reactions, describe all the chemical reactions in which atoms have their oxidation state changed. This can either be a simple redox process like the oxidation of carbon (C) to carbon dioxide (CO_2) or the reduction of C by hydrogen (H) to yield methane (CH_4). However, in biology redox reactions are rather complex and 'redox biology' is fundamental to aerobic life **(Peters et al., 2008; Baliga et al., 2007)**. The simplest example to give is the oxidation of glucose to CO_2 and water in photosynthesis **(Halliwell, 2006)**.

Aerobes are constantly subject to free radicals, but modulate their actions by synthesizing antioxidants. Free radicals are atoms, molecules, or ions with one or more unpaired electrons on an open shell configuration **(Gutteridge & Halliwell, 2000)**. The simplest form is the atomic H. There are many types of free radicals in living systems, but both nitrogen (N) and oxygen (O) radicals are the main concern for the researchers of several fields as they are suspected to be the underlying factors of several conditions and diseases **(Halliwell, 2006)**. O_2 toxicity was suggested to be due to the inactivation of a variety of enzymes (particularly of antioxidant enzymes) by targeting the thiol group of cysteine residues. In the last decades, molecular biology techniques established that the toxic effects of O_2 are directly linked to its reactive forms, the reactive oxygen species (ROS), acting on cellular components. Oxidative stress is a serious imbalance between the generation of ROS and antioxidant protection in favor of the former, causing excessive oxidative damage **(Dröge, 2002; Halliwell, 2011)**. Oxidative stress and ROS can account for changes that may be detrimental to the cells **(Dröge, 2002)**. ROS are shown to contribute to cellular damage, apoptosis and cell death **(Dalton et al., 1999; Finkel, 1998)**. The link between O_2 toxicity and many pathologies, e.g. pulmonary diseases, **(Frankl, 1991)**, and its effect on swelling of the blood–gas barrier **(Drath et al., 1981)**, retina defects **(Geller et al., 2006)**, bowel disease **(Grisham, 1994)** neurodegeneration **(Wang et al., 2006)**, cancer **(Cerutti, 1994)**, diabetes **(Seet et al., 2010)** and ageing **(Irminger-Finger, 2007)** is very well-established. Besides, in the last decade a relationship between obesity and ROS was demonstrated **(Seet et al., 2010; Halliwell, 2011)**.

Antioxidant is a molecule that protects a biological target against oxidative damage **(Halliwell, 2011)**. Accumulating data implicate that both low antioxidant status and genetics may contribute to the risk of several types of malignancies **(Peters et al., 2008;**

Baliga et al., 2007). The field of antioxidants and free radicals is often perceived as focusing around the use of antioxidant supplements to prevent human disease. Currently, there is a growing interest in environmental chemicals that can cause oxidative stress. The genotoxic effects of some compounds are of particular interest for researchers as humans are exposed to these chemicals abundantly. Exposure to such chemicals may result in disturbances of several physiological processes and may lead to wide variety of degenerative diseases including cancer **(Soory, 2009).**

First described by **Östling & Johanson (1984),** and then modified by **Singh et al. in 1988,** the single cell gel electrophoresis assay (also known as Comet assay) is an uncomplicated and sensitive technique for the detection of DNA damage at the level of the individual eukaryotic cell. It has since gained in popularity as a standard technique for evaluation of DNA damage/repair, biomonitoring and genotoxicity testing **(Singh et al., 1988).**

2. Why comet assay is a suitable tool for antioxidant research?

Comet assay can easily detect the *in vitro* toxicity of environmental chemicals on different cell types, as well as *in vivo* toxicity in tissue samples obtained from animals. Besides, it is also a valid technique to evaluate whether antioxidants/micronutrients are able to protect the integrity of the genetic material **(Anderson et al., 1997; Heaton et al., 2002; Novotna et al., 2007).**

The benefits of Comet assay can be summarized as below:

- Sensitivity for detecting low levels of DNA damage: The limit of sensitivity is approximately 50 strand breaks per diploid mammalian cell and will lose sensitivity above about 10,000 breaks per cell **(Olive & Banáth, 2006).**
- Requirement for small number of cells per sample: <10,000 cells are enough to perform the assay.
- Flexibility: Comet assay is applicable to virtually any type of cell, as long as a single cell suspension is obtained. Besides, different combinations of unwinding and electrophoresis conditions and lesion-specific enzymes can be used to detect different types and levels of DNA damage **(Wong et al., 2005).**
- Low cost and ease of application **(Anderson et al., 1997).**
- Studies can be conducted using relatively small amounts of a test substance **(Anderson et al., 1997)**
- A relatively short time is needed to complete an experiment.

The advantages and disadvantages of Comet assay are shown in **Figure 1.**

3. Technical information on comet assay

DNA damage can simply be evaluated using Comet assay that allows the measurement of DNA single- and double-strand breaks (frank strand breaks and incomplete excision repair sites) together with alkali labile sites and crosslinking. By choosing different pH conditions for electrophoresis and the preceding incubation, different levels of damage can be assessed. The degree of DNA migration can be correlated to the extent of DNA damage occurring in each single cell. *In vitro* studies can be performed on virtually with any cell type; however, the cell-type-of-choice in biomonitoring is mostly the lymphocyte because blood is easily

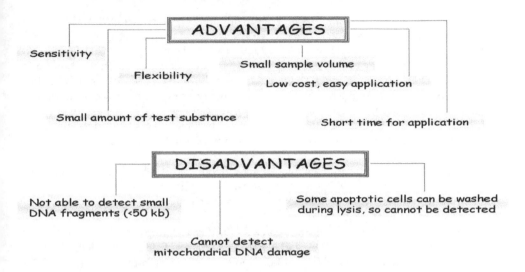

Fig. 1. Advantages and disadvantages of Comet assay

collected and lymphocytes have proved to be good surrogate cells. For example, lymphocytes exhibited genotoxicity caused by anticancer agents targeting several different organs **(Faust et al., 2004)**.

There are differences between laboratories in the isolation of lymphocytes, cells from organs/tissues or other specimens, or in the solutions used for electrophoresis. A simple alkaline Comet assay protocol can be performed in the following steps:

a.   The slides that will be used in the study should be covered with agarose (1%) the day before the experiment.

b.   In the basic alkaline Comet assay, for primary and other cell cultures, after exposing small number of cells to a physical or chemical agent, the cells are trypsinized, centrifuged, washed, and resuspended in PBS. Because of the flexible application of the technique, the cells used can be isolated lymphocytes, cells isolated from bone marrow, cells isolated from solid organs or tissues or cells from primary or other cell cultures. Lymphocytes can be isolated from whole blood using different isolation solutions and centrifugation. Cells from bone marrow can be obtained by perfusing femur in cold mincing solutions and centrifugation. Solid organs or tissues must be minced into fine pieces, later be suspended in cold mincing solutions and centrifugated. Blood-rich organs like liver and kidney have to minced into larger pieces, the mincing solution can be aspirated and fresh mincing solution should be added. Mincing solution can be Hank's Buffered Salt Solution (HBSS, with 20 mM EDTA and 10% DMSO).

c.   Usually 50 µl of the cells obtained from either cell cultures blood or organs/tissues should be mixed with 450 µl solution of low melting point agarose (0.6% in PBS), and 100 µl of the solution is spread on microscope slides covered with agarose.

d.   Cells are lysed (in 2.5 M NaCl, 0.5 MNa$_2$-EDTA, 10 mM Tris, 1% sodium lauryl sulfate, 1% Triton X-100, 10% DMSO, pH 10) at 4°C in dark for 1 h. After lysis, cells were immersed in freshly prepared alkaline electrophoresis buffer (300 mM NaOH, 1 mM Na_2-EDTA, pH 13) for 30 min to allow DNA unwinding.

e.   Electrophoresis is then performed at 25 V/300 mA for 30 min.

f.   After electrophoresis, slides are rinsed three times for 5 min with neutralization buffer (0.4 M Tris-HCl, pH 7.4), and stained with ethidium bromide (20 µg/ml) in PBS. Ethidium bromide is an intercalating agent commonly used as a fluorescent nucleic acid stain in molecular biology. There are a number of alternative stains to ethidium bromide, including acridine orange, propidium iodide, YOYO-1 iodide stain, SYBR Gold nucleic acid gel stain, SbYR Green I stain, TOTO-3 stain and silver (for non-fluorescent staining).

g.   For quantification, a fluorescence microscope can be used which can be connected to a charge-coupled device (CDC) and a computer-based analysis system.

h.   The extent of DNA damage was determined after electrophoretic migration of DNA fragments in the agarose gel.

i.   For each condition randomly selected comets (50/100/200) on each slide can be scored, and % head DNA, % tail DNA, tail length, tail moment and comet length can be determined. Usually, % tail DNA and tail moment are preferred for assessing the DNA damage.

Rather than making use of the cell's own repair enzymes to reveal damage, we can achieve greater specificity and higher sensitivity by treating the DNA with purified repair enzymes which will convert particular lesions into breaks. Thus, Comet assay protocol can also be performed using different base or nucleotide excision repair enzymes (Collins et al., 1997). The most commonly used repair enyme is formamidopyrimidine DNA glycosylase (Fpg) which recognizes and removes 8-oxodeoxyguanosine (8-oxoGua) and other oxidized purines. 8-oxoguanine glycosylase (OGG1) also recognizes 8-oxoGua. Endonuclease III (Endo III) deals with oxidized pyrimidines; and T4 endonuclease V is able to incise at sites of pyrimidine dimers. Digestion with these enzymes is carried out after the initial lysis step. The excision repair pathways act more slowly than strand break rejoining (Collins & Horvathova, 2001), and samples should be taken over a period of a few hours.

Different versions of Comet assay are also used for different puposes. Neutral Comet assay is usually used for assessing double strand DNA breaks in sperm cells. On the other hand, a "Comet Chip" protocol, first introduced by Massachusetts Institute of Technology (MIT) Engelward Lab, is nowadays gaining significant importance as a high throughput DNA damage analysis platform. This new method is also used for evaluating DNA strand breaks, sites of DNA modification and interstrand crosslinks. A limitation of the traditional assay is that each sample requires a separate glass slide and image analysis is laborious and data is intensive, thus reducing throughput. This new technique uses microfabrication technologies to enable analysis of cells within a defined array, resulting in a >200 fold reduction in the area required per condition. Each well of a 96-well plate contains patterned microwells for single cell capture and DNA damage quantification. The "CometChip" can be used to analyze dozens of conditions on a single chip. The newly developed automated image analysis software is used for detection of DNA damage, thus greatly reducing analysis time. This new technology will enable the researchers to conduct both large scale epidemiological and clinical studies (Engelward Lab, 2011).

A new technique "Comet fluorescence in situ hybridization (Comet FISH)" combines two well-established methods. The Comet assay comprises the basis of Comet-FISH and allows separation of fragmented from nonfragmented DNA and quantification of DNA damage and repair. FISH enables detection of specifically labeled DNA sequences of interest, including whole chromosomes. The combined technique of Comet-FISH is a modification of the Comet assay that inserts a hybridization step after unwinding and electrophoresis and permits the labeling of specific gene sequences or telomeres. Comet-FISH has been applied for detection of site-specific breaks in DNA regions that are relevant for development of various diseases, and has also been used to study the distribution of DNA damage and repair in the complete genome. Moreover, DNA sequence modifications can be detected in individual cells using Comet-FISH. The results from the Comet assay alone are only reflections of overall DNA damage. However, the addition of the FISH technique allows the assignment of the probed sequences to the damaged or undamaged part of the comet (tail or head, respectively) **(Schlörmann & Glei, 2009)**.

A spesific illustration for alkaline Comet assay methodology is shown in **Figure 2**. Different protocols of Comet assay in research field are given in **Figure 3**.

In this chapter, I will mainly focus on the genotoxicity of different environmental chemicals and both *in vivo* and *in vitro* protection studies by several selenocompounds, vitamins, and isothiocyanates (ITCs) against the toxicity of these compounds.

4. Protection studies using comet assay

4.1 Prevention of genotoxicity by selenocompounds

There is considerable interest in developing strategies that prevent genotoxicity and cancer with minimal risk or toxicity. Trace elements like selenium (Se) are of particular interest as it is the key component of antioxidant enzyme systems.

The requirement for Se and its beneficial role in human health have been known for several decades. Se is an essential trace element commonly found in grains, nuts, and meats and many years of research showed that that low, non-toxic supplementation with either organic and inorganic forms could reduce cancer incidence following exposure to a wide variety of carcinogens **(El-Bayoumy, 2004)**.

Along with its important role for the cellular antioxidant defense, Se is also essential for the production of normal spermatozoa and thus plays a critical role in testis, sperm, and reproduction **(Flohé, 2007)**. In the physiological dosage range, Se appears to function as an antimutagenic agent, preventing the malignant transformation of normal cells and the activation of oncogenes **(Schrauzer, 2000)**. Although most of its chemopreventive mechanisms still remain unclear, the protective effects of Se seem to be primarily associated with its presence in the glutathione peroxidases (GPxs), which are known to protect DNA and other cellular components from damage by oxygen radicals **(Negro, 2008)**. Low activity of another important peroxidase, GPx4, can lead to reduction in reproduction **(Flohé, 2007)**.

Selenoenzymes are known to play roles in carcinogen metabolism, in the control of cell division, oxygen metabolism, detoxification processes, apoptosis induction and the functioning of the immune system oncogenes **(Schrauzer, 2000)**. Several studies have determined the low activity of Se-containing cytosolic GPx, known as GPx1, as a substantial

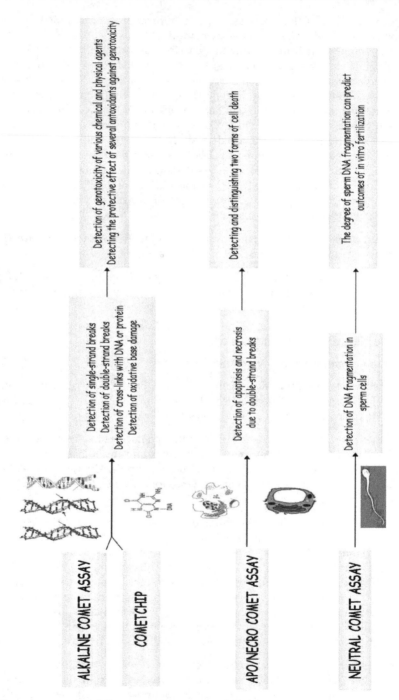

Fig. 2. Different protocols of Comet assay in research field

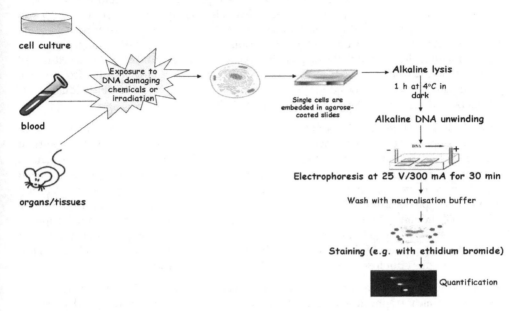

Fig. 3. Alkaline Comet assay methodology

factor in cancer risk **(Esworthy et al., 1985)**. Other modes of action, either direct or indirect, may also be operative, such as the partial retransformation of tumor cells and the inactivation of oncogenes. However, the effects of Se in the physiological dosage range are not attributable to cytotoxicity, allowing Se to be defined as a genuine nutritional cancer-protecting agent **(Yu et al., 1990)**. On the other hand, selenocompounds such as selenodiglutathione, methylselenol, selenomethionine (SM), and Se-methylselenocysteine might affect the metabolism of carcinogens, thus preventing initiation of carcinogenesis **(Gopalakrishna & Gundimeda, 2001)**. These compounds might also restrict cell proliferation by inhibiting protein kinases and by halting phases of the cell cycle that play a central part in cell growth, tumor promotion, and differentiation **(Brinkman et al., 2006)**. A further possible mechanism of action is enhancement of the immune system by stimulating the cytotoxic activities of natural killer cells and lymphokine activated killer cells to act against cancer cells **(Combs, 1998)**. The anticarcinogenic effects of Se are counteracted by Se-antagonistic compounds, and elements **(Schrauzer, 2000)**.

For maximal utilization of its cancer-protective potential, Se supplementation should start early in life and be maintained over the entire lifespan **(Schrauzer & White, 1978; Persson-Moschos et al., 1998; Schrauzer, 2000)**. In addition, exposure to Se antagonists and carcinogenic risk factors should be minimized by appropriate dietary and lifestyle changes **(Schrauzer, 1976; Schrauzer, 1977)**. Because geographical studies done in the 1970s reported a possible inverse association between Se and cancer mortality, epidemiological studies have focused on investigating the anticarcinogenic properties of this nutrient **(Brinkman et al., 2006)**. Two key findings that emerged from these early studies were the inverse association between Se and cancer seemed to be both sex and organ specific **(Li et al., 2004)**.

A larger difference in the reduced death rates was reported for men than for women in regions with high levels of Se, and mortality was significantly lower for some types of cancer **(Shamberger et al., 1976; Clark et al., 1991)**. Higher blood levels of Se have been associated with a lower risk of many types of neoplasia, including prostate, lung, colorectal, and possibly bladder, although the data are inconsistent. A significant 39% decreased risk of bladder cancer associated with high levels of Se by combining results from seven epidemiologic studies, conducted in different populations, which applied individual levels of Se measured in serum or toenails **(Brinkman et al., 2006)**.

Supra-physiological levels of sodium selenite (SS) in the presence of polythiols have oxidative properties that might have an anticancer effect by increasing the vulnerability of cancer cells to destruction. It was stated that Se, independent of type (organic/inorganic), can alter several genes to prevent cancer. High doses of Se might upregulate phase II detoxification enzymes, some Se-binding proteins, and some apoptotic genes, and downregulate phase I activating enzymes and cell proliferation genes **(El Bayoumy & Sinha, 2005)**. Inhibition of carcinogen–DNA adducts formation and induction of apoptosis by high doses of Se suggests that protection occurs at both the initiation and post-initiation phases of carcinogenesis **(El Bayoumy & Sinha, 2005)**. However, at lower physiological doses, Se prevents apoptosis, and induces DNA repair **(Longtin, 2003)**.

The literature agrees on the protective effect of Se evaluated with the Comet assay towards a variety of chemical or physical toxic agents. However, it remains inconclusive which is/are the most suitable Se compound/s to prevent DNA damage and which doses should be used to observe protection. In this chapter, the protective effects of both inorganic and organic selenocompounds, against phthalate and radiation toxicity will be discussed.

4.1.1 Prevention of phthalate genotoxicity by selenocompounds

Phthalate esters are a widespread class of peroxisome proliferators (PPs) and endocrine disruptors. They have attracted substantial attention due to their high production volume and use in a variety of polyvinyl chloride (PVC)-based consumer products **(Akingbemi et al., 2001; Grande et al., 2006)**.

Uses of the various phthalates mainly depend on their molecular weight (MW). Higher MW phthalates, such as di(2-ethylhexyl) phthalate, (DEHP), are used in construction materials and in numerous PVC products including clothing (footwear, raincoats), food packaging, children products (toys, grip bumpers), and medical devices **(Heudorf et al., 2007)**, while relatively lower MW phthalates like di-methyl phthalate (DMP), di-ethyl phthalate (DEP), and di-n-butyl phthalate (DBP) are mainly used as odor/color fixatives or as solvents and in cosmetics, insecticides and pharmaceuticals, but are also used in PVC **(Heudorf et al., 2007)**.

Phthalate migrate out from PVC-containing items into food, air, dust, water, and soils and create human exposure in various ways **(Clark et al., 2003)**. Increasing number of studies on human blood and urine have revealed the ubiquitous phthalate exposure of consumers in industrialized countries **(Wormuth et al., 2006, Frederiksen et al., 2008; Frederiksen et al., 2010; Janjua et al., 2011 , Durmaz et al., 2010)**.

DEHP is the most important phthalate derivative with its high production, use and occurrence in the environment. It is mainly used in PVC plastics in the form of numerous consumer and personal care products and medical devices **(Doull et al., 1999)**. The biological effects of DEHP are hence of major concern but so far elusive. Although, the main mechanism underlying hepatocarcinogenicity of phthalates is not fully elucidated, ROS are thought to be associated with the mechanism of tumorigenesis by PPs, including DEHP. This assumption is based to a fact that various proteins that are induced by DEHP in liver parenchymal cells (peroxisomes, mitochondria and microsomes) are prone to formation of H_2O_2 and other oxidants. Besides, activation of metabolizing enzymes and peroxisome proliferator-activated receptor α (PPARα) might be other substantial factors leading to high intracellular ROS production **(O'Brien et al., 2005; Gazouli et al., 2002)**. However, the mechanisms by which phthalates and particularly DEHP exert toxic effects in reproductive system are not yet fully elucidated. Irreversible and reversible changes in the development of the male reproductive tract like vimentin collapse of Sertoli cells as well as apoptosis of germ cells, effects on sex hormones (mainly on testosterone) as well as follicle stimulating hormone (FSH) and luteinizing hormone (LH), histopathological changes in testis and sperm anomalies were observed with phthalate exposure **(Corton & Lapinskas, 2005; Foster et al., 2001; Erkekoglu et al., 2011a; Erkekoglu et al., 2011b; Kasahara et al, 2002; Noriega et al., 2009)**. Most of the toxic effects were related to its antiandrogenic potential **(Ge et al., 2007)**. A PPARα-mediated pathway based on its peroxisome proliferating (PP) activity **(Gazouli et al., 2002)**, and activation of metabolizing enzymes have also been suggested **(O'Brien et al., 2005)**. While the induction of an oxidative stress may represent a common mechanism in endocrine disruptor–mediated dysfunction, especially on testicular cells **(Latchoumycandane et al., 2002)**, recent studies are also providing supporting evidences for such an effect with DEHP and its major metabolite, mono(2-ethylhexyl)phthalate (MEHP) **(Erkekoglu et al. 2010a; Erkekoglu et al. 2010b; Erkekoglu et al. 2011c; Fan et al. 2010)**. Thus, the primary targets for the DEHP and MEHP are the Sertoli and Leydig cells of testis. In several studies, it was shown that DEHP caused disruption in the function of both cell types. In fact, **Richburg and Boekelheide (1996)** demonstrated histopathological disturbances and alterations of cytoplasmatic distribution of vimentin in Sertoli cells in testis of 28-day-old Fisher rats after a single oral dose of MEHP (2000 mg/kg). Administration of MEHP to Wistar rats at a single oral dose (400 mg/kg bw) was toxic to Sertoli cells and caused detachment of germ cells **(Dalgaard et al., 2000)**. **Tay et al. (2007)** reported vimentin disruption in MEHP-treated C57Bl/6N mice, and gradual disappearance of vimentin in Sertoli cell cultures as time and dose increased. We have also reported that in DEHP-treated rats, significant disruption and collapse of vimentin filaments and disruption of seminiferous epithelium in Sertoli cells was observed **(Erkekoglu et al., 2011b)**. Among several others, an earlier data has demonstrated the increase of ROS generation and depletion in antioxidant defenses by DEHP treatment in rat testis **(Kasahara et al., 2002)**. Our recent studies on MA-10 Leydig **(Erkekoglu et al., 2010b)** and LNCaP human prostate cells **(Erkekoglu et al., 2010a)** have also produced comprehensive data suggesting that at least one of the mechanisms underlying the reproductive toxicity of DEHP is the induction of intracellular ROS. The data of **Fan et al. (2010)** have also suggested oxidative stress as a new mechanism of MEHP action on Leydig cells steroidogenesis *via* CYP1A1-mediated ROS stress. On the other hand, in rats exposed to 1000 mg/kg DEHP for 10 days, we observed that this particular phthalate induced oxidative stress in rat testis, as evidenced by significant decrease in GSH/GSSG redox ratio (>10-fold) and marked increase in TBARS levels **(Erkekoglu et al., 2011d)**.

Several strategies have been attempted to prevent the oxidative stress caused by toxic chemicals and the use of antioxidant vitamins has been the most common approach. **Ishihara et al. (2000)** showed that supplementation of rats with vitamin C and E protected the testes from DEHP-gonadotoxicity. **Fan et al. (2010)** reported that the increase in ROS generation with MEHP exposure in MA-10 cells was inhibited by N-acetylcysteine (NAC). In the above mentioned *in vitro* studies **(Erkekoglu et al., 2010a; Erkekoglu et al., 2010b)**, we demonstrated that Se supplementation in either organic form (SM, 10 μM) or in inorganic form (SS, 30 nM) was highly protective against the cytotoxicity, ROS producing and antioxidant status-modifying effects of DEHP and MEHP in both MA-10 Leydig and LNCaP cells.

Concerning LNCaP cells, we observed that DEHP had a flat dose–cell viability response curve while MEHP showed a very steep dose–response curve and the cytotoxicity of the MEHP was much higher than that of the parent compound. On the other hand, we determined that both organic and inorganic Se supplementation increased resistance to DEHP and MEHP cytotoxicity. From these data, the doses of DEHP and MEHP to be used for the antioxidant status measurements and Comet assay were chosen as close to IC_{50} values and were 3 mM for DEHP and 3 μM for MEHP. We demonstrated that MEHP was the main active form in LnCAP cells with an almost ~1000- fold higher cytotoxicity than the parent compound. Intracellular ROS production showed marked increases with both DEHP and MEHP treatment; however the effect of MEHP was much higher. Both selenocompounds were partially effective in reducing intracellular ROS production. For the antioxidant enzymes, both DEHP and MEHP caused substantial decreases in GPx1 activity (~3-fold, and ~4-fold, respectively) compared to control cells. However, there was no significant difference between the effects of the two phthalate derivatives. Se supplementation with either SS or SM effectively countered the effect of DEHP by completely restoring the activity up to the control level (NT-C) or even higher. In the case of MEHP treatments, both SS and SM supplementations significantly restored the effect of 3 μM MEHP on GPx1 activity, providing ~2-fold increase. For thioredoxin reductase (TrxR) activity, DEHP did not cause a change compared to control; however, MEHP caused a marked increase. Se supplementation in both organic and inorganic forms increased the TrxR activity almost up to the levels of SS and SM supplemented cells alone. However, no changes were observed with both of the phthalates in glutathione S-transferase (GST) activity and total glutathione (GSH) levels. On the other hand, using alkaline Comet assay, we have demonstrated that in LnCAP cells both DEHP and MEHP produced significant DNA damage as evidenced by increased tail % intensity (~2.9-fold and ~3.2-fold, respectively), and tail moment (~2.4-fold and ~2.6-fold, respectively) compared to NT LNCaP cells. The overall difference between the DNA damaging effects of the parent compound and the metabolite was insignificant. Se supplementation itself did not cause any alteration in the steady-state levels of the biomarkers of DNA damage in LNCaP cells, whereas the presence of Se either in SS or SM form reduced the genotoxic effects of DEHP and MEHP as evidenced by significant (~30%) decreases in tail % intensity. These results thus indicated that the Se with the doses and forms used in this study was not genotoxic, but showed antigenotoxic activity against the genotoxicity of DEHP and MEHP. However, the protective effect of Se with the doses used in this study was not complete. Tail intensity remained ~90% and ~80% higher than that of NT-C in SS/DEHP-T and SM/DEHP-T cells, respectively. Similarly, in SS/MEHP-T and SM/ MEHP-T cells, tail intensities were still ~95% and ~120% high compared to NT-C cells. On the other hand, the extent of tail moment increase induced by DEHP was reduced ~30% with SS and ~18% with SM supplementations, and the tail moment induced by MEHP was reduced ~24% with SS supplementation; however, none of these were statistically significant. Only SM supplementation provided a significant (~34%) reduction in the tail moment induced by MEHP. But again, tail moments remained ~64 and ~95% higher than that of NT-C in SS/DEHP-T and SM/DEHP-T cells, respectively; similarly in SS/MEHP-T and SM/MEHP-T cells, tail moments were still ~94 and ~69% high compared to NT-C cells. In all cases, protective effects of SS and SM were not significantly different than each other **(Erkekoglu et al., 2010a)**.

For Leydig MA-10 cells, The IC_{50} values for DEHP and MEHP were again found to be ~3 mM and ~3 µM, respectively. Se supplementation of the cells with either SS (30 nM) or SM (10 µM) was protective against the cytotoxic effects of DEHP, and MEHP. Intracellular ROS production showed substantial increases with both of the phthalates where the effect of MEHP was much more pronounced. SS and SM showed partial protection against the ROS increment for both the phthalates. In cells exposed to DEHP or MEHP, GPx1 and TrxR activities decreased significantly. Se supplementation either with SS or SM in DEHP-exposed cells was able to enhance the both of the selenoenzyme activities. Moreover, GST activity also decreased significantly with both of the phthalates. However, Se supplementation in both of the forms was not effective in restoring GST activity. GSH levels also decreased significantly in DEHP and MEHP treated Leydig cells while Se supplementation in both forms provided significant restoration in both groups. On the other hand, both DEHP and MEHP produced high level of DNA damage as evidenced by significantly increased tail % intensity (~3.4-fold and ~3.8-fold, respectively), and tail moment (~4.2-fold and ~3.8-fold, respectively) compared to non-treated MA-10 cells. The difference between the DNA damaging effects of the parent compound and the metabolite was insignificant. Se supplementation itself did not cause any alteration on the steady state levels of the DNA damage biomarkers of MA-10 cells. But Se was highly effective to decrease the genotoxic effects of phthalate esters. Increased tail % intensities by DEHP and MEHP exposure were lowered ~50–55% with SS supplementation, whereas SM treatment provided ~30–40% protection. SS decreased the tail moments of the DEHP- or MEHP-exposed cells by ~55–65%, whereas the protective effect of SM on tail moments was significantly lower than SS as being ~45% and ~34% for the effects of DEHP and MEHP, respectively. However, both SS and SM reduced the tail moments of the DEHP- and MEHP-exposed cells down to the levels that were not significantly different than that of control cells **(Erkekoglu et al., 2010b)**.

4.1.2 Prevention of radiation genotoxicity by selenocompounds

Ultraviolet (UV) light is electromagnetic radiation with a wavelength shorter than that of visible light, but longer than X-rays, in the range 10-400 nm, and energies from 3 -124 eV. UV light is found in sunlight, can be emitted by electric arcs and specialized lights such as black lights. It can cause chemical reactions, and it causes many substances to glow or fluoresce. Most UV is classified as non-ionizing radiation **(Müller et al., 1998, Griffiths et al., 1998; Grossman et al., 1988)**.

The toxic effects of UV from natural sunlight and therapeutic artificial lamps are a major concern for human health. The major acute effects of UV irradiation on normal human skin comprise sunburn inflammation erythema, tanning, and local or systemic immunosuppression. On the other hand, UV irradiation present in sunlight is an environmental human carcinogen. There is considerable evidence that UV is implicated in skin carcinogenesis and the risk of cutaneous cancers has increased during the last decade due to increase of sun exposure. For a long time, ultraviolet B radiation (UVB: 290-320 nm) have been considered to be the more efficient wavelength in eliciting carcinogenesis in human skin. It is today clear that ultraviolet A (UVA, 320-400 nm), especially UVA$_1$ (340-400 nm) also participate to photo-carcinogenesis. It penetrates deeply, but it does not cause sunburn. One of molecular mechanisms in the biological effects of UV is the induction of ROS directly or through endogenous photosensitization reactions. UVA radiation mainly acts *via* this production of ROS and the subsequent oxidative stress seems to play a crucial role in the deleterious effects of UVA. UVA does not damage DNA directly like UVB and UVC, but it can generate highly reactive chemical intermediates, such as hydroxyl and oxygen radicals, which in turn can damage DNA and lead to the formation of 8-oxoGua **(Ridley et al., 2009)**. UVB light can cause direct DNA damage. The radiation excites DNA molecules in skin cells, causing aberrant covalent bonds to form between adjacent cytosine bases, producing a dimer. When DNA polymerase comes along to replicate this strand of DNA, it reads the dimer as "AA" and not the original "CC". This causes the DNA replication mechanism to add a "TT" on the growing strand. This mutation can result in cancerous growths, and is known as a "classical C-T mutation". The mutations caused by the direct DNA damage carry a UV signature mutation that is commonly seen in skin cancers. The mutagenicity of UV radiation can be easily observed in bacterial cultures. This cancer connection is one reason for concern about ozone depletion and the ozone hole. UVB causes some damage to collagen, but at a very much slower rate than UVA. Fortunately, the skin possesses a wide range of inter-linked antioxidant defense mechanisms to protect itself from damage by UV-induced ROS. However, the capacity of these systems is not unlimited; they can be overwhelmed by excessive exposure to UV and then ROS can reach damaging levels. An interesting strategy to provide photoprotection would be to support or enhance one or more of these endogenous systems **(Béani, 2001)**.

There is limited number of studies in literature concerning the protective effect of selenocompounds on UV-caused genotoxicity. In a study by **Emonet-Piccardi et al. (1998)**, the researchers determined the protective effects of NAC (5 mM), SS (0.6 μM) or zinc chloride (ZnCl$_2$, 100 μM) against UVA radiation in human skin fibroblasts using Comet assay. The cells were incubated with NAC, SS or ZnCl$_2$ and then UVA was applied as 1 to 6 J/cm2 to the cells. The tail moment increased by 45% (1 J/cm^2) to 89% (6 J/cm^2) in non-supplemented cells (p<0.01). DNA damage was significantly prevented by NAC, SS and ZnCl$_2$, with similar efficiency from 1 to 4 J/cm^2. For the highest UVA dose (6 J/cm^2), SS and ZnCl$_2$ were more effective than NAC.

In a study assessing the effects of pretreatment of primary human keratinocytes with Se on UV-induced DNA damage, cells were irradiated with UVB from FS-20 lamps and were subjected to Comet assay. Comet tail length due to UVB-induced T4 endonuclease V-sensitive sites (caused by cyclopyrimidine dimers, CPDs) increased to 100% immediately after irradiation (time 0). After 4 h, 68% of the damage remained and after 24 h, 23% of the damage was still present. Treatment with up to 200 nM SM or 50 nM SS had no effect on CPD formation or rates of repair, or on the number of excision repair sites as measured by cytosine arabino furanoside and hydroxyurea treatment. However, both SS and SM protected against oxidative damage to DNA as measured by formation of formamidopyrimidine (FaPy) glycosylase-sensitive sites, which are indicative of 8-oxoGua photoproduct formation. Preincubation for 18 h with 50 nM SS or with 200 nM SM abolished the UVB-induced increase in comet length. The researchers concluded that both of selenocompounds were protective against UVB-induced oxidative damage in human keratinocytes; however they did not protect from formation of UVB-induced excision repair sites **(Rafferty et al., 2003)**.

Diphenyl diselenide (DPDS) is an electrophilic reagent used in the synthesis of a variety of pharmacologically active organic Se compounds. Studies have shown its antioxidant, hepatoprotective, neuroprotective, anti-inflammatory, and antinociceptive effects. In a study by **Rosa et al. (2007)**, the researchers used a permanent lung fibroblast cell line derived from Chinese hamsters and investigated the antigenotoxic and antimutagenic properties of DPDS. In the clonal survival assay, at concentrations ranging from 1.62 to 12.5 µM, DPDS was not cytotoxic, while at concentrations up to 25 µM, it significantly decreased survival. The treatment with this DPDS at non-cytotoxic dose range increased cell survival after challenge with H_2O_2, methyl-methanesulphonate, and UVC radiation, but did not protect against 8-methoxypsoralen plus UVA-induced cytotoxicity. In addition, the treatment prevented induced DNA damage, as verified in the Comet assay. The mutagenic effect of these genotoxic agents, as measured by the micronucleus test, similarly attenuated or prevented cytotoxicity and DNA damage. Treatment with DPDS also decreased lipid peroxidation levels after exposure to H_2O_2, MMS, and UVC radiation, and increased GPx1 activity in the cells. The results of this study demonstrated that DPDS at low concentrations presents antimutagenic properties, which are most probably due to its antioxidant properties **(Rosa et al., 2007)**.

4.2 Prevention of genotoxicity by vitamins

4.2.1 Ascorbic acid

Diet should include components such as vitamins and flavonoids and the antioxidant capacity of body is directly linked to the diet. Vitamins like ascorbic acid (vitamin C, AA) are important antioxidants. About 90% of AA in the average diet comes from fruits and vegetables **(Vallejo et al., 2002)**.

AA is a water soluble dietary antioxidant that plays an important role in controlling oxidative stress **(Vallejo et al., 2002)**. Most importantly, AA is a mild reducing agent. For this reason, it degrades upon exposure to oxygen, especially in the presence of metal ions and light. It can be oxidized by one electron to a radical state or doubly oxidized to the stable form called "dehydroascorbic acid". Typically it reacts with oxidants such as ROS, such as the •OH formed from H_2O_2. Hydroxyl radical is the most detrimental species, due to its high interaction with nucleic acids, proteins, and lipids. AA can terminate these chain radical reactions by electron transfer. AA is special because it can transfer a single electron, owing to the stability of its own radical ion called "semidehydroascorbate". The oxidized forms of AA are relatively unreactive, and do not cause cellular damage. However, being a good electron donor, high concentrations of AA in the presence of free metal ions can not only promote, but also initiate free radical reactions, thus making it a potentially dangerous pro-oxidative compound in certain metabolic contexts **(Choe and Min, 2006; Blokhina et al., 2003)**.

AA is able to suppress ROS efficiently *in vivo*; thus, reducing DNA damage to tumor suppressor genes which might explain its anticancer properties **(Crott et al., 1999)**. *In vitro*, AA acts in conjunction with vitamin E, present in lipid membranes, to quench free radicals and prevent lipid peroxidation **(Niki et al., 1995)**.

In the Comet assay, evidence of protection was seen against the effects of H_2O_2 when AA was present at low concentrations (up to 1 mM); by contrast, there was exacerbation at higher doses (>5 mM) **(Harréus et al., 2005; Anderson et al., 1994; Anderson and Phillips, 1999)**. After 2-4 h after intake, AA provided significant protection to the DNA of isolated lymphocytes when challenged with H_2O_2 **(Panayiotidis and Collins, 1997)**. Besides, AA was found to be protective against H_2O_2-induced DNA damage (DNA strand breaks and oxidized purines/pyrimidines) in human hepatoma cells (HepG2 cells) **(Arranz et al., 2007a, Arranz et al., 2007b)**. In intervention studies, supplementation of 100 mg/day to 50–59 year-old men led to a decrease in oxidative base damage and enhanced resistance against oxidative damage **(Duthie et al., 1996)**. In a long-term study, the antioxidant effect of AA was studied by measuring oxidative DNA damage and DNA repair in blood cells with the Comet assay. Male smokers were given AA (2 × 250 mg) daily in the form of plain or slow release tablets combined with plain release vitamin E (2 × 91 mg), or placebo for 4 weeks. The results of this study suggested that long-term AA supplementation at a high dose, i.e. 500 mg, together with vitamin E in moderate dose, i.e. 182 mg, decreased the steady-state level of oxidative DNA damage in lymphocytes of smokers **(Møller et al., 2004)**. In a study performed on gastric epithelial cells SGC-7901, both AA and SS were found to be protective against Helicobacter pylori-induced oxidative stress and genotoxicity **(Shi and Zheng, 2006)**.

AA was also tested for its protective effects against the genotoxicity of several toxic chemicals, drugs and metals. Using peripheral blood lymphocytes, AA as well as vitamin E were found to be protective against benzo(a)pyrene [B(a)P]-induced DNA damage **(Gajecka et al., 1999)**. In rats, using Comet assay, the genotoxicity of p-dimethylaminoazobenzene (DAB), a hepatocarcinogen, was found to be decreased by AA administration. Besides, vitamin A, vitamin E and combination of these three vitamins were also found be effective against the toxicity **(Velanganni et al., 2007)**. A significant increase in the levels of protein oxidation, DNA strand breaks, and DNA-protein cross-links was observed in blood, liver, and kidney of rats exposed to arsenic (100 ppm in drinking water) for 30 days. Co-administration of AA and vitamin E in the form of α-tocopherol to arsenic-exposed rats showed a substantial reduction in the levels of arsenic-induced oxidative products of protein and DNA **(Kadirvel et al., 2007)**. For anti-cancer drugs there are inconclusive results. AA was protective against epirubicin- and adriamycin-induced genotoxicity in cancer patients **(Mousseau et al., 2005; Shimpo et al., 1991)**. However, there was no evidence of a protective effect of AA against the damage caused by bleomycin **(Anderson & Phillips, 1999)**. Moreover, results were also inconclusive when oestrogenic compounds were co-incubated with AA (0.5 and 1 mM) in isolated lymphocytes showing no common pattern in the responses **(Anderson et al., 2003)**.

Nitrosamines (NOCs) can be formed endogenously from nitrate and nitrite and secondary amines under certain conditions such as strongly acidic pHs of the human stomach **(Jakszyn and Gonzalez, 2006; Bofetta et al., 2008; Tricker, 1997)**. Humans are exposed to a wide range of NOCs from diet (cured meat products, fried food, smoked preserved foods, foods subjected to drying, pickled and salty preserved foods), tobacco smoking, work place and drinking water **(Bartsch and Spiegelhalder, 1996; Bofetta et al., 2008; Jakszyn & Gonzalez, 2006; Tricker, 1997).**

In several studies, AA was found to be protective against NOC-induced genotoxicity using Comet assay. In a study by **Robichová et al. (2004)**, the researchers used three cell lines (HepG2, V79 and VH10) to determine the genotoxic effect of N-Nitrosomorpholine (NMOR). NMOR was found to induce DNA damage in a dose-dependent manner but the extent of DNA migration in the electric field was unequal in the different cell lines. Although the results obtained by Comet assay confirmed the genotoxicity of NMOR in all cell lines studied, the number of chromosomal aberrations was significantly increased only in HepG2 and V79 cells, while no changes were observed in VH10 cells. In HepG2 cells pre-treated with vitamin A, vitamin E and AA the researchers found a significant decrease of % tail DNA induced by NMOR. The reduction of the clastogenic effects of NMOR was observed only after pretreatment with Vitamins A and E. AA did not alter the frequency of NMOR-induced chromosomal aberrations under the experimental conditions of this study. In a study by **Arranz et al. (2007)**, HepG2 cells were simultaneously treated with AA and the genotoxic effects of the N-nitrosamines, namely, N-nitrosodimethylamine (NDMA), N-nitrosopyrrolidine (NPYR), N-nitrosodibutylamine (NDBA) or N-nitrosopiperidine (NPIP) were reduced in a dose-dependent manner. At concentrations of 1-5 µM AA, the protective effect was higher towards NPYR-induced oxidative DNA damage (78-79%) than against NDMA (39-55%), NDBA (12-14%) and NPIP (3-55%), in presence of Fpg enzyme. However, a concentration of 10 µM AA led to a maximum reduction in NDBA (94%), NPYR (81%), NPIP (80%) and NDMA (61%)-induced oxidative DNA damage, in presence of Fpg enzyme. The greatest protective effect of AA (10 µM) was higher towards NDBA-induced oxidative DNA damage. The authors concluded that one feasible mechanism by which AA exerted its protective effect could be that it might interact with the enzyme systems catalyzing the metabolic activation of the N-nitrosamines, blocking the production of genotoxic intermediates.

In our previous studies performed using Comet assay, we have shown that AA was highly protective in HepG2 cells against the genotoxicity of both nitrite and three important NOC, namely NDMA, Nitrosodiethylamine (NDEA) and NMOR **(Erkekoglu et al., 2010c)**. Nitrite was added as 20 µM, NDMA as 10 mM, NDEA as 10 mM and NMOR as 3 mM to the medium for 30 min with or without AA (10 µM). When compared to untreated cells, nitrite (p>0.05), NDMA (p<0.05), NDEA (p<0.05), and NMOR (p<0.05) raised the tail intensity up to 1.18-, 3.79-, 4.24-, and 4.16-fold, respectively. AA was able to reduce the tail intensity caused by nitrite, NDMA, NDEA, and NMOR to 34%, 59%, 44%, and 44%, respectively, and these reductions were statistically significant when compared to each individual toxic compound applied group (all, p<0.05). Besides, nitrite, NDMA, NDEA, and NMOR increased the tail moment up to 1.94, 6.04, 6.05, and 5.70, respectively. AA (10 µM) enabled a reduction of 27%, 30%, 23%, and 22% in the tail moment in nitrite, NDMA, NDEA, and NMOR-treated cells, respectively, and these reductions were statistically significant when compared to each individual toxic compound applied group (all, p<0.05) **(Erkekoglu et al., 2010c)**.

In an experiment performed on multiple organs of mice, the genotoxicity of endogenously formed N-nitrosamines from secondary amines and sodium nitrite was evaluated in, using Comet assay. Dimethylamine, proline, and morpholine were simultaneously with sodium nitrite and the stomach, colon, liver, kidney, urinary bladder, lung, brain, and bone marrow were sampled 3 and 24 h after these compounds had been ingested. DNA damage was observed mainly in the liver following simultaneous oral ingestion of these compounds **(Ohsawa et al., 2003)**.

4.2.2 Vitamin E

Vitamin E refers to a group of fat-soluble compounds that include both tocopherols and tocotrienols **(Brigelius- Flohé and Traber, 1999)**. Naturally occurring vitamin E exists in eight chemical forms (alpha-, beta-, gamma-, and delta-tocopherol and alpha-, beta-, gamma-, and delta-tocotrienol) that have varying levels of biological activity. Alpha- (or α-) tocopherol is the only form that is recognized to meet human requirements. γ-tocopherol is the most common in the North American diet **(Traber, 1998)**. γ-tocopherol can be found in corn oil, soybean oil, margarine and dressings **(Bieri and Evarts, 1974; Brigelius-Flohé & Traber, 1999)**. The most biologically active form of vitamin E, α-tocopherol, is the second most common form of vitamin E in the North American diet and perhaps the common form in European and Mediterranean diet. This variant of vitamin E can be found most abundantly in wheat germ oil, sunflower, and safflower oils **(Reboul et al., 2006)**. Serum concentrations of α-tocopherol depend on the liver, which takes up the nutrient after the various forms are absorbed from the small intestine. The liver preferentially resecretes only α-tocopherol *via* the hepatic α-tocopherol transfer protein **(Traber, 2006)**. As a result, blood and cellular concentrations of other forms of vitamin E are lower than those of α -tocopherol and have been the subjects of less research **(Sen et al., 2006; Dietrich et al., 2006)**.

Vitamin E is an important vitamin for preventing lipid peroxidation and it has many reported health effects and is recognized as the most important lipid-soluble, chain-breaking antioxidant in the body **(Fenech & Ferguson, 2001)**. This vitamin might have a protective role against chromosomal damage, DNA oxidation and DNA damage. Vitamin E has also been reported to play a regulatory role in cell signaling and gene expression. Epidemiological studies showed that high blood concentrations of vitamin E were associated with a decreased risk of certain cancers. This effect might emerge in part, by enhancing immune function **(Frank, 2005; Claycombe & Meydani, 2001, Salobir et al., 2010)**. Vitamin E might also block the formation of carcinogenic NOCs formed in the stomach from nitrite and secondary amines **(Weitberg and Corvese, 1997)**.

Vitamin E was shown to prevent the genotoxicity of several environmetal chemicals and several drugs. Nitrosamine toxicity was shown to be protected by vitamin E. Hepatocytes freshly isolated from rats fed with a common diet or a vitamin A- or vitamin E-supplemented diet were assayed for sensitivity to DNA breakage and cytogenetic changes induced by several carcinogens including NMOR. NMOR was the only agent that induced DNA breaks, chromosomal aberrations, and micronuclei. Both vitamin A and vitamin E were able to reduce these effects, and the protection by vitamin A was more pronounced **(Slamenová, 2001)**. On the other hand, vitamin E was also found to be protective against the genotoxic properties of one of the most commonly used herbicides, atrazine, in male rats. Atrazine caused a significant increase in tail length of comets from blood and liver cells compared to controls. Co-administration of vitamin E (100 mg/kg bw) along with atrazine resulted in decrease in tail length of comets as compared to the group treated with atrazine alone. Besides, micronucleus assay revealed a significant increase in the frequency of micro-nucleated cells (MNCs) following atrazine administration. In the animals administrated vitamin E along with atrazine, there was a significant decrease in percentage of micronuclei as compared to atrazine treated rats. The increase in frequency of micronuclei in liver cells and tail length of comets confirm genotoxicity induced by atrazine in blood and liver cells. In addition, the findings clearly demonstrated protective effect of vitamin E in attenuating atrazine-induced DNA damage **(Singh et al., 2008)**. In mouse retina, both vitamin E and AA were shown to markedly reduce the cell apoptosis, lipid peroxidation and DNA damage caused by the organophosphorus insecticide chlorpyrifos **(Yu et al, 2008)**. Vitamin E supplementation was also protective against pyrethroid (both cypermethrin and permethrin), induced lymphocyte DNA damage **(Gabbianelli et al., 2004)**.

Vitamin E was also shown to reduce the genotoxic effects of the anti-HIV drug stavudine **(Kaur & Singh, 2007)** and the antibiotic, ciprofloxacin **(Gürbay et al., 2006)**. In a study performed on primary culture of rat astrocytes, the researchers incubated the cultured cells with various concentrations of ciprofloxacin, and DNA damage was monitored by Comet assay. The results showed a concentration-dependent induction of DNA damage by ciprofloxacin. Pretreatment of cells with Vitamin E for 4 h provided partial protection against this effect **(Gürbay et al., 2006)**.

Vitamin E was also found to be protective against the toxicity of anesthesics. In a study performed with sevoflurane on rabbits, vitamin E and SS were administered 15 days before the anesthesia treatment and blood samples were collected after 5 days of treatment with sevoflurane. Both vitamin E and SS administration prevented the sevoflurane induced genotoxicity in the lymphocytes **(Kaymak et al., 2004)**.

Several supplementation studies have also been performed both vitamin E and AA. Supplementation of the diet for 12 weeks with AA and vitamin E resulted in a significant decrease in the DNA damage in diabetic patients **(Sardaş et al., 2001)**. Vitamin E supplementation was also shown to reduce oxidative DNA damage in both hemodialysis and peritoneal dialysis patients **(Domenici et al., 2005)**. In another study performed on 26 healthy subjects, a daily drink including 1.8 mg vitamin E was administered for 26 days and blood samples were obtained. The DNA damage was measured in the lymphocytes subjected to oxidative stress and genotoxicity was found to be significantly lower (42%, $p<0.0001$) **(Porrini et al., 2005)**.

There are few protection studies with vitamin E against radiation toxicity using Comet assay. An *in vitro* study on dermal microvascular endothelial cells by the same research group, gamma- irradiated cells at 3 and 10 Gy, and 0.5 mM of pentoxifylline (PTX) and trolox (Tx, 6-hydroxy-2,5,7,8-tetramethylchroman-2-carboxylic acid, a water-soluble derivative of vitamin E), were added either before (15 min) or after (30 min or 24 h) irradiation. ROS measured by the dichlorodihydrofluorescein diacetate assay, and DNA damage, assessed by the Comet and micronucleus assays, were measured at different times after exposure (0 - 21 days). The PTX/Tx treatment decreased the early and delayed peak of ROS production by a factor of 2.8 in 10 Gy-irradiated cells immediately after irradiation and the basal level by a factor of 2 in non-irradiated control cells. Moreover, the level of DNA strand breaks, as measured by the comet assay, was shown to be reduced by half immediately after irradiation when the PTX/Tx treatment was added 15 min before irradiation. However, unexpectedly, DNA strand breaks was decreased to a similar extent when the drugs were added 30 min after radiation exposure. This reduction was accompanied by a 2.2- and 3.6-fold higher yield in the micronuclei frequency observed on days 10 and 14 post-irradiation, respectively. These results suggest that oxidative stress and DNA damage induced in dermal microvascular endothelial cells by radiation can be modulated by early PTX/Tx treatment. These drugs acted not only as radical scavengers, but they were also responsible for the increased micronuclei frequency in 10 Gy-irradiated cells. Thus, these drugs may possibly interfere with DNA repair processes **(Laurent et al., 2006)**.

In another study, the effects of vitamin E supplementation were evaluated in cultured primary human normal fibroblasts exposed to UVA. Cells were incubated in medium containing α-tocopherol, α-tocopherol acetate or the synthetic analog Trolox for 24 h prior to UVA exposure. DNA damage in the form of frank breaks and alkali-labile sites, collectively termed single-strand breaks (SSB), was assayed by Comet assay, immediately following irradiation or after different repair periods. The generation of H_2O_2 and superoxide ion was measured by flow cytometry through the oxidation of indicators into fluorescent dyes. Pretreatment of cells with any form of vitamin E resulted in an increased susceptibility to the photo-induction of DNA SSB and in a longer persistence of damage, whereas no significant change was observed in the production of H_2O_2 and superoxide, compared to controls. The researchers indicated that in human normal fibroblasts, exogenously added vitamin E exerted a promoting activity on DNA damage upon UVA irradiation and might lead to increased cytotoxic and mutagenic risks **(Nocentini et al., 2001)**.

In an *in vivo* study by **Konopacka at al. (1998),** the modifying effects of treatment with vitamin E, AA and vitamin A in the form of β-carotene on the clastogenic activity of gamma rays were investigated in mice. Damage *in vivo* was measured by the micronucleus assay in bone marrow polychromatic erythrocytes and exfoliated bladder cells. The vitamins were administered orally, either for five consecutive days before or immediately after irradiation with 2 Gy of gamma rays. The results showed that pretreatment with vitamin E (100-200 mg/kg/day) and β-carotene (3-12 mg/kg/day) were effective in protecting against micronucleus induction by gamma rays. AA depending on its concentration enhanced the radiation effect (400 mg/kg/day), or reduced the number of micro-nucleated polychromatic erythrocytes (50-100 mg/kg/day). Such effect was weekly observed in exfoliated bladder cells. The most effective protection in both tissues was noted when a mixture of these vitamins was used as a pretreatment. Administration of the all antioxidant vitamins to mice immediately after irradiation was also effective in reducing the radiation-induced micronucleus frequency. The data from the *in vitro* experiments based on the Comet assay show that the presence of the vitamins in culture medium influences the kinetic of repair of radiation-induced DNA damage in mouse leukocytes.

4.3 Prevention of genotoxicity by thiocyanates

Human cancer can be prevented by changing the dietary habits **(Kelloff , 2000; Vallejo et al., 2002; Hecht, 1996; Milner , 2004; Davis & Milner, 2006)**. Studies show that antioxidant-rich diets are associated with low risk of cancer and whole diet plays a more important role than the individual components. The protective effects of vegetables and fruits may be attributed to the combined effect of various phytochemicals, vitamins, fibers, and allium compounds rather than the effect of a single component **(Lee et al., 2003)**. There is powerful evidence in literature for a cancer-protective effect of the vegetables of the family *Cruciferae* that includes broccoli, watercress, cabbage, kale, horseradish, radish, turnip, and garden cress **(Verhoeven et al., 1996; Hecht, 1999)**. This effect is attributed to ITCs, which occur naturally as thioglucoside conjugates (glucosinolates). They are hydrolysis products of glucosinolates and are generated through catalytic mediation of myrosinase, which is released upon processing (cutting or chewing) of cruciferous vegetables from a compartment separated from glucosinolates. Evidence exists for conversion of glucosinolates to ITCs in the gut. At least 120 different glucosinolates have been identified. ITCs have a common basic skeleton but differ in their terminal R group, which can be an alkyl, an alkenyl, an alkylthioalkyl, an aryl, a β-hydroxyalkyl, or an indolylmethyl group. The widely studied ITCs include phenethyl isothiocyanate (PEITC), benzyl isothiocyanate (BITC), indole-3-carbinol (I_3C) and allyl isothiocyanate (AITC) **(Fahey et al., 2001; Arranz et al., 2006)**.

The most important biological property discovered about ITCs is their ability to inhibit carcinogenesis, induced by several chemicals including nitrosamines in the lung, stomach, colon, liver, esophagus, bladder and mammary glands in animal models **(Hecht, 1999; Zhang et al., 2003; Zhang and Talalay, 1994; Hecht et al., 1995; Munday et al., 2003)**. Two mechanisms can be suggested for the protective effect of ITCs against nitrosamine-induced DNA damage:

a. Blocking the production of genotoxic intermediates by inhibiting Phase I enzymes: PEITC was shown to reduce p-nitrophenol hdroxylase (CYP2E1), ethoxyresorufin O-deethylase (CYP1A1) and coumarin hdroxylase (CYP2A6) activities **(García et al., 2008)**.
b. Enhancement of detoxification pathways through the induction of Phase II enzymes **(Arranz et al., 2006)**.

Furthermore, ITCs may have ROS scavenging capacity, alter cell proliferation, stimulate DNA-repair, and induce NAD(P)H: quinine oxidoreductase activity as also mentioned for AA before **(Gamet-Payrastre et al., 2000; Chaudière and Ferrari-Iliou et al., 1999; Surh, 2002; Surh et al., 2001; Roomi et al., 1998)**.

ITCs were shown to be effective in the inhibition of lung tumorigenesis in mice and rats induced by the tobacco-specific carcinogen 4-(methylnitrosamino)-1-(3-pyridyl)-1-butanone (NNK). Because NNK is believed to play a significant role as a cause of lung cancer in smokers, PEITC is being developed as a chemopreventive agent, which is presently in Phase I a clinical trial in healthy smokers **(Hecht, 1996; Stoner et al., 1991)**. PEITC is a potent inhibitor of rat esophageal tumorigenesis induced by NBMA **(Stoner et al., 1991)**. A comparative study demonstrates that phenylpropyl isothiocyanate (PPITC) is even more potent, whereas BITC and 4-phenylbutyl isothiocyanate (PBITC) have little effect on tumorigenesis **(Wilkinson et al., 1995)**. However, phenylhdroxyl isothiocyanate (PHITC) enhances tumorigenesis in the same model **(Stoner et al., 1995)**. Mechanistic studies clearly show that PEITC inhibits the metabolic activation of NBMA in the rat esophagus, probably through inhibition of a cytochrome P450 (CYP450) enzyme **(Morse et al., 1997)**. Concomitant with this inhibition, inhibition of O^6-methylguanine formation in rat esophageal DNA was observed. The inhibitory effects on tumorigenicity correlate with their inhibitory effects on O^6-methylguanine formation **(Wilkinson et al., 1995; Stoner & Morse, 1997)**. Inhibition of

N'- nitrosonornicotine (NNN) tumorigenicity in the rat esophagus by PEITC also appears to be due to inhibition of its metabolic activation **(Stoner et al., 1998)**.

The antimutagenic properties of ITCs have been reported towards NDMA and NPYR-induced oxidative stress before. In studies performed by **Knasmüller et al. (1996, 2003)** using PEITC as a chemopreventive agent, the researchers observed a reduction in NDMA- and NPYR-induced DNA damage in *Escherichia coli* K-12 and a considerable reduction in NDMA-induced micronuclei in HepG2 cells. The results of several studies demonstrated that ITCs exhibited strong antimutagenic effects against NDMA and NPYR in a dose dependent manner. In a study by **Smerák at al. (2009)**, the researchers investigated the effect of PEITC on the mutagenic activity of indirect-acting mutagens and carcinogens like aflatoxin B1 (AFB1) and 2-amino-3-methylimidazo[4,5-f]quinoline (IQ) using the *Ames* bacterial mutagenicity test, the Comet assay, an *in vivo* micronucleus test, and direct-acting mutagen and carcinogen N-nitroso-N-methylurea (MNU). In the Ames test, the antimutagenic activity of PEITC was studied in the concentration range 0.3-300 μg/plate. PEITC at concentrations of 0.3, 3 and 30 μg/plate reduced dose-dependently mutagenicity of AFB1 and IQ in both *Salmonella typhimurium* TA98 and TA100 strains. In the case of the direct mutagen MNU, the antimutagenic effect of PEITC was detected only at concentration of 30 μg/plate in the strain TA100. The PEITC concentration 300 μg/plate was toxic in the Ames test. The 24 h pre-treatment of HepG2 cells with PEITC at concentration 0.15 μg/ml resulted in a significant decrease of DNA breaks induced by MNU at concentrations 0.25 and 0.5 mM. Although a trend towards reduced strand break level were determined also at PEITC concentrations 0.035 and 0.07 μg/ml, it did not reach the statistical significance. No effect, however, of PEITC on IQ-induced DNA breaks was observed. Chemopreventive effect of PEITC was revealed also *in vivo*. Pretreatment of mice with PEITC concentrations of 25 and 12.5 mg/kg bw administered to mice in three daily doses resulted in reduction of micronucleus formation in mice exposed to all three mutagens under study, with statistically significant effect at concentration of 25 mg/kg. Results of this study indicated that the strong PEITC antimutagenic properties may have an important role in the prevention of carcinogenesis and other chronic degenerative diseases that share some common pathogenetic mechanisms. In a recent study by **Tang et al. (2011),** PEITC was shown to induce a dose-dependent decrease in cell viability through induction of cell apoptosis and cell cycle arrest in the G_2/M phase of DU 145 human prostate cells. Besides, PEITC induced morphological changes and DNA damage in DU 145 cells. The induction of G_2/M phase arrest was mediated by the increase of p53 and Wee1 and it reduced the level of M-phase inducer phosphatase 3 (CDC25C) protein. The induction of apoptosis was mediated by the activation of caspase-8-, caspase-9- and caspase-3-depedent pathways. Results of this study also demonstrated that PEITC caused mitochondrial dysfunction, increasing the release of cytochrome c and Endo G from mitochondria, and led cell apoptosis through a mitochondria-dependent signaling pathway. The researchers concluded that PEITC might exhibit anticancer activity and become a potent agent for human prostate cancer cells in the future.

There are a few studies on ITCs against nitrosamine-induced genotoxicity in literature. In a study by **Arranz et al. (2006)**, the protective effect of three ITCs was tested. ITC were highly protective against NPYR-induced oxidative DNA damage than against NDMA. The greatest protective effect towards NPYR-induced oxidative DNA damage was shown by I_3C (1 μM,

79%) and by PEITC (1 µM, 67%) and I_3C (1 µM, 61%) towards NDMA (in presence of Fpg enzyme). However, in absence of Fpg enzyme, AITC (1 µM, 72%) exerted the most drastic reduction towards NPYR-induced oxidative DNA damage, and PEITC (1 µM, 55%) towards NDMA. These results indicated that ITCs protect human-derived cells against the DNA damaging effect of NPYR and NDMA, two carcinogenic compounds that occur in the environment. Another study performed by **García et al. (2008)** aimed to investigate the protective effect of ITCs alone or in combination with AA towards NDBA or NPIP-induced oxidative DNA damage in HepG2 cells by Comet assay. PEITC and I_3C alone showed a weak protective effect towards NDBA (0.1 µM, 26-27%, respectively) or NPIP (1 µM, 26-28%, respectively)-induced oxidative DNA damage. AITC alone did not attenuate the genotoxic effect provoked by NDBA or NPIP. In contrast, HepG2 cells simultaneously treated with PEITC, I_3C and AITC in combination with AA showed a stronger inhibition of oxidative DNA-damage induced by NDBA (0.1 µM, 67%, 42%, 32%, respectively) or NPIP (1 µM, 50%, 73%, 63%, respectively) than ITCs alone. One feasible mechanism by which ITCs alone or in combination with AA exert their protective effects towards N-nitrosamine-induced oxidative DNA damage could be by the inhibition of their CYP450 dependent bioactivation. PEITC and I_3C strongly inhibited the p-nitrophenol hydroxylation (CYP2E1) activity (0.1 µM, 66-50%, respectively), while the coumarin hydroxylase (CYP2A6) activity was slightly reduced (0.1 µM, 25-37%, respectively). However, the ethoxyresorufin O-deethylation (CYP1A1) activity was only inhibited by PEITC (1 µM, 55%). The results indicated that PEITC and I_3C alone or PEITC, I_3C and AITC in combination with AA protect human-derived cells against the oxidative DNA damaging effects of NDBA and NPIP.

In our study performed on HepG2 cells, we tested AITC (0.5 µM) against the nitrite and nitrosamine toxicity. Nitrite was added as 20 µM, NDMA as 10 mM, NDEA as 10 mM and NMOR as 3 mM to the medium for 30 min with or without AITC. When compared to untreated cells, nitrite, NDMA, NDEA and NMOR raised the tail intensity up to 17 %, 279 %, 324 % and 288 %, respectively (all, $p<0.05$). AITC was able to reduce the tail intensity caused by nitrite 36 %, by NDMA 36 %, by NDEA 49 % and by NMOR 32 %, respectively. These reductions were statistically significant when compared to each individual toxic compound applied group (all, $p<0.05$). Besides, when compared to untreated cells, nitrite, NDMA, NDEA and NMOR raised the tail intensity up to 94%, 126%, 157% and 207%, respectively (all, $p<0.05$). AITC was able to reduce the tail moment caused by nitrite 16 %, by NDMA 32 %, by NDEA 41 % and by NMOR 19 %, respectively and these reductions were statistically significant when compared to each individual toxic compound applied group **(Erkekoglu & Baydar, 2010d)**.

5. Conclusion

The protective effect of antioxidants is universally accepted. However, as also seen in AA, the mode of action of antioxidants particularly with dual behavior (prooxidant and antioxidant) remain unclear and more research must be conducted on these compounds. For instance, the elucidation of how antioxidant properties operate *in vitro* can provide a better understanding of the *in vivo* situation. On the other hand, Comet assay can be an important tool for the determining of the genotoxic effect of several environmental chemicals, as well as the antioxidant properties of several compounds.

Most of these chemicals exert their toxicity over their ability of producing ROS. ROS can be balanced by the antioxidant action of non-enzymatic antioxidants as well as antioxidant enzymes and it was shown that the genotoxicity of several environmental chemicals can be reversed by proper doses of antioxidants *in vitro*. More *in vitro* studies are needed to prove the beneficial antioxidant effects of trace elements and vitamins. Medicine might benefit from current investigations demonstrating the properties of a vast number of antioxidants as well as studying the effects of different diets. Modest antioxidant supplementation might help prevent chemical-induced carcinogenesis in healthy individuals. On the other hand, antioxidant applications might be beneficial in individuals who may have polymorphisms in genes, including those for antioxidant enzyme. Additionally, populations deficient in several trace elements and vitamins might exhibit modest DNA-repair defects that could be functionally rescued by dietary antioxidants. The future interest of several researchers as well as ours is to understand the pathways underlying the genotoxicity of several agents, particularly phthalates and to determine the antioxidant effect of trace elements and vitamins against the toxic effects of such agents *in vitro* and *in vivo* systems.

6. References

Akingbemi, B.T.; Youker, R.T.; Sottas C.M.; Ge R.; Katz E.; Klinefelter G.R.; Zikrin, B.R. & Hardy, M.P. (2001). Modulation of rat Leydig cell steroidogenic function by di(2-ethylhexyl)phthalate. *Biology of Reproduction*, Vol. 65, No.4, (October 2001), pp. 1252-1259.

Anderson, D. & Phillips, B.J. (1999). Comparative in vitro and in vivo effects of antioxidants. *Food and Chemical Toxicology*, Vol. 37, No. 9-10, (September-October 1999), pp. 1015-1025.

Anderson, D.; Schmid, T.E.; Baumgartner, A.; Cemeli-Carratala, E.; Brinkworth, M.H. & Wood, J.M. (2003). Oestrogenic compounds and oxidative stress (in human sperm and lymphocytes in the Comet assay). *Mutation Research*, Vol. 544, No. 2-3 (November 2003), pp. 173-178.

Anderson, D.; Phillips B.J. & Schmezer P. (1994) The effect of various antioxidants and other modifying agents on oxygen-radical-generated DNA damage in human lymphocytes in the COMET assay. *Mutation Research*, Vol. 307, No. 1, (May 1994), pp. 261-271.

Arranz, N.;, Haza, A.I.; García, A.; Rafter, J. & Morales P. (2007a). Protective effect of vitamin C towards N-nitrosamine-induced DNA damage in the single-cell gel electrophoresis (SCGE)/HepG2 assay. *Toxicology In Vitro*, Vol. 27, No. 1, (October 2007), pp. 1311-1317.

Arranz, N.; Haza, A.I.; García, A.; Delgado, E.; Rafter, J. & Morales, P. (2007b). Effects of organosulfurs, isothiocyanates and vitamin C towards hydrogen peroxide-induced oxidative DNA damage (strand breaks and oxidized purines/pyrimidines) in human hepatoma cells. *Chemical and Biological Interactions*, Vol. 169, No. 1, (August 2007), pp. 3-71.

Novotna, B.; J. Topinka, I.; Solansky, I.; Chvatalova, Z. & Lnenickova, R.J. (2007). Impact of air pollution and genotype variability on DNA damage in Prague policemen, *Toxicology Letters*, Vol.172, No. 1-2, (July 2007), pp. 37–47.

Baliga, M.S.; Wang, H.; Zhuo, P.; Schwartz, J.L. & Diamond, AM. (2007). Selenium and GPx-1 overexpression protect mammalian cells against UV-induced DNA damage. *Biological Trace Element Research*, Vol. 115, No. 3, (March 2007), pp. 227-242.

Béani, J.C. (2001). Enhancement of endogenous antioxidant defenses: a promising strategy for prevention of skin cancers. *Bulletin de l'Académie Nationale de Médecine*, Vol. 185, No. 8, pp. 507-525.

Bieri, J.G. & Evarts, R.P. (1974). Gamma tocopherol: metabolism, biological activity and significance in human vitamin E nutrition. *American Jotrnal of Clinical Nutrition*, Vol. 27, No. 9, (September 1974), pp. 980-986.

Blokhina, O.; Virolainen, E.; & Fagerstedt, K.V. (2003). Antioxidants, oxidative damage and oxygen deprivation stress: a review. *Annuals of Botanics*, (January 2003), pp. 179-194.

Boffetta, P., Hecht, S., Gray, N., Gupta, P. & Straif, K. (2008). Smokeless tobacco and cancer. *Lancet Oncology*, Vol. 9, No. 7, (July 2008), pp. 667-675.

Brigelius-Flohé, R. & Traber, M.G. (1999). Vitamin E: function and metabolism. *FASEB Journal*, Vol. 13, No. 10, (July 1999), pp. 1145-1155.

Brinkman, M., Buntinx, F., Muls, E. & Zeegers MP. Use of selenium in chemoprevention of bladder cancer. *Lancet Oncology*, Vol. 7, No. 9, (September 2006), pp. 766-774.

Chaudière, J. & Ferrari-Iliou, R. (1999). Intracellular antioxidants: from chemical to biochemical mechanisms, *Food and Chemical Toxicology*, Vol. 37, No. 9-10, (September-October 1999), pp.949-962.

Choe, E. &Min, D.B. (2006). Chemistry and reactions of reactive oxygen species in foods. *Critical Reviews in Food Science and Nutrition*, Vol. 46, No. 1, pp. 1-22.

Clark, L.C.; Cantor, K.P.; & Allaway, W.H. (1991). Selenium in forage crops and cancer mortality in U.S. counties. *Archieves of Environmental Health*. Vo. 46, No. 1, (January-February 1991), pp. 37-42.

Claycombe, K.J. & Meydani SN. (2001).Vitamin E and genome stability. *Mutation Research*, Vol. 475, No. 1-2, (April 2001), pp. 37-44.

Collins, A.R.; Dobson, V.L.; Dusinska, M.; Kennedy, G.; & Stĕtina, R. (1997). The comet assay: what can it really tell us? *Mutat Research*, Vol. 375, No. 2, (April 1997), pp. 183-193.

Collins, A.R. & Horváthová, E. (2001). Oxidative DNA damage, antioxidants and DNA repair:applications of the comet assay. *Biochemical Society Transactions*, Vol. 29, No. Pt 2, (May 2001), pp. 337-341.

Combs, G.F. Jr. & Gray, W.P. (1998). Chemopreventive agents: selenium. *PharmacologicalTherapy*, Vol. 79, No. 3, (September 1998), pp. 179-192.

Corton, J.C. & Lapinskas, P.J. (2005). Peroxisome proliferator-activated receptors: mediators of phthalate ester-induced effects in the male reproductive tract? *Toxicological Sciences*, Vo. 83, No. 1, (January 2005), pp. 4-17.

Crott, J.W. & Fenech, M. (1999). Effect of vitamin C supplementation on chromosome damage, apoptosis and necrosis ex vivo. *Carcinogenesis*, Vol. 20, No. 6, (June 1999), pp. 1035-1041.

Drath, D.B.; Shorey, J.M. & Huber, G.L. (1981). Functional and metabolic properties of alveolar macrophages in response to the gas phase of tobacco smoke, *Infection and Immunity*, Vol. 34, No. 9-10, (September-October1981), pp. 11–15.

Anderson, D.; Phillips B.J.; Yu T.W.; Edwards A.J.; Ayesh R. & Butterworth K.R. (1997). The effects of vitamin C supplementation on biomarkers of oxygen radical generated damage in human volunteers with low or high cholesterol levels. *Environmental and Molecular Mutagenesis*, Vol. 30, No. 2, pp. 161–174.

Dalgaard, M.; Ostergaard, G.; Lam, H.R.; Hansen, E.V. & Ladefoged, O. (2000). Toxicity study of di(2-ethylhexyl)phthalate (DEHP) in combination with acetone in rats. *Pharmacological Toxicology*, Vol. 86, No. 2, (February 2000), pp. 92-100.

Dalton, T.P.; Shertzer, H.G. & Puga, A. (1999). Regulation of gene expression by reactive oxygen. *Annual Review of Pharmacology and Toxicology*, Vol. 39, pp. 67-101.

Davis, C.D. & Milner, J.A. (2006). Nutritional health: strategies for disease prevention In: *Diet and Cancer Prevention*, N.J.Temple; T. Wilson & D.V. Jacobs, (Eds). 151-171, Humana Press; Totowa, NJ, USA.

Dietrich, M.; Traber, M.G.; Jacques, P.F.; Cross, C.E.; Hu, Y & Block, G. (2006). Does gamma-tocopherol play a role in the primary prevention of heart disease and cancer? A review. *Journal of American College of Nutrition*, Vol. 25, No. 4, (August 2006), pp. 292-299.

Domenici, F.A.; Vannucchi, M.T.; Jordão, A.A. Jr.; Meirelles, M.S. & Vannucchi, H. (2005). DNA oxidative damage in patients with dialysis treatment. *Renal Failure*. Vol. 27, No. 6, pp. 689-694.

Doull, J.; Cattley, R.; Elcombe, C.; Lake, B.G.; Swenberg, J.; Wilkinson, C.; Williams, G. & van Gemert, M. A cancer risk assessment of di(2-ethylhexyl)phthalate: application of the new U.S. EPA Risk Assessment Guidelines. *Regulatory Toxicology and Pharmacology*, Vol. 29, No. 3, (June 1999), pp. 327-357.

Dröge, W. (2002). The plasma redox state and ageing. *Ageing Research Reviews*, Vol.1, No.2, (April 2022), pp. 257-78.

Durmaz, E.; Ozmert, E.N.; Erkekoglu, P.; Giray, B.; Derman, O.; Hincal, F. & Yurdakök, K. (2010). Plasma phthalate levels in pubertal gynecomastia. *Pediatrics*, Vol. 125, No. 1, (January 2010), pp. e122-e129.

Duthie, S.J.; Ma, A.; Ross, M.A. & Collins, A.R. (1996). Antioxidant supplementation decreases oxidative DNA damage in human lymphocytes. *Cancer Research*, Vol. 56, No. 6, (March 1996), pp. 1291-1295.

El-Bayoumy, K. & Sinha, R. Mechanisms of mammary cancer chemoprevention by organoselenium compounds. (2004). *Mutation Research*, Vol. 551, No. 1-2, (July 2004), pp. 181-197.

El-Bayoumy, K. & Sinha, R. (2005). Molecular chemoprevention by selenium: a genomic approach. *Mutation Research*, Vol. 591, No. 1-2, (December 2005), pp. 224-236.

Emonet-Piccardi, N.; Richard, M.J.; Ravanat, J.L.; Signorini, N.; Cadet, J. & Béani JC. (1998). Protective effects of antioxidants against UVA-induced DNA damage in human skin fibroblasts in culture. *Free Radical Research*, Vol. 29, No. 4, (October 1998), pp. 307-313.

Engelward Lab. (2011). High throughput DNA damage analysis. Available at:http://web.mit.edu/engelward-lab/comet.htm.

Erkekoglu, P.; Rachidi, W.; De Rosa, V.; Giray, B.; Favier, A & Hincal, F. (2010a). Protective effect of selenium supplementation on the genotoxicity of di(2-ethylhexyl)phthalate and mono(2-ethylhexyl)phthalate treatment in LNCaP cells. *Free Radical Biology anf Medicine*, Vol. 49, No. 4, (August 2010), pp. 559-566.

Erkekoglu, P.; Rachidi, W.; Yuzugullu, O.G.; Giray, B.; Favier, A.; Ozturk, M. & Hincal F. (2010b). Evaluation of cytotoxicity and oxidative DNA damaging effects of di(2-ethylhexyl)-phthalate (DEHP) and mono(2-ethylhexyl)-phthalate (MEHP) on MA-10 Leydig cells and protection by selenium. *Toxicology and Applied Pharmacology*, Vol. 248, No. 1, (October 2010), pp. 52-62.

Erkekoglu, P. & Baydar, T. (2010c). Evaluation of the protective effect of ascorbic acid on nitrite- and nitrosamine-induced cytotoxicity and genotoxicity in human hepatoma line. *Toxicology Mechanisms and Methods*, Vol. 20, No. 2, (February 2010), pp. 45-52.

Erkekoglu, P. & Baydar, T. (2010d). Effect of allyl isothiocyanate (AITC) in both nitrite- and nitrosamine-induced cell death, production of reactive oxygen species, and DNA damage by the single-cell gel electrophoresis (SCGE): does it have any protective effect on HepG2 cells? *International Journal of Toxicology*, Vol. 29, No. 3, (May-June 2010), pp. 305-312.

Erkekoglu, P.; Zeybek, N.D.; Giray, B.; Asan, E.; Arnaud, J. & Hincal F. (2011a) Reproductive toxicity of di(2-ethylhexyl) phthalate in selenium-supplemented and selenium-deficient rats. *Drug and Chemical Toxicology*, Vol. 34, No. 4, (October 2011), pp. 379-389.

Erkekoglu, P.; Zeybek, N.D.; Giray, B.; Asan, E. & Hincal F. (2011b). The Effects of Di(2-Ethylhexyl)Phthalate Exposure and Selenium Nutrition on Sertoli Cell Vimentin Structure and Germ-Cell Apoptosis in Rat Testis. *Archives of Environmental Contamination and Toxicology*. doi 10.1007/s00244-011-9712-9.

Erkekoglu, P.; Rachidi, W.; Yüzügüllü, O.G.; Giray, B.; Oztürk, M.; Favier, A & Hıncal F. (2011c). Induction of ROS, p53, p21 in DEHP- and MEHP-exposed LNCaP cells-protection by selenium compounds. *Food and Chemical Toxicolgy*, Vol. 49, No. 7, (July 2011), pp. 1565-1571.

Erkekoglu, P.; Giray, B.; Rachidi, W; Hininger-Favier, I.; Roussel, A.M.; Favier, A. & Hincal F. (2011d). Effects of di(2-ethylhexyl)phthalate on testicular oxidant/antioxidant status in selenium-deficient and selenium-supplemented rats. *Environmental Toxicology*, doi: 10.1002/tox.20776.

Esworthy, R.S.; Baker, M.A. & Chu FF. (1995). Expression of selenium-dependent glutathione peroxidase in human breast tumor cell lines. *Cancer Research* Vol. 55, No. 4, (February 1995), pp. 957-962.

Faust, F.; Kassie, F.; Knasmuller S.; Boedecker, R.H.; Mann M. & Mersch-Sundermann, V. (2004). The use of the alkaline comet assay with lymphocytes in human biomonitoring studies, *Mutation Research*, Vol. 566, No. 3, (May 2004),pp. 209-229.

Fahey, J.W.; Zalcmann, A.T. & Talalay P. (2001) The chemical diversity and distribution of glucosinolates and isothiocyanates among plants. *Phytochemistry*. Vol. 56, No 1, pp. 5-51.

Fan, J.; Traore, K.; Li, W.; Amri, H.; Huang, H.; Wu, C.; Chen, H.; Zikrin, B. & Papadopoulos, V. (2010). Molecular mechanisms mediating the effect of mono-(2-ethylhexyl) phthalate on hormone-stimulated steroidogenesis in MA-10 mouse tumor Leydig cells. *Endocrinology*. Vol. 151, No. 7, (July 2010), pp. 3348-3362.

Fenech, M. & Ferguson, L.R. (2001). Vitamins/minerals and genomic stability in humans. *Mutation Researh*, Vol, 475, No. 1-2, (April 2001), pp. 1-6.

Finkel, T. (1998). Oxygen radicals and signaling. *Current Opinions in Cell Biology*, Vol. 10, No. 2, (April 1998), pp. 248-253.

Flohé, L. (2007). Selenium in mammalian spermiogenesis. *Biological Chemistry*, Vol. 288, No. 10, (October 2007), pp. 987-995.

Foster, P.M.; Mylchreest, E.; Gaido, K.W. & Sar M. (2001). Effects of phthalate esters on the developing reproductive tract of male rats. *Human Reproduction Update*, Vol. 7, No. 3, (May-June2001), pp. 231-235.

Frank J. (2005). Beyond vitamin E supplementation: an alternative strategy to improve vitamin E status. *Journal of Plant Physiology*, Vol. 162, No. 7, (July 2005), pp. 834-843.

Frederiksen, H.; Aksglaede, L.; Sorensen, K.; Skakkebaek, N.E.; Juul, A. &Andersson AM. (2011). Urinary excretion of phthalate metabolites in 129 healthy Danish children and adolescents: estimation of daily phthalate intake. *Environmental Research*, Vol. 111, No. 5, (July 2011), pp. 656-663.

Frederiksen, H.; Jørgensen, N. & Andersson, A.M. (2010). Correlations between phthalate metabolites in urine, serum, and seminal plasma from young Danish men determined by isotope dilution liquid chromatography tandem mass spectrometry. *Journal Analytical Toxicology*, Vol. 34, No. 7, (September 2010), pp. 400-410.

Gabbianelli, R.; Nasuti, C.; Falcioni, G. & Cantalamessa F. (2004). Lymphocyte DNA damage in rats exposed to pyrethroids: effect of supplementation with Vitamins E and C. *Toxicology*. Vol. 203, No. 1-3, (October 2004), pp. 17-26.

Gajecka, M.; Kujawski, L.M.; Gawecki, J. & Szyfter K. (1999). The protective effect of vitamins C and E against B(a)P-induced genotoxicity in human lymphocytes. *Journal of Environmental Pathology, Toxicology and Oncology*, Vol. 18, No. 3, pp. 159-167.

Gamet-Payrastre, L.; Li, P.; Lumeau, S.; Cassar, G.; Dupont, M.A.; Chevolleau, S.; Gasc, N.; Tulliez, J. &Tercé, F. (2000). Sulforaphane, a naturally occurring isothiocyanate, induces cell cycle arrest and apoptosis in HT29 human colon cancer cells, *Cancer Research*, Vol 60, No. 5, (March 2000), pp. 1426-1433.

García, A.; Haza, A.I.; Arranz, N.; Rafter, J. & Morales, P. (2008). Protective effects of isothiocyanates alone or in combination with vitamin C towards N-nitrosodibutylamine or N-nitrosopiperidine-induced oxidative DNA damage in the single-cell gel electrophoresis (SCGE)/HepG2 assay. *Journal of Applied Toxicology*, Vol. 28, No. 2, (March 2008), pp. 196-204.

Gazouli, M.; Yao, Z.X.; Boujrad, N.; Corton, J.C.; Culty, M. & Papadopoulos V. (2002). Effect of peroxisome proliferators on Leydig cell peripheral-type benzodiazepine receptor gene expression, hormone-stimulated cholesterol transport, and steroidogenesis: role of the peroxisome proliferator-activator receptor alpha. *Endocrinology*, Vol. 143, No. 7, (July 2002),pp. 2571-2583.

Gopalakrishna, R. & Gundimeda, U. (2001). Protein kinase C as a molecular target for cancer prevention by selenocompounds. *Nutrition and Cancer*. Vol. 40, No. 1, pp. 55-63.

Grande, S.W.; Andrade, A.J.; Talsness, C.E.; Grote, K.; Golombiewski, A.; Sterner-Kock, A. & Chahoud, I. (2007). A dose-response study following in utero and lactational exposure to di-(2-ethylhexyl) phthalate (DEHP): reproductive effects on adult female offspring rats. *Toxicology*, Vol. 229, No. 1-2, (January 2007), pp. 114-122.

Griffiths, H.R.; Mistry, P.; Herbert, K.E. & Lunec, J. (1998). Molecular and cellular effects of ultraviolet light-induced genotoxicity. *Critical Reviews in Clinical Laboratory Sciences*, Vol. 35, No. 3, (June 1998), pp. 189-237.

Grossman, L.; Caron, P.R.; Mazur, S.J. & Oh, E.Y. (1998). Repair of DNA-containing pyrimidine dimers. *FASEB Journal*, Vol. 2, No. 11, (August 1988), pp. 2696-2701.

Gutteridge, J.M. & Halliwell, B. (2000). Free radicals and antioxidants in the year 2000. A historical look to the future. *Annuals of New York Academy of Sciences*, Vol. 899, pp. 136-147.

Gürbay, A.; Gonthier, B.; Signorini-Allibe, N.; Baret, L.; Favier, A. & Hincal, F. (2006). Ciprofloxacin-induced DNA damage in primary culture of rat astrocytes and protection by Vitamin E. *Neurotoxicology*, Vol. 27, No. 1, (January 2006), pp. 6-10.

Halliwell, B. (2006). Reactive species and antioxidants. Redox biology is a fundamental theme of aerobic life. *Plant Physiology*, Vol. 141, No. 2, (June 2006), pp. 312-322.

Halliwell B. Free radicals and antioxidants – quo vadis? (2011). *Trends in Pharmacological Sciences*, Vol. 32, No. 3, (March 2011), pp. 125-130.

Harréus, U.; Baumeister, P.; Zieger, S. & Matthias C. (2005). The influence of high doses of vitamin C and zinc on oxidative DNA damage. *Anticancer Research*, Vol. 25, No. 5, (September-October 2005), pp. 3197-3201.

Hecht, S.S. (1996). Recent studies on mechanisms of bioactivation and detoxification of 4-(methylnitrosamino)-1-(3-pyridyl)-1-butanone (NNK), a tobacco-specific lung carcinogen. *Critical Reviews in Toxicology*, Vol. 26, No. 2, pp. 163-181.

Hecht, S.S. (1999). Chemoprevention of cancer by isothiocyanates, modifiers of carcinogen metabolism. *Journal of Nutrition*, Vol. 129, No. 3, pp. 768-774.

Hecht, S.S.; Chung, F.L.; Richie, J.P. Jr; Akerkar, S.A.; Borukhova, A.; Skowronski, L. & Carmella, S.G. (1995). Effects of watercress consumption on metabolism of a tobacco-specific lung carcinogen in smokers. *Cancer Epidemiology Biomarkers and Prevention*, Vol. 4, No. 8, (December 1995), pp. 877-884.

Heudorf, U.; Mersch-Sundermann, V. & Angerer, J. (2007). Phthalates: toxicology and exposure. *International Journal of Hygiene and Environmental Health*, Vol. 210, No. 5, (October 2007), pp. 623-634.

Irminger-Finger, I. (2007). Science of cancer and aging, *Journal of Clinical Oncology*, Vol. 25, No. 14, (May 2007), pp. 1844–1851.

IARC (International Agency for Research on Cancer). (2000). Some industrial chemicals. Vol. 77. Available at: http://monographs.iarc.fr.

Ishihara, M.; Itoh, M.; Miyamoto, K.; Suna, S.; Takeuchi, Y.; Takenaka, I. & Jitsunari F. (2000). Spermatogenic disturbance induced by di-(2-ethylhexyl) phthalate is significantly prevented by treatment with antioxidant vitamins in the rat. *International Journal of Andrology*, Vol. 23, No. 2, (April 2000), pp. 85-94.

Wang, J.Y.; Wen, L.L.; Huang, Y.N.; Chen, Y.T. & Ku M.C. (2006). Dual effects of antioxidants in neurodegeneration: direct neuroprotection against oxidative stress and indirect protection via suppression of glia-mediated inflammation, *Current Pharmaceutical Design*, Vol. 12, No. 27, pp. 3521-3533.

Jakszyn, P. & Gonzalez, C.A. (2006). Nitrosamine and related food intake and gastric and oesophageal cancer risk: a systematic review of the epidemiological evidence. *World Journal of Gastroenterology*, Vol. 12, No. 27, (July 2006), pp. 4296-4303.

Janjua, N.R.; Frederiksen, H.; Skakkebaek, N.E.; Wulf, H.C. Andersson, A.M. (2008). Urinary excretion of phthalates and paraben after repeated whole-body topical application in humans. *Internatonal Journal of Andrology*, Vol. 31, No. 2, (April 2008), pp. 118-30.

Kadirvel, R; Sundaram, K.; Mani, S.; Samuel, S.; Elango, N. & Panneerselvam C. (2007). Supplementation of ascorbic acid and alpha-tocopherol prevents arsenic-induced protein oxidation and DNA damage induced by arsenic in rats. *Human and Experimental Toxicology*, Vol. 26, No. 12, (December 2007), pp. 939-946.

Kasahara, E.; Sato, E.F.; Miyoshi, M.; Konaka, R.; Hiramoto, K.; Sasaki, J.; Tokuda, M.; Nakano, Y. & Inoue M. (2002). Role of oxidative stress in germ cell apoptosis induced by di(2-ethylhexyl)phthalate. *Biochemical Journal*, Vol. 365, No. Pt 3, pp. 849-856.

Kaur, P. & Singh, R. (2007). In vivo interactive effect of garlic oil and vitamin E against stavudine induced genotoxicity in Mus musculus. *Indian Journal of Experimental Biology*, Vol. 45, No. 9, (September 2007), pp. 807-811.

Kaymak, C.; Kadioglu, E.;Basar, H. & Sardas, S. (2004). Genoprotective role of vitamin E and selenium in rabbits anaesthetized with sevoflurane. *Human and Experimental Toxicology*, Vol. 23, No. 8, (August 2004), pp. 413-419.

Kelloff, G.J. (2000). Perspectives on cancer chemoprevention research and drug development. *Advances in Cancer Research*, Vol. 78, pp. 199-334.

Knasmüller, S.; Huber, W.W.; Kienzl, H.; & Schulte-Hermann, R. (1992). Inhibition of repairable DNA-damage in Escherichia coli K-12 cells recovered from various organs of nitrosamine-treated mice by vitamin A, phenethylisothiocyanate, oleic acid and triolein. *Carcinogenesis*, Vol. 13, No. 9, (September 1992), pp. 1643-1650.

Knasmüller, S.; Mersch-Sundermann, V.; Kevekordes, S.; Darroudi, F.; Huber, W.W., Hoelzl, C.; Bichler, J. & Majer, B.J. (2004). Use of human-derived liver cell lines for the detection of environmental and dietary genotoxicants; current state of knowledge. *Toxicology*, Vol. 198, No. 1-3, pp. 315-328.

Konopacka, M.; Widel M. & Rzeszowska-Wolny J. (1998). Modifying effect of vitamins C, E and beta-carotene against gamma-ray-induced DNA damage in mouse cells. *Mutation Research*, Vol. 417, No. 2-3, (September 1998), pp. 85-94.

Frankl, L. (1991). Developmental aspects of experimental pulmonary oxygen toxicity. *Free Radical Biology and Medicine*, Vol. 11, No. 5, pp. 463-494.

Laurent, C.; Pouget, J.P. & Voisin P. (2005). Modulation of DNA damage by pentoxifylline and alpha-tocopherol in skin fibroblasts exposed to Gamma rays. *Radiation Research*, Vol. 164, No. 1, (July 2005), pp. 63-72.

Laurent, C.; Voisin, P. & Pouget, J.P. (2006). DNA damage in cultured skin microvascular endothelial cells exposed to gamma rays and treated by the combination pentoxifylline and alpha-tocopherol. *International Journal of Radiation Biology*, Vol. 82, No. 5, (May 2006), pp. 309-321.

Lee, K.W.; Lee, H.J.; Surh, Y.J. & Lee, C.Y. (2003). Vitamin C and cancer chemoprevention: reappraisal. *American Journal of Clinical Nutrition*, Vol. 78, No. 6, pp. 1074-1078.

Li, H.; Stampfer, M.J.; Giovannucci, E.L.; Morris, J.S.; Willett, W.C.; Gaziano, J.M. & Ma, J. (2004). A prospective study of plasma selenium levels and prostate cancer risk. *Journal of National Cancer Institute*, Vol. 96, No. 9, (May 2004), pp. 696-703.

Longtin, R. (2003). Selenium for prevention: eating your way to better DNA repair? *Journal of National Cancer Institute* Vol. 95, No. 2, (January 2003), pp. 98-100.

Grisham, M.B. (1994). Oxidants and free radicals in inflammatory bowel disease, *Lancet*, Vol. 344, No. 8926, (September 1994), pp. 859–861.

Milner, J.A. (2004). Molecular targets for bioactive food components. *Journal of Nutrition*, Vol. 134, No. 9, pp. 2492S-2498S.

Møller, P.; Viscovich, M.; Lykkesfeldt, J.; Loft, S.; Jensen, A. & Poulsen H.E. (2004). Vitamin C supplementation decreases oxidative DNA damage in mononuclear blood cells of smokers. *European Journal of Nutrition*, Vol. 43, No. 5, (October 2004), pp. 267-274.

Morse, M.A.; Lu, J.; Gopalakrishnan, R.; Peterson, L.A.; Wani, G.; Stoner, G.D.; Huh, N. & Takahashi, M. (1997). Mechanism of enhancement of esophageal tumorigenesis by 6-phenylhexyl isothiocyanate. *Cancer Letters*, Vol 12, No. 1, (January 1997), pp. 119–125.

Mousseau, M.; Faure, H.; Hininger, I.; Bayet-Robert, M. & Favier, A. (2005). Leukocyte 8-oxo-7,8-dihydro-2'-deoxyguanosine and comet assay in epirubicin-treated patients. *Free Radic Research*, Vol. 39, No. 8, (August 2005), pp. 837-843.

Munday, R.; Munday, J.S. & Munday, CM. (2003). Comparative effects of mono-, di-, tri-, and tetrasulfides derived from plants of the Allium family: redox cycling in vitro and hemolytic activity and Phase 2 enzyme induction in vivo. *Free Radical Biology and Medicine*, Vol. 34, No. 9, pp. 1200-1211.

Müler, L.; Kasper, P.; Kertsen, B. & Zhang, J. (1998). Photochemical genotoxicity and photochemical carcinogenesis--two sides of a coin? *Toxicology Letters*, Vol. 102-103, (December 19989, pp. 383-387.

Negro, R. (20089. Selenium and thyroid autoimmunity. *Biologics*. Vol. 2, No. 2, (June 20089, pp. 265-273.

Niki, E.; Noguchi, N.; Tsuchihashi, H. & Gotoh, N. (1995). Interaction among vitamin C, vitamin E, and beta-carotene. *American Journal of Clinical Nutrition*, Vol. 62, No. 6 Supplement, (December 1995), pp. 1322S-1326S.

Nocentini, S.; Guggiari, M.; Rouillard, D. & Surgis, S. (2001). Exacerbating effect of vitamin E supplementation on DNA damage induced in cultured human normal fibroblasts by UVA radiation. *Photochemical Photobiology*, vol. 73, No.4, (April 2001), pp. 370-377.

Noriega, N.C.; Howdeshell, K.L.; Furr, J.; Lambright, C.R.; Wilson, V.S. & Gray, L,E, Jr. (2009). Pubertal administration of DEHP delays puberty, suppresses testosterone production, and inhibits reproductive tract development in male Sprague-Dawley and Long-Evans rats. *Toxicological Sciences*, Vol. 111, No. 1, (September 2009), pp. 163-178.

O'Brien, M.L.; Spear, B.T. & Glauert HP. (2005). Role of oxidative stress in peroxisome proliferator-mediated carcinogenesis. *Critical Reviews in Toxicology*, Vol. 35, No. 1, (January 2005), pp. 61-88.

Ohsawa, K.; Nakagawa, S.Y.; Kimura, M.; Shimada, C.; Tsuda, S.; Kabasawa, K.; Kawaguchi, S. & Sasaki, Y.F. (2003). Detection of in vivo genotoxicity of endogenously formed N-nitroso compounds and suppression by ascorbic acid, teas and fruit juices. *Mutation Research*, Vol. 539, No.1-2, (August 2003), pp. 65-76.

Olive, P.L.; Banáth, J.P. (2006). The comet assay: a method to measure DNA damage in individual cells. *Nature Protocols*, Vol. 1, No. 1, pp. 23-29.

Östling, O. & Johanson, K.J. (1984). Microelectrophoretic study of radiation-induced DNA damages in individual mammalian cells, *Biochemical and Biophysical Research Communications*, Vol. 123, No. 1, (August 1984), pp. 291–298.

Cerutti, P.A. (1994). Oxy-radicals and cancer. *Lancet*, Vol. 344, No. 8926 , (September 1994), pp. 862–863.

Heaton, P.R.; Reed, C.F.; Mann, S.J.; Ransley R.; Stevenson J.; Charlton C.J.; Smith, B.H.; Harper, E.J. & Rawlings, J.M. (2002). Role of dietary antioxidants to protect against DNA damage in adult dogs. *Journal of Nutrition*, Vol. 132, No. 6 Supplement 2 , (June 2002), pp. 1720S–1724S.

Panayiotidis, M. & Collins, A.R. (1997). Ex vivo assessment of lymphocyte antioxidant status using the comet assay. *Free Radical Research*, Vol. 27, No. 5, (November 1997), pp. 533-537.

Persson-Moschos, M.; Alfthan, G. & Akesson, B. (1998). Plasma selenoprotein P levels of healthy males in different selenium status after oral supplementation with different forms of selenium. *European Journal of Clinical Nutrition*, Vol. 52, No. 5, (May 1998), pp. 363-367.

Peters, U. & Takata, Y. (2008). Selenium and the prevention of prostate and colorectal cancer. *Molecular Nutrition and Food Research*, Vol. 52, No. 11, (November 2008), pp. 1261-1272.

Porrini, M.; Riso, P.; Brusamolino, A.; Berti, C.; Guarnieri, S. & Visioli, F. (2005). Daily intake of a formulated tomato drink affects carotenoid plasma and lymphocyte concentrations and improves cellular antioxidant protection. *British Journal of Nutrition*, Vol. 93, No. 1, (January 2005), pp. 93-99.

Rafferty, T.S.; Green, M.H.; Lowe, J.E.; Arlett, C.; Hunter, J.A.; Beckett, G.J. & McKenzie, R.C. (2003). Effects of selenium compounds on induction of DNA damage by broadband ultraviolet radiation in human keratinocytes. British *Journal of Dermatology*, Vol. 148, No. 5, (May 2003), pp. 1001-1009.

Reboul, E.; Richelle, M.; Perrot, E.; Desmoulins-Malezet, C.; Pirisi, V. & Borel, P. (2006). Bioaccessibility of carotenoids and vitamin E from their main dietary sources. *Journal of Agricultural Food Chemistry*, Vol. 54, No. 23, (November 2006), pp. 8749-8755.

Richburg, J.H. & Boekelheide, K. (1996). Mono-(2-ethylhexyl) phthalate rapidly alters both Sertoli cell vimentin filaments and germ cell apoptosis in young rat testes. *Toxicology and Applied Pharmacolgy*, Vol. 137, No. 1, (March 1996), pp. 42-50.

Ridley, A.J.; Whiteside, J.R.; McMillan, T.J. & Allinson, S.L. (2009). Cellular and sub-cellular responses to UVA in relation to carcinogenesis. *International Journal of Radiation Biology*, 85, No. 3, (March 2009), pp. 177-195.

Robichová, S.; Slamenová, D.; Chalupa, I. & Sebová, L. (2004). DNA lesions and cytogenetic changes induced by N-nitrosomorpholine in HepG2, V79 and VH10 cells: the protective effects of Vitamins A, C and E. *Mutation Research*, Vol. 560, no. 2, (June 2004), pp. 91-99.

Roomi, M.W.; Netke, S. & Tsao, C. (1998). Modulation of drug metabolizing enzymes in guinea pig liver by high intakes of ascorbic acid. *International Journal for Vitamin and Nutrition Research*, Vol. 68, No. 1, pp. 42-47.

Rosa, R.M.; Moura, D.J.; Romano, E.; Silva, A.C.; Safi, J. & Pêgas Henriques, J.A. (2007). Antioxidant activity of diphenyl diselenide prevents the genotoxicity of several mutagens in Chinese hamster V79 cells. *Mutation Research*, Vol. 631, No. 1, (July 2007), pp. 44-54.

Geller, S.; Krowka, R.; Valter, K. & Stone, J. (2006). Toxicity of hyperoxia to the retina: evidence from the mouse. *Advances in Experimental Medicine and Biology*, Vol. 572, 425–437.

Salobir, J.; Zontar, T.P.; Levart, A. & Rezar, V. (2010). The comparison of black currant juice and vitamin E for the prevention of oxidative stress. *International Journal for Vitamin and Nutrition Research*, Vol. 80, No. 1, (January 2010), pp. 5-11.

Sardaş, S.; Yilmaz, M.; Oztok, U.; Cakir, N. & Karakaya, A.E. (2001). Assessment of DNA strand breakage by comet assay in diabetic patients and the role of antioxidant supplementation. *Mutation Research*, Vol. 490, No. 2, (February 2001), pp. 123-129.

Schlörmann, W. & Glei, M. (2009). Comet fluorescence in situ hybridization (Comet-FISH):detection of DNA damage. *Cold Spring Harbor Protocols*, No.5, (May 2009), pdb.prot5220.

Schrauzer, G.N. (2000). Anticarcinogenic effects of selenium. *Cellular and Molecular Life Sciences*, Vol. 57, No. 13-14, (December 2000), pp. 1864-1873.

Schrauzer, G.N. (1976). Selenium and cancer: a review. *Bioinorgic Chemistry*, Vol. 5, No. 3, pp. 275-281.

Schrauzer, GN. (1977). Trace elements, nutrition and cancer: perspectives of prevention. *Advances in Experimental Medicine and Biology*, Vol. 91, pp. 323-344.

Schrauzer, G.N. & White, D.A. (1978). Selenium in human nutrition: dietary intakes and effects of supplementation. *Bioinorgic Chemistry*, Vol. 8, No.,4, (April 1978), pp. 303-318.

Seet, R.C.; Lee, C.Y.; Lim, E.C.; Quek, A.M.; Huang, S.H.; Khoo, C.M. & Halliwell, B. (2010) Markers of oxidative damage are not elevated in otherwise healthy individuals with the metabolic syndrome. *Diabetes Care*, Vol. 33, No. 5, (May 2010), pp. 1140-1142.

Sen, C.K.; Khanna, S. &Roy, S. (2006). Tocotrienols: Vitamin E beyond tocopherols. *Life Sciences*, Vol. 78, No. 18, (March 2006), pp. 2088-2098.

Shamberger, R.J.; Tytko, S.A. & Willis, CE. (1976). Antioxidants and cancer. Part VI. Selenium and age-adjusted human cancer mortality. *Archieves of Environmental Health*, Vol. 31, No. 5, (September-October 1976), pp. 231-235.

Shi, L.Q. & Zheng, R.L. (2006). DNA damage and oxidative stress induced by Helicobacter pylori in gastric epithelial cells: protection by vitamin C and sodium selenite. *Pharmazie*, Vol. 61, No. 7, (July 2006), pp. 631-637.

Shimpo, K.; Nagatsu, T.; Yamada, K.; Sato, T.; Nimi, H; Shamoto, M.; Takeuchi, T.; Umezawa, H. & Fujita, K. (1991). Ascorbic acid and adriamycin toxicity. *American Journal of Clinical Nutrition*, Vol. 54, No. 6 Supplement, (December 1991), pp. 1298S-1301S.

Singh, M.; Kaur, P.; Sandhir, R. & Kiran, R. (2008). Protective effects of vitamin E against atrazine-induced genotoxicity in rats. *Mutation Research*, Vol. 654, No. 2, (July 2008), pp. 145-149.

Singh, N.P.; McCoy, M.T.; Tice, R.R. & Schneider, E.L. (1988). A simple technique for quantitation of low levels of DNA damage in individual cells. *Experimental Cell Research*, Vol. 175, No. 1, (March 1988), pp. 184-191.

Slamenová, D. (2001). Contemporary trends in in vivo and in vitro testing of chemical carcinogens. *Neoplasma*, Vol. 48, No. 6, pp. 425-434.

Smerák, P.; Polívková, Z.; Stetina, R. & Bártová, J. Bárta I. (2009). Antimutagenic effect of phenethyl isothiocyanate. *Central European Journal of Public Health*, Vol. 17, No.2, (June 2009), pp. 86-92.

Soory, M. (2009). Redox status in periodontal and systemic inflammatory conditions including associated neoplasias: antioxidants as adjunctive therapy? *Infectious Disorders Drug Targets*, Vol. 9, No. 4, (August 2009), pp. 415-427.

Spiegelhalder, B. & Bartsch H. (1996). Tobacco-specific nitrosamines. *European Journal of Cancer Prevention*, Vol. No. 5 Supplement 1, (September 1996), pp. 33-38.

Stoner, G.D.; Morrissey, D.T.; Heur, Y.H.; Daniel, E.M.; Galati, A.J. & Wagner, S.A. (1991). Inhibitory effects of phenethyl isothiocyanate on N-nitrosobenzylmethylamine carcinogenesis in the rat esophagus. *Cancer Research*, Vol. 51, No. 8, (April 1991), pp. 2063-2068.

Stoner, G.D.; Siglin, J.C.; Morse, M.A.; Desai, D.H.; Amin, S.G.; Kresty, L.A.; Toburen, A.L.; Heffner, E.M&. Francis, D.J. (1995). Enhancement of esophageal carcinogenesis in male F344 rats by dietary phenylhexyl isothiocyanate. *Carcinogenesis* Vol. 16, No. 10, (October 1995), pp. 2473-2476.

Stoner, G.D. & Morse, M.A. (1997). Isothiocyanates and plant polyphenols as inhibitors of lung and esophageal cancer. *Cancer Letters*, Vol. 114, No. 1-2, (March 1997), pp. 113-119.

Stoner, G.D.; Adams, C.; Kresty, L.A.; Hecht, S.S.; Murphy, S.E. & Morse, M.A. (1998). Inhibition of N-nitrosonornicotine-induced esophageal tumorgenesis by 3-phenylpropyl isothiocyanate. *Carcinogenesis*, Vol. 19, No. 12, (December 1998), pp. 2139-2143.

Surh, Y.J. (2002). Anti-tumor promoting potential of selected spice ingredients with antioxidative and anti-inflammatory activities: a short review. *Food and Chemical Toxicology*, Vol. 40, No. 8, (August 2002), pp. 1091-1097.

Surh, Y.J., Chun, K.S., Cha, H.H., Han, S.S., Keum, Y.S., Park, K.K., Lee, S.S. (2001). Molecular mechanisms underlying chemopreventive activities of anti-inflammatory phytochemicals: down-regulation of COX-2 and iNOS through suppression of NF-kappa B activation. *Mutation Research*, Vol. 480-481, pp. 243-268.

Svensson, K.; Hernández-Ramírez, R.U.; Burguete-García, A.; Cebrián, M.E.; Calafat, A.M.; Needham, L.L.; Claudio, L. & López-Carrillo L. (2011). Phthalate exposure associated with self-reported diabetes among Mexican women. *Environmental Research*, Vol. 111, No. 6 (August 2011), pp. 792-796.

Tay, T.W.; Andriana, B.B.; Ishii, M.; Tsunekawa, N.; Kanai, Y. & Kurohmaru, M. (2007). Disappearance of vimentin in Sertoli cells: a mono(2-ethylhexyl) phthalate effect. *International Journal of Toxicology*, Vol. 26, No. 4, (July-August), pp. 289-295.

Traber, M.G. (1998). The Biological Activity of Vitamin E. http://lpi.oregonstate.edu/sp-su98/vitamine.html. Last updated May, 1998.

Traber, M.G. (2006). Vitamin E. In: *Modern Nutrition in Health and Disease*, M.E. Shils; M. Shike; A.C. Ross; B. Caballero; R. Cousins, (Eds), 396-411, Lippincott Williams & Wilkins, Baltimore, MD, USA.

Traber, M.G. (2007). Vitamin E regulatory mechanisms. *Annual Review of Nutrition*, Vol.27, pp. 347-362.

Tricker, A.R. (1997). N-nitroso compounds and man: sources of exposure, endogenous formation and occurrence in body fluids. *European Journal of Cancer Previews*, Vol. 6, No. 3, (June 1997), pp. 226-268.

Vallejo, F.; Tomas-Barberan, F.A. & Garcia-Viguera, C. (2002). Glucosinolates and vitamin C content in edible parts of broccoli florets after domestic cooking. *European Food Research and Technology*, Vol. 215, No. 4, pp. 310-316.

Vallejo, F.; Tomas-Barberan, F.A. & Garcia-Viguera, C. (2002). Potential bioactive compounds in health promotion from broccoli cultivars grown in Spain. *Journal of the Science of Food and Agriculture*, Vol. 82, No. 11, pp. 1293-1297.

Velanganni, A.A.; Dharaneedharan, S.; Geraldine, P. & Balasundram, C. (2007). Dietary supplementation of vitamin A, C and E prevents p-dimethylaminoazobenzene induced hepatic DNA damage in rats. *Indian Journal of Biochemistry and Biophysics*, Vol. 44, No. 3, (June 2007), pp. 157-163.

Verhoeven, D.T.H.; Goldbohm, R.A.; van Poppel, G.; Verhagen, H. & van den Brandt PA. (1996). Epidemiological studies on Brassica vegetables and cancer risk. *Cancer Epidemiology, Biomarkers and Prevention*, Vol. 5, No.9, pp. 733-748.

Weitberg, A.B. & Corvese, D. (1997). Effect of vitamin E and beta-carotene on DNA strand breakage induced by tobacco-specific nitrosamines and stimulated human phagocytes. *Journal of Exprimental and Clinical Cancer Research*, Vol. 16, No. 1, (March 1997), pp. 11-14.

Wilkinson, J.T., Morse, M.A., Kresty, L.A. Stoner, G.D. (1995). Effect of alkyl chain length on inhibition of N-nitrosomethylbenzylamine-induced esophageal tumorigenesis and DNA methylation by isothiocyanates. *Carcinogenesis*, Vol. 16, No. 5, (May 1995), pp. 1011–1115.

Wong, V.W.C.; Szeto, Y.T.; Collins, A.R. & Benzie, I.F.F. (2005). The comet assay: a biomonitoring tool for nutraceutical research. Current Topics in Nutraceutical Research Vol. 3, No. 1, pp. 1-14.

Wormuth, M.; Scheringer, M.; Vollenweider, M. & Hungerbühler K. (2006). What are the sources of exposure to eight frequently used phthalic acid esters in Europeans? *Risk Analaysis*, Vol. 26, No. 3, (June 2006), pp. 803-824.

Yu, F.; Wang, Z.; Ju, B.; Wang, Y.; Wang, J. & Bai, D. (2008). Apoptotic effect of organophosphorus insecticide chlorpyrifos on mouse retina in vivo via oxidative stress and protection of combination of vitamins C and E. *Experimental Toxicology and Pathology*, Vol. 59, No. 6, (April 2008), pp. 415-423.

Zhang, Y. & Talalay, P. (1994). Anticarcinogenic activities of organic isothiocyanates: chemistry and mechanisms. *Cancer Research,* Vol. 54, No. 7S, pp. 1976s-1981s.

Zhang, Y.; Tang, L. & Gonzalez, V. (2003). Selected isothiocyanates rapidly induce growth inhibition of cancer cells. *Molecular Cancer Therapy,* Vol. 2, No. 10, pp. 1045-1052.

9

Gel Electrophoresis as a Tool to Study Polymorphism and Nutritive Value of the Seed Storage Proteins in the Grain Sorghum

Lev Elkonin, Julia Italianskaya and Irina Fadeeva
Agricultural Research Institute for South-East Region
Russia

1. Introduction

Seed storage proteins of cereals constitute the basis of mankind nutrition. However, climate changes, especially, increased droughts that are distinctly observed in many regions all over the globe, hamper sustainable production of traditional cereals, such as wheat, maize and barley, and dictates necessity to cultivate drought resistant and heat tolerant crops. Among these crops, the grain sorghum, owing to its ability for sustainable grain production in conditions of minimal level of precipitation, takes one of the leading places. However, application of sorghum grain for food and feed purposes is limited by its relatively low nutritive value in comparison with other cereals.

One of the reasons of poor nutritive value of sorghum grain is the resistance of its seed storage proteins (kafirins) to protease digestion. Kafirins are alcohol-soluble prolamin proteins making up to 80% of endosperm sorghum proteins (Hamaker et al., 1995). As well as other prolamins, sorghum kafirins contain high levels of proline and glutamine and are deposited in protein bodies of endosperm cells during kernel development. According to differences in solubility in aqueous *tert*-butanol solutions, molecular weight, structure and immunochemical similarity to zeins (maize prolamins) the kafirins were classified into α-, β- and γ-kafirins (Shull et al., 1991; for review, see: Belton et al., 2006). The α-kafirins are highly hydrophobic prolamin proteins (soluble in 40-90% aqueous *tert*-butanol solutions), they comprise 66-84% of total kafirins, depending on the endosperm type (vitreous or opaque). By SDS-PAGE the α-kafirins usually are resolved into two proteins, 25 kDa and 23 kDa. The γ-kafirin accounts for 9-21% of total kafirins depending on the endosperm type (Waterson et al., 1993). According to immunochemical data, the γ-kafirin is a protein with molecular mass of 28 kDa (Shull et al., 1991) although the sequence of the γ-kafirin gene corresponds to the protein with molecular mass of about 20 kDa (De Barros et al., 1991). The β-kafirin, in different endosperm types, accounts for about 7-13% of the total kafirins, and is resolved by the SDS-PAGE into three bands of 20, 18 and 16 kDa (Shull et al., 1991; 1992) or produced one band of 20kDa (El Nour et al., 1998); such variability, perhaps, is due to genotype differences.

One of the main characteristic features of kafirin proteins is their ability to form olygo- or polymers of high molecular weight. These oligomers comprise α- and γ-kafirins linked together by disulphide (S-S) bonds, which are formed by sulphur-containing amino acids (Nunes et al., 2005). In the native state, both mono- and oligomers are present, while in 'reduced' extracts (i.e. with addition of 5% 2-mercaptoethanol that destroys S-S bonds) only monomers were detected (El Nour et al., 1998).

The causes of the poor kafirin digestibility appear to be multi-factorial (Duodu et al., 2003). Among these factors are chemical structure of kafirin molecules, some of which (γ- and β-kafirins) are abundant with sulfur-containing amino acids that are capable to form S-S bonds, resistant to protease digestion; interactions of kafirins with non-protein components such as polyphenols and polysaccharides; and spatial organization of different kafirins in the protein bodies of endosperm cells.

Among the methods that were developed for investigation of sorghum protein digestibility (Pedersen & Eggum, 1983; Mertz et al., 1984; Aboubacar et al., 2003), pepsin digestion of the flour proteins with subsequent gel electrophoresis is the most informative. This method, originally applied by B. Hamaker and co-workers (Weaver et al., 1998; Aboubacar et al., 2001) has been used in a number of studies (Nunes et al., 2004; Wong et al., 2010). Application of this method allowed to isolate sorghum lines with high protein digestibility (Weaver et al., 1998) and to find out that γ-kafirin plays an important role in resistance of sorghum seed storage proteins to protease digestion, namely, γ-kafirin forms a disulfide-bound enzyme-resistant layer at the periphery of protein bodies that restricts access of proteases to the inferior-located and more easily digested α-kafirins (Oria et al., 2000).

In our investigations (Italianskaya et al., 2009), we studied the protein digestibility in different sorghum lines and hybrids using this method and revealed significant polymorphism for *in vitro* kafirin digestibility as well as the strong genetic bases of this trait. In this paper, we summarize the results of these studies, which allowed isolating sorghum lines and F_1 hybrids with increased nutritive value. In addition, we demonstrate that kafirin polymorphism may be used in genetic experiments, namely, in determination of genetic structure of endosperm in sorghum.

2. Material and methods

In vitro protein digestibility was studied in 10 lines and seven F_1 hybrids of the grain sorghum (*Sorghum bicolor* (L.) Moench) (Table 1).

To study *in vitro* protein digestibility the modified method of whole-grain flour pepsin treatment was used (Oria et al. 1995). For each variety 25 mg of flour was treated with 5 ml of 0.15% pepsin solution (P7000 Sigma-Aldrich) in the 0.1 M potassium-phosphate buffer (pH 2.0) for 120 min at 37 °C with repeated shaking. Analysis of seed storage protein (kafirin) spectra was performed before and after pepsin treatment by SDS-PAG electrophoresis (SDS-PAGE) in reducing conditions. SDS-PAGE was carried out in the 12.5% (w/v) acrylamide separating gel (0.375 M TRIS·HCl, pH 8.8) and 4% stacking gel (0.125 M TRIS, pH 6.8) according to modified Laemmli method (Laemmli, 1970). SDS-reducing buffer: 62.5 mM TRIS·HCl, pH 6.8, 20% glycerol, 2% SDS, 5% β-mercaptoethanol; running buffer: 25.0 mM TRIS·HCl, 192 mM glycine, 0.1% SDS, pH 8.3; spacer thikness 1.00 mm. Gels were electrophoresed at 20-23 ma for about 5 hr. Gels were stained with Coomassie Brilliant Blue G-250 or R-250 (Diezel et al, 1972).

Gel Electrophoresis as a Tool to Study Polymorphism and Nutritive Value of the Seed Storage
Proteins in the Grain Sorghum
205

Line, F₁ hybrid[1]	Grain color	Endosperm type
VIR-120	white	floury
Pishchevoe-614 (P-614)	light-brown	semi-vitreous
Volzhskoe-4	light-brown	floury
Volzhskoe-4 waxy (V-4w)	pink	semi-vitreous
Karlikovoe beloe (KB)	white	semi-vitreous
Milo-10	yellow	floury
KVV-45	white	semi-vitreous
KVV-97	white	vitreous
KVV-3	white	semi-vitreous
KP-70	creamy	semi-vitreous
Topaz	creamy	semi-vitreous
O-1237	white	semi-vitreous
Sudzern svetlyi (Sud)	creamy	semi-vitreous
F₅ [M35-1A] Pishchevoe-614/KVV-45	white	semi-vitreous
A2 Karlikovoe beloe/Pishchevoe-614 (A2 KB/P-614)	light-brown	semi-vitreous
A2 Karlikovoe beloe/KP-70 (A2 KB/P-614)	white-yellowish	semi-vitreous
M35-1A Karlikovoe beloe /KVV-45 (M35-1A KB/KVV-45)	white	semi-vitreous
A2 KVV-97/Pishchevoe-614	light-brown	semi-vitreous
A2 Sudzern svetlyi/Topaz (A2 Sud/Topaz)	creamy	semi-vitreous
A2 O-1237/ Pishchevoe-614 (A2 O-1237/P-614)	light-brown	semi-vitreous

In parenthesis: brief designation used in the paper. F₁ hybrids were obtained using male-sterile counterparts of fertile lines; they are designated as A2 or M35-1A depending on the type of male sterility-inducing cytoplasm.

Table 1. The grain sorghum entries used in this investigation

For quantitative estimation of kafirin digestibility the SDS-PAGE banding patterns were scanned by laser densitometer ULTROSCAN XL (LKB-Pharmacia) with wavelength 633 nm. The protein quantity in each fraction was expressed as the area (mm²) of the appropriate peak on densitogram, which was calculated by Software LKB 2222 (Version 3.00). In some experiments, the SDS-PAGE banding patterns were analyzed by Scangel program (developed by Dr. A.F. Ravich). The protein quantity in each fraction and in each lane of electrophoregramm was expressed as the amount of dots in the appropriate protein band. Experiments were performed in two replications. The data on digestibility of kafirins (the ratio of protein peak area before and after pepsin digestion) were subjected to variance analysis using the program Agros (Version 2.09; Dr. S. Martynov, Wheat Genetic Resources Department, N.I. Vavilov Institute of Plant Production, St. Petersburg, Russia).

In some lines and hybrids, the dependence of in vitro protein digestibility from in vitro starch digestibility was studied. In this experiment, the flour, firstly, was subjected to amylolitic enzyme treatment according to the method of B.V. McCleary (McCleary et al., 2002) using Megazyme Resistant Starch Kit (Megazyme Co, Ireland). The pellet remained after removal of solubilised starch was used for pepsin treatment according to the method described above, and, after that, the protein spectrum of the sample was studied by SDS-PAGE.

In order to use kafirins as markers of genetic structure of endosperm the modified technique of SDS-PAGE was applied. In these experiments, AS-1a line of the grain sorghum, which is characterized by a low frequency of parthenogenic embryo formation (Elkonin et al., 2012)

was used. Emasculated panicles of this line were pollinated with the pollen of the line Volzhskoe-4w homozygous for dominant gene *Rs*, conditioning purple color of coleoptiles, seedling leaves and stem. To study the origin of the kernels (apomictic or sexual) with the aid of the kafirin polymorphism, the kernels were split into two parts. The part with an embryo was put in a tray on a moisture filter paper to study the phenotypic traits of a seedling (expression of the *Rs* gene). Another part was used in SDS-PAGE to study its kafirin spectrum. In these experiments, gels were electrophoresed at constant voltage (70 V) for about 15 hr. Gels were stained with $AgNO_3$ solution.

3. *In vitro* kafirin digestibility

SDS-PAGE spectra of the seed storage proteins of a number of lines used in our investigations, before and after pepsin digestion, are shown on Figures 1 and 2.

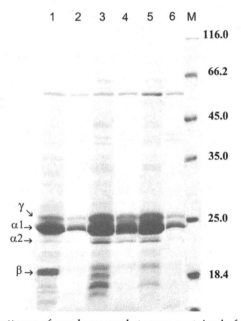

Fig. 1. Electrophoretic patterns of sorghum seed storage proteins before (1, 3, 5) and after (2, 4, 6) pepsin digestion. Lanes 1, 2 – Volzhskoe-4; 3,4 – Pishchevoe-614; 5,6 – F₅ [M35-1A] Pishchevoe-614/KVV-45; M – molecular weight markers (kDa).

α, β, γ – individual kafirin fractions. Gels were stained with Coomassie Brilliant Blue R-250.

In electrophoretic spectra of sorghum lines subjected to pepsin digestion, one could clearly distinguish the γ- (28 kDa), α1 (25 kDa) and α2 (23 kDa) kafirins and one or several bands of β-kafirin fractions (Fig. 1). These electrophoretic patterns correspond to kafirin spectra previously described in the literature (Shull et al., 1991; El Nour et al., 1998; Nunes et al., 2004). In our previous investigations (Table 2) we determined the relative content of different kafirin fractions and observed significant variation among different cultivars. The α1 and γ-kafirins were the most abundant in all lines and hybrids tested: 24-37% and 10-13% of all endosperm proteins, respectively; β-kafirins represent relatively small fractions

Gel Electrophoresis as a Tool to Study Polymorphism and Nutritive Value of the Seed Storage
Proteins in the Grain Sorghum

207

(4-10%) that is in concordance with the literature data (Shull et al., 1991; Waterson et al., 1993).

Line, F_1 hybrid	Protein fraction, % [1]			
	γ	α1	α2	β
KVV-45	13.2	37.3	2.0	3.9
Milo-10	12.8	30.7	5.3	7.2
A2 KVV-97	13.1	24.3	3.2	5.7
A2 KVV-97/P-614	13.3	31.4	4.4	5.3
P-614	10.7	26.9	5.5	5.5
A2 KVV-114	10.8	26.0	4.1	6.9
A2 KVV-114/V-4w	9.5	24.9	4.6	8.9
V-4w	10.3	33.1	3.4	10.2

[1] Relative content of each fraction is expressed as percentage of its peak area from the total endosperm proteins peak area sum. Mean data of two replications.

Table 2. Relative content of different kafirin fractions in some sorghum lines and F_1 hybrids (Italianskaya et al., 2009)

After pepsin digestion the amount of protein in kafirin fractions substantially reduced (Figs. 1; 2). Different sorghum lines and cultivars differed significantly by this trait. For example, among the entries presented in Figure 2 the highest digestibility level had VIR-120 – 90.8% (lanes 1 and 2), while the kafirins of line KVV-3 (lanes 9 and 10) were the most resistant to pepsin digestion (54.5% digestibility level) (Table 3).

In our previous study (Italianskaya et al., 2009), we observed significantly higher variation among the lines. For example, in the cultivar Volzhskoe-4 (V-4, registered standard), the amount of undigested γ- and α-kafirins after pepsin digestion was 80% and 73% from their initial contents, respectively. The total amount of undigested kafirins in cv. V-4 was 70% (digestibility level was 30%). At the same time, in the line KVV-45, the total amount of undigested proteins was 37% (digestibility level was 63%). Percentage of undigested α1 and γ-kafirins in the line KVV-45 was only 25% and 30%, respectively. The differences in kafirin spectra between this line and cv. V-4 before and after pepsin treatment are clearly seen in the Figure 3. Further investigation confirmed a high level of protein digestibility in this line (78.4%) (Table 3). Perhaps, the line KVV-45 contains mutation(s) in the genes encoding structure or deposition of kafirin molecules and, therefore, is of a great interest for future experiments.

Remarkably, in subsequent investigation it was found that in the line Topaz the digestibility level was even higher than in the KVV-45 and reached 89% (see chapter 4). This value is sufficiently high; it corresponds to digestibility level of whole grain flour protein of the best condenced-tannin-free sorghum entries (Axtell et al., 1981, and other reports, as cited in Duodu et al., 2003). One should expect that this line would have high nutritive value.

One should note high digestibility of the β-kafirin fractions in majority of lines. This fact contradicts to hypothesis that explains poor kafirin digestibility by formation of S-S bonds because β-kafirins as well as γ-kafirins contain a high amount of cystein, a sulfur-containing amino acid (Belton et al, 2006). In addition, in all lines, the polypeptides with molecular weight approx. 42 and 46 kDa were prominent in electrophoretic spectra after pepsin digestion. These polypeptides, perhaps, represent kafirin dimers, which were formed as a result of association of kafirin monomers. Earlier, the formation of similar polypeptides (45 kDa) was observed after the cooking process (Duodu et al., 2003; Nunes et al., 2004).

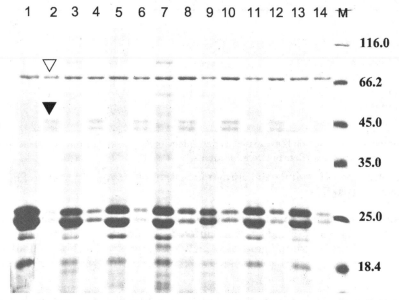

Fig. 2. Electrophoretic patterns of sorghum seed storage proteins before (1, 3, 5, 7, 9, 11, 13) and after (2, 4, 6, 8, 10, 12, 14) pepsin digestion. Lanes 1, 2 – VIR-120; 3, 4 – Volzhskoe-4w; 5, 6 – KVV-45; 7, 8 – KVV-97; 9, 10 – KVV-3; 11, 12 – Karlikovoe beloe; 13, 14 – KP-70; M – molecular weight markers (kDa). di- and trimers of kafirins are indicated by arrows, ◀ and ◁, respectively. Gels were stained with Coomassie Brilliant Blue R-250.

Lane number	Line	Total amount of dots in the lanes		Amount of undigested protein, %	Digestibility, %
		control	after pepsin digestion		
1,2	VIR-120	9769124	897710	9.2	90.8
3,4	Volzhskoe-4w	7285338	2692241	37.0	63.0
5,6	KVV-45	16465667	3554046	21.6	78.4
7,8	KVV-97	26995517	9483915	35.1	64.9
9,10	KVV-3	12242662	5571704	45.5	54.5
11,12	Karlikovoe beloe	13897393	4335642	31.2	68.8
13,14	KP-70	14462063	3651537	25.2	74.8

Table 3. Densitometry of electrophoretic patterns of seed storage proteins shown in Figure 2. The SDS-PAGE banding patterns were scanned and analyzed by Scangel program (developed by Dr. A.F. Ravich)

Gel Electrophoresis as a Tool to Study Polymorphism and Nutritive Value of the Seed Storage
Proteins in the Grain Sorghum

209

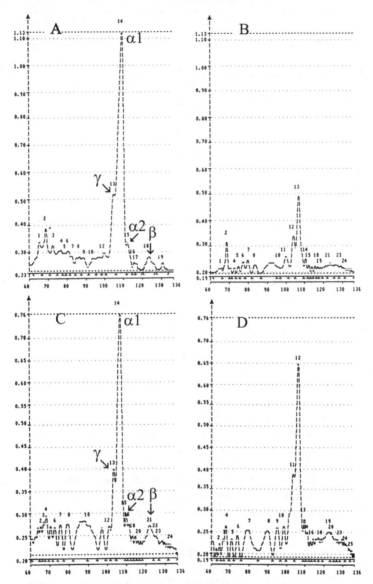

Fig. 3. Densitograms of electrophoretic spectra of endosperm proteins of sorghum line KVV-45 (a, b) and cultivar Volzhskoe-4 (c, d) before (a, c) and after (b, d) pepsin digestion. α1, α2, β and γ-kafirin fractions are indicated.

In order to explore the genetic basis of kafirin digestibility, we studied the expression of this trait in the F_1 hybrids between parental lines differing by resistance to pepsin digestion. Comparison of kafirin digestibility in the F_1 hybrids and their parental lines showed that different hybrid combinations had different mode of inheritance of resistance to pepsin affect (Table 4).

Line, F₁ hybrid[1]	Amount of undigested protein, percent from untreated sample[1]			
	γ	α1	β	Total proteins
KVV-45	24.4	24.6	32.2	24.5 a
M35-1A Karlikovoe beloe /KVV-45	36.2	33.9	34.2	26.8 ab
Karlikovoe beloe	21.3	37.2	26.0	32.1 bcd
A2 Karlikovoe beloe /KP-70	39.5	51.5	42.3	41.6 g
KP-70	22.4	29.5	22.3	26.1 a
A2 Karlikovoe beloe/Pishchevoe-614	41.1	51.3	42.3	40.4 efg
Pishchevoe-614	53.4	64.5	34.7	33.7 cd
A2 KVV-97/Pishchevoe-614	48.9	55.7	44.3	40.5 fg
KVV-97	40.4	30.3	20.4	34.2 d
$F_{0.05}$				14.76*
$LSD_{0.05}$				5.4

[1] Mean from two replications. Data followed by the same letter did not differ significantly ($p<0.05$) according to Duncan Multiple Range Test.
* Significant at $p<0.05$.

Table 4. *In vitro* protein digestibility of endosperm proteins in F₁ sorghum hybrids and their parental lines

The F₁ hybrids A2 KB/P-614, A2 KB/KP-70 and A2 KVV-97/P-614 had significantly lower kafirin digestibility than parental lines, which were characterized by its relatively high level. The reasons of such negative heterosis are unclear. Perhaps, genetic factors conditioning relatively high kafirin digestibility of KP-70, KB and P-614 are recessive and locate in different loci. At the same time, the F₁ hybrid M35-1A KB/KVV-45 did not differ from parental lines and retained high level of kafirin digestibility of the line KVV-45. Perhaps, high digestibility of KVV-45 contrary to other lines may be controlled by any dominant gene(s). This hybrid as well as the line KVV-45, is of great importance for fundamental investigation of factors influencing seed storage protein digestibility in sorghum (kafirin gene structure, structural organization of protein bodies and others) and for practical breeding.

Strong effect of genotype was also found on spectrum of high-molecular weight kafirins that were observed after pepsin digestion (Fig. 4). In some lines and F₁ hybrids two peaks di- and trimers) were found (Fig. 4, A-C), while in others only one peak (trimers) was seen (Fig. 4, D-F). Remarkably, densitograms of the F₁ hybrids in the peak area clearly resembled parental ones. One should note that while the peaks corresponding to trimers were observed in electrophoretic spectra already before pepsin treatment and their amount usually reduced after that, the dimers (45 kDa) were observed only after pepsin action. In some entries kafirin polymers were highly resistant to pepsin digestion, as in the KVV-45, while in others, as in the line P-614 and F₁ hybrid A2 KVV-97/P-614 (Fig. 5, A,B), these peaks were faint or almost absent. These data point on the genetic bases of formation of these molecules, which affect nutritive value of sorghum grain.

Gel Electrophoresis as a Tool to Study Polymorphism and Nutritive Value of the Seed Storage
Proteins in the Grain Sorghum

211

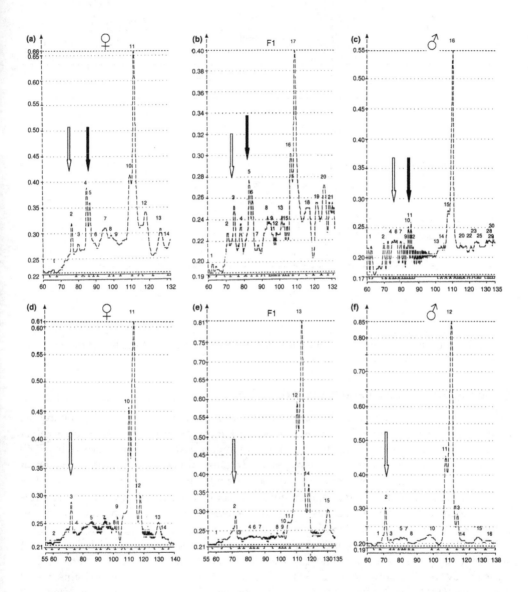

Fig. 4. Densitograms of endosperm proteins electrophoretic spectra of F₁ hybrids and their
parental lines after pepsin digestion: a – A2 KVV-114, b – F₁ A2 KVV-114/V-4w, c – V-4w, d
– A2 KB, e – F₁ A2 KB/KP-70, f – KP-70. Fractions of di- and trimers of kafirin proteins
(45kDa and 66 kDa) are shown by arrows, ➡ and ⇨, respectively.

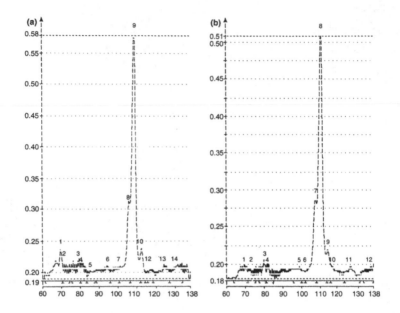

Fig. 5. Densitograms of endosperm proteins electrophoretic spectra of the line P-614 (a) and F_1 hybrid A2 KVV-97/P-614 (b) after pepsin digestion.

4. Interaction of starch and protein digestibility

In order to found out dependence of sorghum protein digestibility on starch digestibility the flour of several lines and F_1 hybrids was subjected to pepsin action after removal of digestible starch by the amylolytic enzymes treatment, and then was studied by SDS-electrophoresis for the presence of undigested proteins. It was found that after action of amylolitic enzymes the amount of protein in the kafirin fractions significantly increases (Fig. 6): in the lanes 3, 7 and 11 (samples after amylolitic enzyme action) almost all the protein is concentrated in the kafirin fractions, in comparison with the lanes 1, 5 and 9 (samples without amylolitic enzyme action). However, contrary to expectation that removal of starch will favor to kafirin digestion, the pepsin treatment of the samples treated before it with amylolitic enzymes (lanes 4, 8 and 12) were digested significantly fewer than samples digested by pepsin only (lanes 2, 6 and 10). Gel densitometry confirmed this visual conclusion (Table 5). Such phenomenon was observed in all F_1 hybrids studied (A2 Sud/Topaz, A2 O-1237/P-614, M35-1A KB/KVV-45) and their parental lines. Perhaps, partially digested starch molecules may interact with kafirin molecules by any physical or, probably, chemical way and prevent their protease digestion. One should not exclude that similar process might take place in *in vivo* conditions and thus decrease sorghum protein digestibility and reduce its nutritive value.

Gel Electrophoresis as a Tool to Study Polymorphism and Nutritive Value of the Seed Storage
Proteins in the Grain Sorghum

213

In addition, it was found that after amylolitic enzyme treatment the amount of di- and trimer fractions significantly reduced in comparison with the non-fermented control samples. In the F_1 hybrid A2 Sud/Topaz their amount was significantly fewer even in comparison with pepsin treatment only. Such a reduction of kafirin oligomers may be also responsible in increase of the level of kafirin monomers. These data testify that starch molecules might participate in formation of kafirin oligomer molecules. They are important for understanding the factors influencing kafirin and starch interactions in sorghum endosperm and their digestibility.

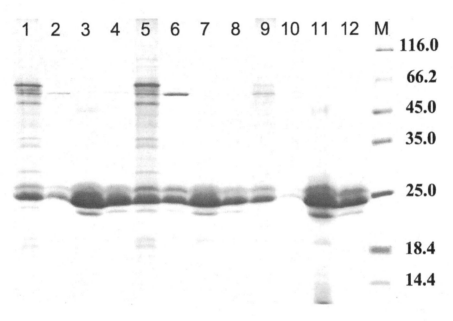

Fig. 6. Electrophoretic patterns of sorghum seed storage proteins from the flour before (1, 3, 5, 7, 9, 11) and after (2, 4, 6, 8, 10, 12) pepsin digestion; lanes 3, 4, 7, 8, 11, 12 – after removal of soluble starch by amylolitic enzymes before pepsin digestion; lanes 1, 2, 5, 6, 9, 10 – without this procedure. Lanes 1-4 – Sudzern svetlyi; 5-8 – F_1 A2 Sudzern svetlyi/Topaz; 9-12 – Topaz; M – molecular weight markers (kDa). Gels were stained with Coomassie Brilliant Blue G-250.

5. Kafirins as the markers of endosperm genetic structure

In addition to variation of a number of β-kafirin fractions in different sorghum entries described above, we have revealed polymorphism of the α-kafirins. The line Volzhskoe-4w (V-4w) that is used as a tester line to distinguish the hybrid seedlings from the maternal ones, possessed specific kafirin spectrum, which was rarely observed in other sorghum lines and cultivars. The α1 fraction was composed from three polypeptides: α1-1, α1-2, and α1-3; α2 fraction was composed from two polypeptides: α2-1 and α2-2 (Fig. 7, lanes 1-3). We hypothesized that this polymorphism could be used in studies of genetic structure of endosperm in apomixis research in sorghum.

To test this possibility we used the AS-1a line, which is characterized by ability for development of aposporous embryo sacs and parthenogenetic embryos (Elkonin et al., 2012). Gel electrophoresis showed that kafirin spectrum of this line differs from V-4w (Fig. 7). Two polypeptides were observed in the α1 fraction (α1 and α1-2), the α1-2 was in trace amount, and α1-3 was absent; the α2 fraction did not subdivide into two polypeptides (Fig. 7, lanes 4-6).

Genotype	Experimental treatment	Amount of protein in different kafirin fractions		Total proteins percent to the control
		Individual fractions (α+β+γ), percent to the control	Olygomers percent to the control	
Sudzern svetlyi	Control	100.0	100.0	100.0
	Pepsin	48.2	7.4	23.0
	Amylolitic enzymes	177.0	8.9	76.5
	Amylolitic enzymes, pepsin	102.9	5.5	44.2
A2 Sudzern / Topaz	Control	100.0	100.0	100.0
	Pepsin	66.5	19.1	35.6
	Amylolitic enzymes	111.2	4.5	52.4
	Amylolitic enzymes, pepsin	60.9	1.2	27.1
Topaz	Control	100.0	100.0	100.0
	Pepsin	18.1	10.2	11.1
	Amylolitic enzymes	225.6	24.9	139.4
	Amylolitic enzymes, pepsin	124.7	6.4	64.7
F_A (genotypes)		0.858	3.834	1.338
F_B (treatment)		6.340*	836.245***	7.945*
F_{AB}		0.995	5.964*	1.245
In average for treatment				
	Control	100.0 a	100.0 c	100.0 b
	Pepsin	44.3 a	12.2 b	23.2 a
	Amylolitic enzymes	171.2 b	12.8 b	89.4 b
	Amylolitic enzymes + pepsin	96.1 a	4.4 a	45.3 a

Mean data of two replications; data followed by the same letter did not differ significantly ($p<0.05$) according to Duncan Multiple Range Test;
*, and *** significant at $p<0.05$, and $p<0.001$, respectively.

Table 5. Densitometry of seed storage proteins electrophoretic patterns of F_1 A2 Sudzern/Topaz and its parents after treatment with pepsin and/or α-amylase and amyloglucosidase

In order to use this polymorphism for identification of seeds formed via apomixis, the kernels obtained by pollination of emasculated panicles of AS-1a with the pollen of V-4w were split into two parts. The part with an embryo was used to study the phenotypic traits of a seedling. Another part was used in SDS-PAGE to study its kafirin spectrum. In the case of autonomous endosperm development, no V-4w proteins should be found in the kafirin spectra of the kernels yielded maternal seedlings, while in the case of pseudogamous endosperm development, in the electrophretic spectra of these kernels, the SDS-PAGE must reveal V-4w proteins. It was found that kafirin spectra of kernels, which yielded maternal seedlings (Fig. 7, lanes 11,12) did not differ from the spectrum of AS-1a line (Fig. 7, lanes 4-6), while in the spectra of the kernels, which yielded hybrid seedlings the α1-3 protein was clearly distinguished (Fig. 7, lanes 7-10). These data support the results of our cyto-embryological observations of autonomous endosperm development in the AS-1a line (Elkonin et al., 2012) and are in accordance with the literature data on other sorghum lines with apomictic potentials (Rao et al., 1978; Wu et al., 1994; Ping et al., 2004).

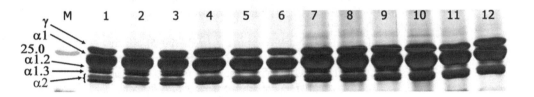

Fig. 7. Kafirin spectra of Volzhskoe-4w (lanes 1-3), AS-1a (4-6) and of the kernels, which were set on emasculated panicles of AS-1a pollinated with the Volzhskoe-4w pollen and yielded the F_1 hybrid seedlings (7-10) and maternal plants (11-12); M – molecular weight marker (kDa). Gels were stained with $AgNO_3$.

6. Conclusion

Summarizing, the results of our investigation demonstrate that gel electrophoresis of the seed storage proteins is a powerful instrument in researches on sorghum genetics and breeding that have both fundamental and applied orientation. It allowed to isolate of sorghum lines with individual kafirin fractions more sensitive to protease action, and, therefore, with increased protein digestibility – one of the main trait characterizing the nutritive value of sorghum grain. These lines may be used in breeding programs for developing new CMS-lines and F_1 hybrids. In addition, these lines (for example, KVV-45) may be used in future investigations on molecular organization of genes encoding structure and/or deposition of kafirins, their cloning and transfer into other sorghum lines by methods of classical genetics or genetic engineering.

Gel electrophoresis of the flour subjected to amylolitic enzyme action has demonstrated that starch digestion decreases content of kafirin polymers and reduces subsequent kafirin digestion by pepsin. This finding may explain the reduced nutrient value of sorghum grain, in comparison with other cereals. These data point on the complex mode of interactions of storage proteins and starch in sorghum endosperm.

Gel electrophoresis of the seed storage proteins allowed to determine genetic structure of endosperm in sorghum kernels with parthenogenic embryos developing in the line AS-1a with apomictic potentials and may be used in development of sorghum lines with high frequency and stable expression of this trait.

7. Acknowledgement

Authors are grateful to Dr. Alexander Ravich for the Software Scangel. This work was funded partly by the Russian Foundation for Basic Researches, grant 10-04-00475.

8. References

Aboubacar, A.; Axtell, J. D.; Huang, C.P. & Hamaker, B.R. (2001). A rapid protein digestibility assay for identifying highly digestible sorghum lines. *Cereal Chemistry*, Vol. 78, pp. 160-165, ISSN 0009-0352

Aboubacar, A.; Axtell, J. D., Nduulu, L. & Hamaker, B. R. (2003). Turbidity assay for efficient identification of high protein digestibility sorghum lines. *Cereal Chemistry*, Vol. 80, pp. 40-44, ISSN 0009-0352.

Belton, P.S.; Delgadillo, I.; Halford, N.G. & Shewry, P.R. (2006). Kafirin structure and functionality. *J Cereal Science*, Vol. 44, pp. 272-286, ISSN 0733-5210

De Barros, E.G.; Takasaki, K.; Kirleis, A.W. & Larkins, B.A. (1991). Nucleotide sequence of a cDNA clone encoding γ-kafirin protein from *Sorghum bicolor*. *Plant Physiolology*, Vol. 97, pp. 1606–1607, ISSN 0032-0889

Diezel, W.; Kopperschlader, G. & Hofman, E. (1972). An improved procedure for protein staining in polyacrylamide gels with a new type of Coomassie Brilliant Blue. *Annals of Biochemistry*, Vol. 48, pp. 617-620, ISSN 0003-2697

Duodu, K.G.; Taylor, J.R.N.; Belton, P.S. & Hamaker, B.R. (2003). Factors affecting sorghum protein digestibility. *J Cereal Science*, Vol. 38, pp. 117-131, ISSN 0733-5210

El Nour, I.N.A.; Peruffo, A.D.B. & Curioni, A. (1998). Characterisation of sorghum kafirins in relation to their cross-linking behaviour. *J Cereal Science*, Vol. 28, pp. 197–207, ISSN 0733-5210

Elkonin, L.A.; Belyaeva, E.V. & Fadeeva, I.Yu. (2012). Expression of apomictic potentials and selection for apomixis in sorghum line AS-1a. *Russian Journal of Genetics*, Vol. 48, No 1, pp. 32–40, ISSN 1022-7954

Hamaker, B.R.; Mohamed, A.A. & Habben, J.E. (1995). Efficient procedure for extracting maize and sorghum kernel proteins reveals higher prolamin contents than the conventional method. *Cereal Chemistry*, Vol. 72, pp. 583–588, ISSN 0009-0352

Italianskaya, J.V.; Elkonin, L.A. & Kozhemyakin, V.V. (2009). Characterization of kafirin composition and *in vitro* digestibility in CMS-lines, fertility restorers and F₁ hybrids with the new types of CMS-inducing cytoplasms of sorghum. *Plant Breeding*, Vol. 128, pp. 624-630, ISSN 0179-9541

Laemmli, U.K. (1970). Cleavage of structural proteins during the assembly of the head of bacteriophage T4. *Nature*, Vol. 227, pp. 680-685, ISSN 0028-0836

Gel Electrophoresis as a Tool to Study Polymorphism and Nutritive Value of the Seed Storage
Proteins in the Grain Sorghum

217

Mertz, E.T.; Hassen, M.M.; Cairnms-Wittern, C.; Kirleis, A.W.; Tu, L. & Axtell, J.D. (1984). Pepsin digestibility of proteins in sorghum and other major cereals. *Proceedings of National Academy of Science of USA*, Vol. 81, pp. 1-2, ISSN 0027-8424

McCleary, B.V.; McNally, M. & Rossiter, P. (2002). Measurement of resistant starch by enzymic digestion in starch samples and selected plant materials: Collaborative Study. *Journal of AOAC International*, Vol. 85, №5, pp. 1103-1111, ISSN 1060-3271

Nunes, A.; Correia, I.; Barros, A. & Delgadillo, I. (2004). Sequential *in vitro* pepsin digestion of uncooked and cooked sorghum and maize samples. *Journal of Agricultural and Food Chemistry*, Vol. 52, pp. 2052-2058, ISSN 0021-8561

Nunes A.; Correia, I.; Barros A. & Delgadillo, I. (2005). Characterization of kafirin and zein oligomers by preparative sodium dodecyl sulfate-polyacrylamide gel electrophoresis. *Journal of Agricultural and Food Chemistry*, Vol. 53, pp. 639-643, ISSN 0021-8561

Oria, M.P.; Hamaker, B.R. & Shull, J.M. (1995). Resistance of Sorghum α-, β- and γ-kafirins to pepsin digestion. *Journal of Agricultural and Food Chemistry*, Vol. 43, pp. 2148-2153, ISSN 0021-8561

Oria, M.P.; Hamaker, B.R.; Axtell, J.D. & Huang, C.P. (2000). A highly digestible sorghum mutant cultivar exhibits a unique folded structure of endosperm protein bodies. *Proceedings of National Academy of Science of USA*, Vol. 97, pp. 5065-5070, ISSN 0027-8424

Pedersen, B. & Eggum, B.O. (1983). Prediction of protein digestibility by an *in vitro* enzymatic pH-stat procedure. *Zeitschrift fur Tierphysiologie, Tierernaehrung und Futtermittelkunde*, Vol. 49, pp.265-277, ISSN 1439-0396

Ping, J.-A.; Zhang, F.-Y.; Cui, G.-M.; Cheng Q.-J.; Du Z.-H.; Zhang Y.-X. (2004). A study of the properties of autonomous seed setting and embryology in Sorghum apomictic line 2083. *Acta Agronomica Sinica*. Vol. 30, № 7, pp. 714-718, ISSN 0496-3490

Rao, N.G.P.; Narayana, L.L. & Reddy, V.R. (1978). Apomixis and its utilization in grain Sorghum. I. Embryology of two apomictic parents. *Caryologia*. Vol. 31, № 4, pp. 427-433, ISSN 0008-7114

Shull, J.M.; Watterson, J.J. & Kirleis, A.W. (1991). Proposed nomenclature for the alcohol-soluble proteins (kafirins) of *Sorghum bicolor* (L. Moench) based on molecular weight, solubility and structure. *Journal of Agricultural and Food Chemistry*, Vol. 39, pp. 83–87 ISSN 0021-8561

Shull, J.M.; Watterson, J.J. & Kirleis, A.W. (1992). Purification and immunocytochemical localization of kafirins in *Sorghum bicolor* (L. Moench) endosperm. *Protoplasma*, Vol. 171, pp. 64–74, ISSN: 0033-183X

Watterson, J.J.; Shull, J.M., & Kirleis, A.W. (1993). Quantitation of α-, β- and γ-kafirins in vitreous and opaque endosperm of *Sorghum bicolor*. *Cereal Chemistry*, Vol. 70, pp. 452–457, ISSN 0009-0352

Weaver, C.A.; Hamaker, B.R. & Axtell, J.D. (1998). Discovery of grain sorghum germplasm with high uncooked and cooked *in vitro* protein digestibility. *Cereal Chemistry*, Vol. 75, pp. 665-670, ISSN 0009-0352

Wong, J.H.; Marx, D.B.; Wilson, J.D.; Buchanan, B.B.; Lemaux, P.G. & Pedersen, J.F. (2010). Principal component analysis and biochemical characterization of protein and starch reveal primary targets for improving sorghum grain. *Plant Science*, Vol. 179, pp. 598-611, ISSN 0168-9452

Wu, S.-B.; Shang, Y.-J. & Han, X.-M. (1994). Embryological study on apomixis in a sorghum line SSA-1. *Acta Botanica Sinica.*, Vol. 36, №11, pp. 833-837, ISSN 0577-7496

10

Gel Electrophoresis as Quality Control Method of the Radiolabeled Monoclonal Antibodies

Veronika Kocurová

Nuclear Physics Institute, Academy of Sciences of the Czech Republic,
Řež near Prague
Czech Republic

1. Introduction

Neurodegeneration is the leading term for the progressive loss of the neuron structure, including death of neurons. Many neurodegenerative diseases including the specific diseases - such as Parkinson's, Alzheimer's, and Huntington's occur as a result of the neurodegenerative processes. As research progresses, many similarities appear which relate these diseases to one another on a sub-cellular level. Discovering these similarities offers hope for therapeutic advances that could ameliorate many diseases simultaneously. There are many parallels between different neurodegenerative disorders including atypical protein assemblies as well as induced cell death followed by an apoptosis. Apoptosis is a form of the programmed cell death in the multicellular organisms. It is one of the main types of the programmed cell death, and, last but not least, involves a series of the biochemical reactions leading to a characteristic cell morphology changes, and, finally, death. In according to the previously mentioned knowledge, there is a necessity to develop an imaging method which describes these cellular changes. The principal goal of the investigation monoclonal antibodies and their fragments is to examine the possibility of developing of an imaging radiotracer that would be specific for cytoskeleton of destructed dendrites and neuronal bodies. One of the suitable fitting marker, specific for neuronal tissue, performs anti III β-tubulin (bTcIII) antibody - TU-20 with molecular weight 150 kDa and its scFv fragment with molecular weight 27.7 kDa. The scFv fragment of TU-20 was synthesized for its higher mobility through tissue and vascular barriers. Biochemical characteristics (especially immunoaffinity) of the specific binding substance - anti III β-tubulin scFv fragment - is preserved, and, moreover, the biological availability is much better than in case of the whole antibody. See the structure in the Fig. 1.

To examine this hypothesis, it is necessary to radiolabel both substances with ^{125}I and ^{123}I. The next step is chemical analysis and, furthermore, biochemical properties are extensively investigated. The quality control, performed by gel filtration, electrophoresis, ELISA testing determines adequate properties of the radiolabeled substances for further studies.

Affinity coupling and RIA analytic methods occur under development with focusing on specifics of the antibody and its fragment behavior. *In vitro* experiment shows an extent of the preserved binding specifity of the species by incubation of the both radiolabeled substances with mice brain slices followed by an autoradiography.

The *in vivo* biodistribution confirmes behavior of elimination of the radiolabeled TU-20 and scFv from mice. The bi-exponential model for two-phase clearance to determine short phase half-life $t_{1/2\alpha}$ and long phase half-life $t_{1/2\beta}$ values is used. For comparative study, a transgene population G93A1 Gur was chosen to show different behavior of the substances in normal mouse and in modified organism with amyotrophic lateral sclerosis (ALS).

The main objective of this work is to develop a method for direct imaging of the structural degradation of peripheral neurones by various types of neuropathies.

2. Methods and materials

The monoclonal antibody TU-20 and its scFv was purchased from Exbio, CZ. The antibody recognizes the peptide sequence ESESQGPK. ScFv TU-20 is a recombinant protein expressed in *E. coli*. (Dráberová et al., 1998)

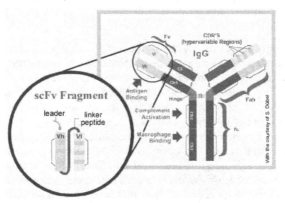

Fig. 1. The structure of the monoclonal antibody on the base of IgG and its scFv fragment.

2.1 Radioiodination of the antibody

^{125}I ($T_{1/2}$ = 59,4 h) radioiodination of TU-20 and scFv TU-20 was performed via chloramine-T with or without stopping reaction with sodium thiosulfate agent. The ratio of an amount of TU-20 to radioactivity was 1 μg to 5.5-7.0 MBq of ^{125}I. The ratio of an amount of the fragment to radioactivity was 1 μg 1.5-2.0 MBq of ^{125}I. ^{123}I ($T_{1/2}$ = 13,3 h) radioiodination of the fragment scFv TU-20 was performed via chloramine-T with stopping reaction with sodium thiosulfate. The ratio of an amount of the fragment to radioactivity was 1 μg to 3-5 MBq ^{123}I (Švecová et al., 2008). The structure of the radiolabeled antibody is shown in the Fig. 2.

The monoclonal antibody TU-20 was radioiodinated by using either chloramine-T or iodogen as an oxidizing agent. Iodination via chloramine-T was provided in two alternative ways: either with or without stopping a reaction by a reducing agent (Dráberová et al., 1998).

The reaction was performed under following conditions: 10 μl of TU-20 (1 mg/ml) was transferred to 10 μl phosphate buffer (PBS, 0,01 M, pH 7,4) in a reaction vessel and ^{125}I radioactivity (approximately 5,36 MBq) was added. Finally, the solution of chloramine-T in PBS (0,1 mg/ml) was added to the reaction vessel. The amount of chloramine-T ranged from 0,5 to 6 μg per 10 μg of the antibody. After the reaction time (60 seconds), during which the

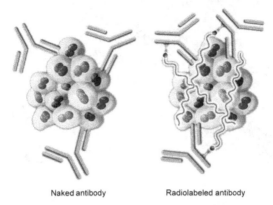

Naked antibody   Radiolabeled antibody

Fig. 2. Radiolabeled monoclonal antibody. Radiotracer is bound to the antibody structure via -OH group of the tyrosine.

reaction mixture was gently agitated, the reaction alternatively might be or not stopped with 100 μl of the solution sodium thiosulfate in water (4 mg/ml) (Chizzonite et al., 1991).

Iodination tubes, for both methods, were prepared in the same way. 100 μl of iodogen dissolved in chloroform (10 - 500 μg/ml) was given in a glass tube and chloroform was evaporated under a slow stream of nitrogen. The prepared iodination tubes were used immediately. The procedure for the direct method consisted in adding 10 μl of TU-20 (1 mg/ml) into the reaction tube with 50 μl of phosphate buffer (PB, 0,05 M, pH 8,5) and an equal amount of Na^{125}I around 5,4 MBq. Reaction time was 15 minutes.

The indirect method was performed in two steps. Firstly, radioactivity in PB was added into the tube coated with iodogen. After 15 minutes an activated iodide was withdrawn, transferred into the vessel containing 10 μl of the antibody and the mixture was agitated for 20 minutes (Švecová et al., 2008).

Radioiodination of the fragment scFv TU-20 was performed via chloramine-T without stopping reaction with thiosulfate as described previously for TU-20. In both cases, at the end of labeling, the reaction mixture was loaded on the top of a BSA-blocked polyacrylamide desalting column with an exclusion limit 6 kDa. Fractions were eluted with 0,1 % BSA in PBS and measured for radioactivity. (Hamilton, 2002), (Katsetos, 2003).

2.2 Immunoreactivity testing by enzyme linked immunosobent assay (ELISA)

The immunoreactivity of the radiolabeled monoclonal antibody TU-20 was determined by an enzyme linked immunosorbent assay (ELISA) using the commercial set for detection of mouse anti - β III tubulin antibodies from VIDIA, CZ. One of the most useful of the immunoassays is the two antibody sandwich ELISA. This assay is used to determine the antigen concentration in unknown samples. This ELISA is fast and accurate, and if a purified antigen standard is available, the assay can determine the absolute amount of antigen in an unknown sample. The principle of ELISA testing is shown in the Fig. 3.

The sandwich ELISA requires two antibodies that bind to epitopes that do not overlap on the antigen. This can be accomplished with either two monoclonal antibodies that recognize discrete sites or one batch of affinity-purified polyclonal antibodies. To utilize this assay, one

ELISA

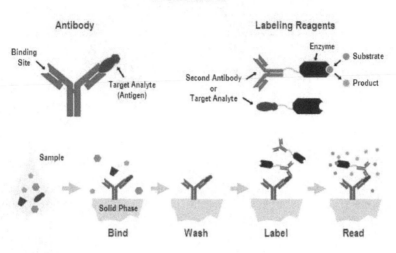

Fig. 3. ELISA principle. A specific antigen (an antibody plays the role of the "antigen" in the case of the antibody ELISA detection) is bound to the specific antibody coated on the solid carrier (microtitration plate). Subsequently, another specific antibody (labeled by an apropriate enzyme which catalyzes the coloured and easily detectable reaction) is added to the previously bound antigen.

antibody (the 'capture' antibody) is purified and bound to a solid phase typically attached to the bottom of a plate well.

Afterwards, an antigen is added, and, allowed to complex with the bound antibody. Unbound products are then removed with a wash, and a labeled second antibody (the 'detection' antibody) is allowed to bind to the antigen, and, therefore, the setting is described as the sandwich. The assay is then quantified by measuring the amount of labeled second antibody bound to the matrix, through the use of a colorimetric substrate.

Major advantages of this technique are that the antigen does not need to be purified prior to use, and that these assays are very specific. However, one disadvantage is that not all antibodies can be used. Monoclonal antibody combinations must be qualified as "matched pairs", meaning that they can recognize separate epitopes on the antigen so they do not hinder each other's binding.

ELISA procedures utilize substrates that produce soluble products. Ideally the enzyme substrates should be stable, safe and inexpensive. Popular enzymes are those that convert a colorless substrate to a colored product, e.g., pnitrophenylphosphate (pNPP), which is converted to the yellow p-nitrophenol by alkaline phosphatase. Substrates used with peroxidase include 2,2'-azo-bis(3-ethylbenzthiazoline-6-sulfonic acid) (ABTS), o-phenylenediamine (OPD) and 3,3'5,5'-tetramethylbenzidine base (TMB), which yield green, orange and blue colors, respectively. In our case, TMB was used for colorimetric visualization. The settlement of the procedure see in the Fig. 4.

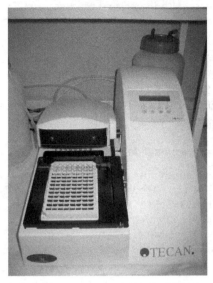

Fig. 4. Equipment for optical density measurement in the ELISA settings.

2.3 Immunoaffinity testing by radioimmunoassay (RIA)

In radioimmunoassay, a fixed concentration of radio-labeled antigen in trace amounts is incubated with a constant amount of antiserum such that the total antigen binding sites on the antibody are limited such that the only 30–50 % of the total radio-labeled antigen may be bound in the absence of the antigen. When unlabeled antigen, either as standard or test sample, is added to this system, there is competition between radio-labeled antigen and unlabeled antigen for the limited constant number of binding sites on the antibody.

The amount of radio-labeled antigen bound to antibody decreases as the concentration of unlabeled antigen increases. Following optimal incubation condition e.g. buffer, pH, time and temperature, radio-labeled antigen bound to antibody is separated from unbound radio-labeled antigen.

RIA analytic method was developed in two modifications of surface of the reactive vessel.

2.4 Immunoaffinity separation affinity coupling (AC)

Affinity coupling was develop by use the basic matrix activated Sepharose 4 Fast Flow by Pierce which was modified specific binding octapeptide (Vijayalakshmi, 1992). Activated media enable successful, convenient immobilization of ligands without the need for complex chemical syntheses or special equipment. The Sepharose matrix provides a wide range of high-capacity media with a variety of coupling chemistries for fast, easy, and safe immobilization through a chosen functional group. The principle is to immobilize the antibodies or other large proteins containing -NH2 groups by coupling them to the matrix without the need for an intermediate spacer arm.

The correct choice of an activated medium is dictated by both the group available in the ligand molecule, and by the nature of the binding reaction with the substance to be purified. To ensure minimal interference with the normal binding reaction, immobilization should be attempted through the least critical region of the ligand (Haugland, 1995).

2.5 Stability testing by electrophoresis

Mostly used variation of the electrophoresis for the intention of the quality control of the radiolabeled substances is SDS-PAGE formation of the electrophoresis. It concerns of zone electrophoresis in gel in surface placement. The mixture of the substances is analyzed by division in accordance to the molecular weight.

2.5.1 Polymerization of the polyacrylamide gel

Polyacrylamide gel is prepared to the form by polymerization of the basic monomer acrylamide (CH2=CH-CO-NH2; abbrev.   AA) and N,N'-methylen-bis-akrylamid (CH2=CH-CO-NH-CH2-NH-CO=CH-CH2; abbrev.   BIS) which is implemented to the polymere randomly and might covalently bind two linear chains of the polyacrylamide. Ammonium persulfate (abbrev. APS) is used as the initiative reactant and N,N'-tetramethylendiamine (abbrev. TEMED) as the catalyzer, see the Fig. 5.

The inhibitor of the reaction is oxygen, and, therefore, the gel must be protected against the oxygen atmosphere. The polymerization has the radical and exothermic process, and, therefore, the cooling is neccessary during the whole polymerization. The ratio of AA:BIS is crucial for the gel mechanic and separation characteristics. The suggested ratio is ranging of about 40:1 (from 20:1 up to 100:1) (Jones, 2004).

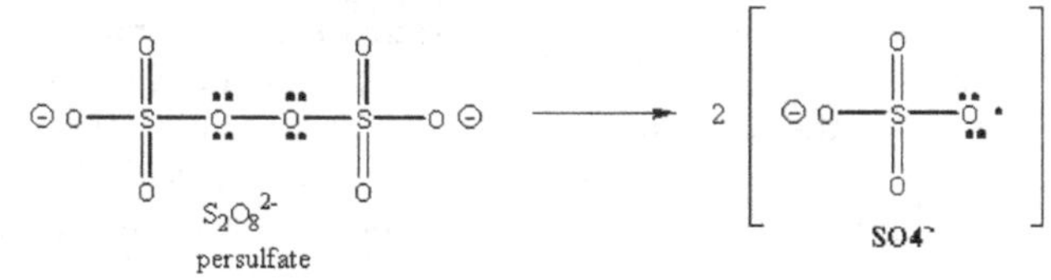

Fig. 5. The Free radical Polymerization of the Acrylamide Initiated on the Addition of the Ammonium Persulfate which Forms the Free Reactive Radicals in the Water.

2.5.2 PAGE electrophoresis

PAGE separation could be conducted in the gel with the same content of the acrylamide in two different following gels, so called Laemmli electrophoresis, when the first gel contents lower percentage of the acrylamide and it is intended to the concentration of the sample at the begining of the separation (so called the concentration gel). The bigger sharpness of the zones in the gel is provided by means of the lower pH (of about two degree) against the surrounding setting. The itself separation takes place in the following part of the gel with the higher density (so called the separation gel). (Laemmli, 1970) The structure of the polymerization process see in the Fig. 6.

Other variation performs the creation of the gradient gel, where the concentration gradient of the polyacrylamide (from the part with lowe density to the part of higher density, in the direction of the separation) is created. The bigger sharpness of the gel zones of the molecules of the similar size is ensured in this arrangement.

The choice of pH of the used buffer by polymerization process, and, also the division of the molecules by the classical PAGE, because, the suitable buffer ensures the sufficient differences in the specific charge of the assorted parts of the protein mixture. The acid proteins require slightly alkaline or neutral pH (the molecules moves to the anode) and alkaline proteins require a slightly acid pH (the molecules migrate to the cathode) (Bernard et al., 1979).

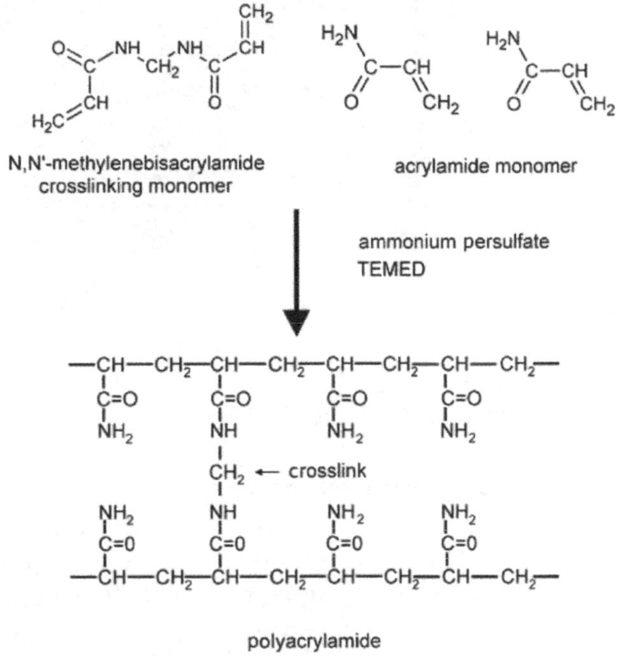

N,N'-methylenebisacrylamide
crosslinking monomer

acrylamide monomer

ammonium persulfate
TEMED

polyacrylamide

Fig. 6. Polymerization Process of the Structure Networking.

2.5.3 SDS-PAGE electrophoresis

The perfectly suitable modification of the PAGE electrophoresis is an arrangement in the –sodium dodecyl sulfate (abbrev. SDS, or NaDS), which makes an ability of the proteins to bind the SDS in amount of abouti 1,4 mg per 1 mg of the protein by means of the hydrofobic reaction. SDS carries a huge negative charge which enables to equalize the charge of the molecules, and, those, move in one direction in the electrophoretic gel in accordance of the molecular size. The complex SDS-protein unifies either the charge density, or, conformation on the surface of the complex, see the structure in the Fig. 7.

The mobility of the SDS-protein complex in the polyacrylamide gel is proportional to the logarithm of the molecular weight of an appropriate protein, which enables the gel calibration (Rédei, 2008). It is quiet convenient that the examined samples are adjusted before the whole process.

First, an appropriate buffer is added (e.g. TrisHCL) and SDS so that we have the same homogennous reaction setting.

Second, the glycerol is added, because it makes the settings in the gel more dense, so that the samples fill the sample holes properly and do not swirl. Glycerol also decreases the electroendoosmosis and makes the movement and distribution of the proteins even better.

Third, the bromophenol blue is added as the protein movement indicator. Fourth, dithiothreitol (abbrev. DTT) could be added to cleave the proteins to make an analysis more suitable. The samples could be also denatured in the hot water by the temperature of about 65 °C.

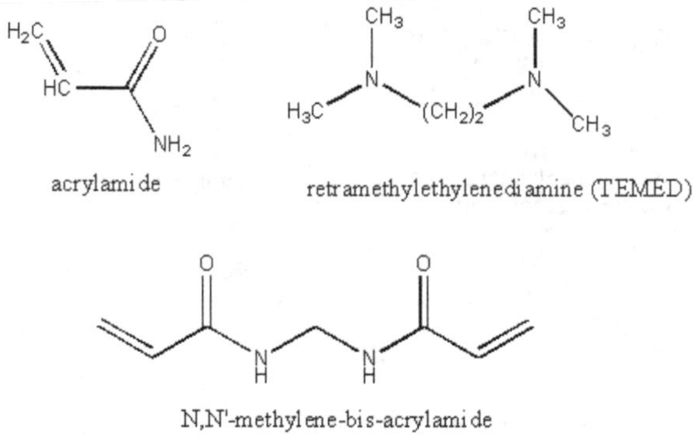

acrylamide   retramethylethylenediamine (TEMED)

N,N'-methylene-bis-acrylamide

Fig. 7. The Structural Formulae of the Substances in the SDS-PAGE.

2.5.4 Visualization and radiodetection in electrophoresis

The proteins can be visualized directly in gel after electrophoresis proceeding, or, subsequently Western Blot technique could be processed and detection is performed in the membrane where the proteins are transferred from gel. Adsorption of the pigment is used for visualization.

2.5.5 Staining in electrophoresis

A Silver Staining shows another alternative for dying of the proteins in gel. The silver ion is insoluble and colourless, and, distinguishes the places with protein and without proteins in the polyacrylamide gel (formation of the silver complexes with alkaline or sulphuric proteins).

After this procedure, the silver ions are reduced by formaldehyde into the form of the metal silver which is perfectly visible and insoluble. The amount of proteins, which could be visualized by this procedure, ranges from the hundreds of picograms to the units nanograms.

Another staining, which si possible for this purpose of detection, is dying by means of Coomassie Blue which is less sensitive (of about 50 times), but it has another advantage that Coomassie Blue is bound to the protein in the stechiometry ratio, and, therefore, it represents a quantitative densitometry detection (maximum absorbance ranges from 560 nm to 575 nm), see the Fig. 8. An autoradiography may be used as an alternative for detection in gel of the radiolabeled compounds. The differences between electrophoresis by non-reductive (see Fig. 9 and Tab. 1) and reductive conditions (see Fig. 10 and Tab. 2) are shown below.

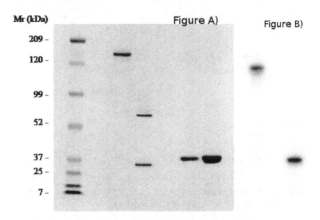

Fig. 8. Electrophoresis of the TU-20 and scFv TU-20 in the Gradient Gel by the Non-Reductive (–) and Reductive (+) Conditions Figure A) Gel Coloured by Coomassie Blue: 1. Molecular Marker; 2. TU-20 (–); 3. TU-20 (+); 4. scFv TU-20 (–); 5. scFv TU-20 (+). Figure B) Autoradiography: 1. [^{125}I]TU-20 (–); 2. [^{125}I]scFv TU-20 (–). .

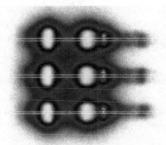

Fig. 9. Autoradiography of electrophoresis SDS-PAGE ^{125}I-TU-20 (all lines) by non-reductive conditions

Peak	Integral Density in PSL	% Ratio of Peak
^{125}I-TU-20	59237,2	46,7
BSA(I)	53228,4	42,0
BSA(II)	14319,4	11,3

Table 1. Autoradiographical interpretation of SDS-PAGE of ^{125}I TU-20 by non-reductive conditions. An autoradiographical visualization of the SDS-PAGE gel (which contains the radiolabeled antibody by non-reductive conditions) after developing on the luminiscent plate by means of the AIDA software.

2.5.6 Immunoblotting

Western Blot transfers the proteins, closed into the gel matrix, into the nitrocellulose membrane for further purposes of investigation after finishing of electrophoresis. Western Blot (Immunoblotting), used for the protein detection, transferes the proteins from the gel into the membrane by means of electrophoresis.

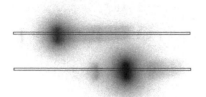

Fig. 10. Autoradiography of electrophoresis SDS-PAGE of ^{125}I-TU-20 and ^{125}I-scFv TU-20 by reductive conditions. An autoradiographical visualization of the SDS-PAGE gel (which contains the radiolabeled antibody and fragment by reductive conditions) after developing on the luminiscent plate by means of the AIDA software.

Peak	Integral Density in PSL	% Ratio of Peak
^{125}I-TU-20	63445,2	98,1
BSA(I)	1228,8	1,9
^{125}I-scFv TU-20	77988,6	97,6
BSA(I)	1909,7	2,4

Table 2. Autoradiographical interpretation of SDS-PAGE of ^{125}I-scFv TU-20 (upper line) and ^{125}I TU-20 (bottom line) by reductive conditions.

The particular proteins are subsequently indentified by the appropriate radiolabbeled antibodies (labeled by enzymatic reaction, or, by the radiolabeling reaction with ^{125}I). The proteins bound into the membrane could be submitted to the non-specific staining, or, as an alternative, to the autoradiography. After drying, the membrane is stored with much better results than dried gel.

When the electrophoresis with all its instruments and alternatives is used as a quality control method of the radiolabeled antibodies, the following parametres were proved and chosen for this setting as the most suitable. Stability of the radiolabeled TU-20 and its scFv TU-20 was investigated on 4 - 12 % Bis-Tris gel electrophoresis.

Protein bands were visualized by staining the gels with Silver Stain Plus. ^{125}I-labeled scFv fragment was processed by autoradiography exposing plate BAS-SR 2025, and finally developed by BAS-1800II. Autoradiographs were evaluated by AIDA 2.0 software, see the Fig. 11.

2.6 Immunohistochemistry testing

Preserved binding properties of the radiolabeled MAb or scFv for neuronal tissue were confirmed by the method of double labeling. It is based on the immunohistochemistry and autoradiography of the brain tissue slices. The 50 μm thick brain slices from the wild type mouse (C57B/6/J) were incubated with the radiolabeled TU-20. The second incubation was performed with anti-mouse IgG polyclonal antibody conjugated with horseradish peroxidase (Sigma-Aldrich, USA). Afterwards, the immunohistochemistry was finalized by staining with 3,3' – diaminobenzidine (DAB) that revealed the neuronal structure, see the Fig. 12 and 13.

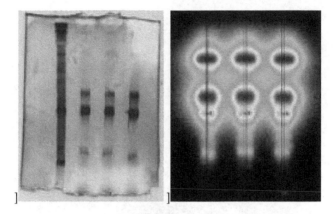

Fig. 11. Gel electrophoresis analysis of [^{125}I]TU-20 – autoradiography (a) and silver staining (b).

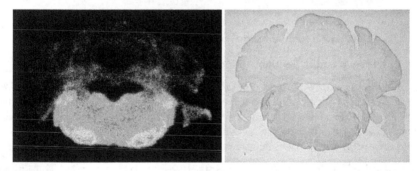

Fig. 12. [^{125}I]TU-20 autoradiographical - Figure A) ,and, immunohistochemical visualization of the bound radiolabeled antibody in the mice brain slice - Figure B) image of the coronal mice brain slice.

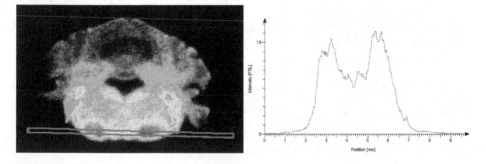

Fig. 13. Autoradiography visualization - Figure A), and, visualization of the 1D-interpretation of the bound radiolabeled antibody in the mice brain slice - Figure B) of the labeled mice brain slices by means of the software AIDA.

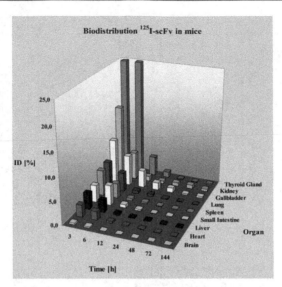

Fig. 14. [^{125}I]scFv biodistribution in normal mice.

2.7 In vivo preparative biodistribution testing in normal mice

The in vivo biodistribution was carried out with the male normal mice - wild type C57B/6/J. Biodistribution studies were performed following an i.v. injection. The main focus is intended for scFv fragment due to its better mobility in organism. ^{125}I-labeled scFv fragment, for comparison with the biodistribution of Na^{125}I, was applied in amount of 50 kBq/50 µl. ^{123}I-labeled scFv fragment was injected in amount of 200 kBq/50 µl.

Mice were sacrificed at designated times points in groups by 3 animals. The kinetic time intervals were: 3, 6, 12, 24, 48, 72, 144 hours for ^{125}I-labeled scFv TU-20 fragment and 0,5, 1, 2, 3, 6, 12 hours for ^{123}I-labeled scFv fragment.

Blood and major organs (included thyroid gland, kidneys, lung, heart, brain, spleen, muscle, fat, skin, gallbladder, testicles, stomach, liver, small intestine, and colon) were removed, weighed, and counted in a gamma scintillation counter to determine the % ID/g (percentage of injected dose per gram) for each radiolabeled substance.

The biodistribution figures are shown below, see the Fig. 14 and 15.

Blood clearance data for ^{125}I-labeled scFv fragment were obtained by analyzing blood samples by using a bi-exponential model for two-phase clearance to determine short phase half-life $t_{1/2\alpha}$ and long phase half-life $t_{1/2\beta}$ values.

2.8 In vivo SPECT imaging biodistribution testing

[^{123}I]scFv TU-20 and [^{123}I]TU-20 behavior in mice (wild type C57B/6/J) was observed by use of the SPECT camera. Kinetic intervals were 0.5, 1, 2, 3 h by [^{123}I]scFv TU-20 - see the Fig. 16 and 1, 2, 3, 6 h by [^{123}I]TU-20 - see the Fig. 17.

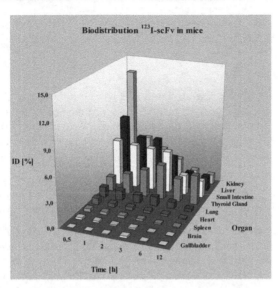

Fig. 15. [^{123}I]scFv biodistribution in normal mice.

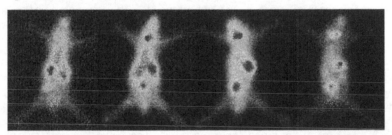

Fig. 16. [^{123}I]scFv TU-20 SPECT camera images –biodistribution study in kinetic intervals 0,5, 1, 2 and 3 h.

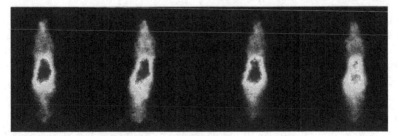

Fig. 17. ^{123}I]-TU-20 SPECT camera images –biodistribution study in kinetic intervals 1, 2, 3, 6 h.

2.9 In vivo biodistribution testing in genetically modified mice

Transgene population G93A1 Gur was used for comparative study to show different behavior of the substances in normal mouse and in modified organism with amyotrophic lateral sclerosis (ALS). Biodistribution kinetic intervals were 3 h (^{125}I-scFv) and 6 h (^{125}I-TU-20). (Heiman-Patterson T.D., 2005) (Naini A., 2007)

3. Conclusion

TU-20 and its scFv were labeled with ^{125}I and ^{123}I by chloramine-T (with average yield 0,72 and 0,50, resp.). Radiochemical purity and stability was revealed by gel filtration (decrease to 80 % and 50 % in two months, resp.) Fragmentation of the labeled antibody and its fragment was estimated by bis-tris gel electrophoresis followed by silver staining and autoradiography (over 95 % of radioactivity bound in the substances).

Affinity coupling and RIA adaptation for the specific conditions showed 10-30 % preserved immunoreactivity of the labeled compounds. Otherwise, these methods carry out quite high discrepancy and it will be neccessary to provide further optimising search.

In vitro studies performed on mice brain slices confirmed several important assumptions. The antibody is preferentially bound in the layer of Purkinje cells in the cerebellum. SPECT camera in vivo experiment deals with these results: activity bound in scFv is primarily distributed to the thyroid gland and digestive tract, then passes quickly through kidneys.

Distribution images of the labeled TU-20 provides ambigous because the substance is accumulated in the chest and ventral part and image resolution do not afford more detailed biodistribution identification. However, it is known from previous biodistribution preparative study that activity is distributed in lung, heart, liver, stomach and colon in first 6 h.

In vivo experiments were focused on investigation of the blood clearance and organ distribution of the radiolabeled TU-20 and scFv fragment in mice. Let´s show especially the results from scFv biodistribution study in preference. It was verified that the major part of activity, according to the amount of the labeled scFv fragment, was eliminated from blood during 2-3 hours. Minor part of activity, according to the amount of the labeled scFv fragment (0,5 - 1,0 %), was kept in the blood for some days. The value $t_{1/2\alpha}$ for ^{125}I-labeled scFv fragment was calculated as 2,3 h and the $t_{1/2\beta}$ was estimated as 62,4 h. The half-life for overall elimination of Na^{125}I from blood was 4,5 h.

In comparison, we found that the ^{125}I-labeled scFv fragment uptake in thyroid gland appeared much lower than for Na^{125}I, as expected. The $t_{1/2\alpha}$ value for ^{123}I-labeled scFv fragment was calculated as 1,4 h, but the long phase elimination half-life $t_{1/2\beta}$ was not estimated due to short half-life of the isotope ^{123}I. The radiolabeled scFv fragment passed in general through the digestive tract (stomach and intestine) and finally was eliminated through kidneys in preference.

TU-20 and ScFv TU-20 showed suitable properties for further investigation in animals which are genetically modified mutants with the ALS (Amyotrophic Lateral Sclerosis). Comparing biodistribution experiments in modified organism confirmed expected behavior. The most significant biodistribution differences occured in the area of the limbs and caudal part of spinal cord and spine.

Finally, as I can summarize, TU-20 and its scFv fragment were successfully labeled with radioiodine ^{123}I and ^{125}I, and, subsequently, the biochemical and analytical characteristics were investigated. Biological properties of the radiolabeled TU-20 and its scFv were evaluated in vivo by biodistribution studies.

The expected behavior of biomolecules during their elimination was observed. Furthermore, the elimination parametres were calculated. ^{125}I-labeling of the TU-20 and its scFv is very suitable for investigation of the radiolabeled antibody fragment behavior and properties due to the long ^{125}I half-life. On the other hand, ^{123}I-labeling of the scFv fragment TU-20 is intended for practical imaging at SPECT camera.

In summary, TU-20 shows better immunospecific behavior in organism together with slower kinetics, on the other hand, scFv TU-20 reveales worse immunospecific characteristics in combination with much faster kinetics.

4. Acknowledgement

This work was supported by the projects No. E!3177 - DIAGIM (1P040E167) and E!2510 - NEUROTUB (0E91) of EUREKA, and by the project No. IBS1048301 of Grant Agency CAS.

5. References

Bernard, A., Goret, A., Buchet, J. P., Roels, H. & Lauwerys, R. (1979). Comparison of sodium dodecyl sulfate-polyacrylamide gel electrophoresis with quantitative methods for the analysis of cadmium-induced proteinuria, *International Archives of Occupational and Environmental Health* 44: 139–148. 10.1007/BF00381129.
    URL: *http://dx.doi.org/10.1007/BF00381129*
Chizzonite, R., Truitt, T., Podlaski, F., Wolitzky, A., Quinn, P., Nunes, P., Stern, A. & Gately, M. (1991). Il-12: monoclonal antibodies specific for the 40-kda subunit block receptor binding and biologic activity on activated human lymphoblasts, *The Journal of Immunology* 147(5): 1548–1556.
    URL: *http://jimmunol.org/content/147/5/1548.abstract*
Dráberová, E., Lukáš, Z., Ivanyi, D., Viklický, V. & Dráber, P. (1998). Expression of class iii β-tubulin in normal and neoplastic human tissues, *Histochemistry and Cell Biology* 109: 231–239. 10.1007/s004180050222.
    URL: *http://dx.doi.org/10.1007/s004180050222*
Hamilton, S., O. J. W. G. K. J. (2002). Reducing renal accumulation of single-chain fv against melanoma-associated proteoglycan by coadministration of l-lysine, *Melanoma Res.* 12: 373–379.
Haugland, R. P. (1995). Coupling of monoclonal antibodies with enzymes, *in* W. C. Davis & J. M. Walker (eds), *Monoclonal Antibody Protocols*, Vol. 45 of *Methods in Molecular Biology*, Humana Press, pp. 235–243. 10.1385/0-89603-308-2:235.
    URL: *http://dx.doi.org/10.1385/0-89603-308-2:235*
Heiman-Patterson T.D., Deitch J.S., B. E. (2005). Background and gender effects on survival in the tgn(sod1-g93a)1gur mouse model of als, *J. Neurol. Sci.* 236: 1–7.
Holliger, P., H. P. (2005). Engineered antibody fragments and the rise of single domains, *Nat. Biotechnol* 23: 1126–1136.
Jones, N. A. (2004). Two-dimensional polyarcylamide gel electrophoresis for proteome analyses, *in* P. Cutler (ed.), *Protein Purification Protocols*, Vol. 244 of *Methods in Molecular Biology*, Humana Press, pp. 353–359. 10.1385/1-59259-655-X:353.
    URL: *http://dx.doi.org/10.1385/1-59259-655-X:353*
Katsetos, C.D., H. M. M. S. (2003). Class iii β-tubulin in human development and cancer. cell. motil. cytoskeleton, *Cell. Motil.* 55: 77–96.

Laemmli, U. (1970). Cleavage of structural proteins during the assembly of the head of bacteriophage t4, *Nature* 227: 680–685.

Naini A., Mehrazin M., L. J. G. P. M. H. (2007). Identification of a novel d109y mutation in cu-zn superoxide dismutase (sod1) gene associated with amyotrophic lateral sclerosis, *J. Neurol. Sci.* 254: 17–21.

Rédei, G. P. (2008). Electrophoresis, *Encyclopedia of Genetics, Genomics, Proteomics and Informatics*, Springer Netherlands, pp. 591–592. 10.1007/978-1-4020-6754-99_5192.
    URL: *http://dx.doi.org/10.1007/978-1-4020-6754-9_5192*

Robles, A. M., B. H. O. P. W. M. P. E. (2001). Improved radioiodination of biomolecules using exhaustive chloramine-t oxidation, *Nucl. Med. Biol.* 28: 999–1008.

Švecová, H., Kleinová, V., Seifert, D., Chaloupková, H., Bäurle, J., Kranda, K., Král, V. & Fišer, M. (2008). Radioiodination of mouse anti-iii β-tubulin antibodies and their evaluation with respect to their use as diagnostic agents for peripheral neuropathies, *Applied Radiation and Isotopes* 66(3): 310 – 316.
    URL: *http://www.sciencedirect.com/science/article/pii/S0969804307002813*

Vijayalakshmi, M. A. (1992). Histidine ligand affinity chromatography, *in* A. Kenney, S. Fowell & J. M. Walker (eds), *Practical Protein Chromatography*, Vol. 11 of *Methods in Molecular Biology*, Humana Press, pp. 33–44. 10.1385/0-89603-213-2:33.
    URL: *http://dx.doi.org/10.1385/0-89603-213-2:33*

Extraction and Electrophoresis of DNA from the Remains of Mexican Ancient Populations

Maria de Lourdes Muñoz et al*

Department of Genetics and Molecular Biology, Centro de Investigación y de Estudios
Avanzados del Instituto Politécnico Nacional, Mexico D. F.
Mexico

1. Introduction

Ten years ago, the first reports of human genome sequencing were published in Nature and Science (Venter et al., 2001; Sachidanandam et al., 2001; Lander, 2011). This was very exciting and expectations for the application of genome sequencing technology were high. In the past decade, the cost of sequencing has gone down several orders of magnitude, making it a more accessible technology for research studies. The medical value of comprehensive genome sequencing is now becoming apparent: for example, the genetic cause of a rare and debilitating vascular disorder was solved by genome sequencing at NIH (Jasny and Zahn, 2011; Lander, 2011). It is also possible to solve the genetics of individual Mendelian disorders thereby relating phenotype to genotype. In addition, better treatments for diseases such as cancer, metabolic disorders, inflammation, neurodegeneration or diabetes are expected to be found through studies involving genome sequencing (Lander, 2011). Sequencing also has been used to query variation in populations worldwide, and sequences are now available from extinct hominids as well as from thousands of other species (Rasmussen et al., 2010; Krause et al., 2010; Reich et al., 2010; Balter, 2010; Rasmussen et al., 2010). We expect to know very soon what variation exists among individuals at almost all sites in the genome. This is a great opportunity for population genetics to reconstruct the entire genealogical and mutational history of humans (Callaway, 2011), to understand the evolutionary and genetic forces that affected every region of the genome, to determine disease mutations present in human populations, to elucidate the genetic bases of cognitive and physiological adaptations, and/or to determine the demographic events that led to the colonisation of the earth.

The question remains: what is the relationship between morphological features and ancient deoxyribonucleic acid (aDNA)? The evolutionary processes that generated modern species

* Mauro Lopez-Armenta[1,2], Miguel Moreno-Galeana[1], Alvaro Díaz-Badillo[1], Gerardo Pérez-Ramirez[1],
Alma Herrera-Salazar[1], Elizabeth Mejia-Pérez-Campos[3], Sergio Gómez-Chávez[4]
and Adrián Martínez-Meza[5]
[1]*Department of Genetics and Molecular Biology, Centro de Investigación y de Estudios Avanzados del Instituto Politécnico Nacional; Mexico*
[2]*Posgrado en Ciencias Genómicas, Universidad Autónoma de la Ciudad de México, Mexico*
[3]*Instituto Nacional de Antropología e Historia, Querétaro, Mexico D.F., Mexico*
[4]*Teotihuacan, Mexico*
[5]*Mexico City, Mexico*

and populations are commonly inferred through the analysis of morphological and genetic markers in addition to analyses of contemporary organisms to create tentative reconstructions. To confirm this indirect evidence, it is necessary to check the reconstructions against the fossil records. Nevertheless, the comparison has been made possible now by analysing morphological characters, and the application of recent advances in deoxyribonucleic acid (DNA) sequencing technologies for aDNA are now allowing the genetic record to be generated. This new technology let us focus not only on single genetic loci, such as mitochondrial DNA (mtDNA), but it made possible to obtain whole genome sequences of extinct species and populations (Lander et al., 2011), our closest extinct relatives the Neanderthal (Green et al., 2010), and the extinct hominid group from Siberia, the Denisovans (Reich et al., 2010).

The field of aDNA was initiated more than twenty years ago (Higuchi et al., 1984; Cooper et al., 1992; Greenwood et al., 1999) and research efforts continue to grow and expand into new areas (Stoneking and Krause, 2011). The first aDNA studies demonstrated the inefficiency of bacterial cloning to amplify small sequences recovered from the skins of animals and human mummies (Higuchi et al. 1984; Pääbo, 1985) and showed that DNA was at very low concentrations of short damaged fragments. However, these studies are considered very important because they will elucidate population origins, migrations, relationships, admixture and changes in population size, essentially revealing the demographic history of the human population.

It is now accepted that DNA is preserved in ancient samples under a wide range of depositional environments (Willerslev and Cooper, 2005). Although the DNA of a deceased organism degrades rapidly, part of it may survive for more than 100,000 years under favourable conditions, such as cold, stable temperatures and a dry environment (Pääbo et al., 2004). Fortunately, the development of new technologies has made possible the recovery and manipulation of these molecules as well as the genetic characterisation of these samples. Because this DNA is degraded the analysis is complicated, nevertheless, the new sequencing technology makes it possible to obtain historical information. In addition, the presence of polymerase inhibitors makes DNA amplification exceedingly difficult. Research in this area shares a common problem with forensics and other approaches requiring analyses of museum and non-invasively collected specimens; the amount of endogenous DNA available in the samples is limited. In addition, when working with human samples it is also possible to have contamination from contemporary human DNA. Careful adherence to currently established procedures is necessary to avoid such contamination (Deguilloux et al., 2011).

Because aDNA contains the information of our past its analysis is of high importance. Here, we will review a variety of methods for extraction, purification, amplification and sequencing of aDNA segments informative for genetic population studies. Future prospects for the potential direction of ancient DNA research will be discussed. Furthermore, contributions to migratory theories will also be analysed based on population diversity, taking into account ancient mtDNA studies.

Although there is new technology to determine the sequence of nuclear DNA, we will focus on mtDNA analysis. mtDNA analysis has been very useful to extensively examine human population history throughout the world because of its relatively rapid rate of mutation, lack of recombination and maternal inheritance. Mitochondrial DNA sequence variations at the hypervariable regions HVI and HVII will be described and their importance in population genetic studies will be discussed. Technical differences between DNA extraction procedures for ancient bones and mummy tissue will also be described. Molecular phylogenetic analysis, haplotype and haplogroup determination through software will also be defined and examined.

2. Procedures to study ancient DNA

There have been several aDNA extraction protocols suggested over the years. The first method was purification based on phenol/chloroform extraction, alcohol precipitation (Kalmár et al., 2000; Munoz et al., 2003; Hagelberg and Clegg, 1991; Hänni et al., 1995) and silica binding (Höss and Pääbo, 1993; Yang et al., 1998). In addition, other methods have been suggested, such as using Chelex (Faerman et al., 1995), centricon filters (Anzai et al., 1999), Dextran Blue (Kalmár et al., 2000), decalcifying bone with EDTA (Hagelberg and Clegg, 1991; Hänni et al., 1995; Yang et al., 1998) and hybridisation and magnetic separation (Anderung et al., 2008). The methods most commonly used now combine EDTA decalcification and silica purification (Yang et al., 1998; Krings et al., 1997; Anzai et al., 1999). It is evident that many different techniques have been used, demonstrating that no single procedure has clear advantages. Based on our experience, the selected method is a function of the sample characteristics, including considerations for the origin of the sample, from the skeleton or a mummy.

2.1 Samples

Samples from this study include bones pertaining to pre-Hispanic populations from different periods of time (200 to 1500 years before present). Bone samples of two individuals from Monte Albán, Oaxaca, one from Teotihuacán and a tissue portion from the mummy Pepita were used in the examples presented in this study. To work with the ancient Mexican samples, we made a written agreement with the "Intituto Nacional de Antropologia e Historia" (Mexico). Research on ancient unidentifiable human remains is excluded from the requirement of ethics review by the Research Ethics Boards.

Sampling should be conducted as soon as the bones appear in excavation, and gloves, mask and coat must be used to prevent contamination from excavators. This is not always possible because some samples were collected before these studies were initiated. Samples also have to be deposited directly in hermetic sterile tubes and frozen at -70°C. These practices prevent the introduction of contaminant DNA during the sample collection. In addition, it is also very important to manipulate the sample in a sterile clean room, to use bleach and ultraviolet light to degrade potential contaminants and to keep strict physical separation of modern DNA work from aDNA (Miller et al., 2008, Cooper and Poinar, 2000).

2.2 Ancient DNA extraction

All DNA purification and PCR experiments were carried out under sterile conditions in separate dedicated rooms. Samples were handled wearing protective clothing from collection to DNA isolation, and the laboratory equipment and reagents are maintained DNA-free. The laboratory managing the ancient samples has a high-pressure system to filter the incoming air and a laminar flow hood as well as UV light irradiation and bleach were used to clean of every surface to avoid contamination (Knapp et al., 2011).

DNA purification was preceded by a decontamination step to eliminate surface exogenous DNA when samples were collected and manipulated by unknown people. Each sample was washed with bleach followed by a water rinse and UV light irradiation for 30 min on each face. Some authors suggest removing the surface of the bone, however this procedure may also contaminate the inside of the bone if it has some kind of porosity. When bones were collected as soon as they appeared during the exaction with the necessary equipment to prevent contamination from excavators (i.e., gloves, mask and coat), it was not necessary to treat the sample with the decontamination steps (Deguilloux et al., 2011) and the potential to damage template or impede the efficiency of PCR was therefore avoided. Overall, there is no way to guarantee complete removal of contaminant DNA through decontamination procedures, but these practices are used to eliminate as much contamination as possible.

Bone powder was generated by grinding in a mortar with pestle until a fine powder was obtained, when bone quantities were around 1 g. When the weight of the samples was ≤ 0.5 mg, the bone sample was ground under liquid nitrogen with a sterile screw cylinder modified from those suggested by Thomas M. G. and Moore L. J., (1997). The powder (0.250-0.500 g) was transferred into a sterile 15 ml tube and was suspended in 2 ml of extraction buffer (0.01 M Tris-HCl, 0.1 M EDTA and 0.2% SDS pH 8.0), and the tubes were capped and sealed with Parafilm. After incubation with gentle agitation for 1 h at 37°C, 1 mg/ml proteinase K was added, and the sample was incubated at 50°C for 2 h. A blank extraction treated identically to the experimental samples throughout the procedure was included to monitor for contamination during the DNA extraction process. Finally, the samples were centrifuged at 5,000×g for 5 min, and the supernatants were extracted using phenol-chloroform-isoamyl alcohol (24:24:1) organic extraction (Maniatis, et al., 1989; Munoz et al., 2003; Hughes et al., 2006). Subsequently, the aqueous phase was concentrated by precipitation by the addition of 0.1 volumes of 3 M sodium acetate at pH 5.0 and 2.5 volumes of ethanol. After mixing, the sample was incubated at -78°C overnight and centrifuged at 15,000 rpm for 10 min at 4°C. The supernatant was decanted, and the precipitate was rinsed with 70% ethanol. After drying the pellet at ambient temperature in a sterile area, the pellet was resuspended in 100 µl of high quality sterile water. Alternatively, the aqueous phase can be concentrated using Amicon® Ultra-0.5 30 kDa columns (Millipore, Billerica, USA), in a final volume of 40 µl.

Another method to extract the aDNA is by binding to silica: the powered sample (0.250 g) was suspended in 1 ml of extraction buffer (0.01 M Tris-HCl, 0.5 M EDTA pH 8.0) and after incubation at 37°C for 16 h, the suspension was incubated at 56°C for 3 h and centrifuged at 5,000xg for 2 min. The supernatant was transferred into 3 ml of binding buffer (5 M GuSCN, 0.025 M NaCl, 0.010 M Tris-HCl pH 8.0) in a 15 ml sterile conical tube and adjusted to pH 4.0 by adding 30% HCl in 25 µl aliquots. Then, the solution is passed through a QIAquick (Qiagen) silica column. The column was rinsed twice with the washing buffer (50% ethanol, 0.125 M NaCl and 0.010 M Tris and 0.001 M EDTA, pH 8.0) and dried for 15 min. Finally, aDNA was eluted from the column with 100 µl of TE buffer (0.01 M Tris-HCl, 0.001 M EDTA, pH 8).

Extracted DNA was kept in aliquots of 25 µl at -70 °C.

Ancient DNA can also be extracted by the Chelex-100 method: Extraction of DNA using Chelex1-100 (Bio-Rad Laboratories, CA, USA) was performed with 5% Chelex-100 in sterile H_2O using the protocol described by Walsh et al. (1991). Briefly, 200 µl of DNA extracted by phenol-chloroform-isoamyl alcohol (24:24:1) was boiled at 94 °C for 10 min with 5% Chelex-100 and centrifuged, and an aliquot of the supernatant was taken as the template for the PCR experiment.

2.3 Amplification of DNA from pre-Hispanic samples

Analysis in Native Americans of mtDNA by PCR amplification and high-resolution restriction analysis with 14 endonucleases (Torroni, et al., 1992; Torroni, et al., 1993; Torroni, et al., 1994a,b; Richards et al., 1996) identified four major mtDNA lineages or haplogroups (A-D). These haplogroups of Asian ancestry, each defined by specific polymorphisms, together encompass 96.9 % of the mtDNA observed in modern Native Americans. Each lineage is characterized by specific mtDNA marker: the 9-bp deletion in the COII/tRNAlys region (haplogroup B); a HaeIII restriction site gain at nucleotide position 663 of the reference sequence (haplogroup A) (Anderson et al., 1981); a HincII restriction site loss at nucleotide 13259 (haplogroup C); and an AluI restriction site loss at nucleotide 5176 (haplogroup D) (Wallace et al., 1985; Schurr et al., 1990; Torroni et al., 1992; Wallace and Torroni, 1992). Sequence data indicate a correspondence between each marker and particular hypervariable region I (HVI) mutations (Horai et al., 1993; Bailliet et al. 1994). Consequently, the mtDNA amplification of the specific region has to be performed to characterize the Native Americans (ancient and contemporary) populations. Primers to amplify HVRII were also included, although we did not included any example, because analysing the HVRII region is not as informative as the HVRI.

Enzymatic amplification by PCR was performed as described previously (Munoz et al., 2003; Campos, et al., 2011) using heat-resistant *Thermus aquaticus* (Taq) DNA polymerase (FINNZYMES), or Platinum® Taq High Fidelity (Invitrogen). The PCR parameters were as follows: 2.5 U of hot start DNA polymerase, 1X buffer, 2.5 mM $MgCl_2$, 200 μM of each dNTP, 0.25 mg/ml bovine serum albumin (BSA) and 0.2 μM of each primer in a total volume of 25 μl, and 5 μl of the aDNA template. The primers used to amplify and sequence human mitochondrial DNA were as follow:

HVR I:
L15975-15996 5′-CTCCACCATTAGCACCCAAAGC-3′;
H16401-16420 5′-TGATTTCACGGAGGATGGTG-3′ (Vigilant et al., 1989);
L16140-16159 5′-TACTTGACCACCTGTAGTAC-3′;
H16236-16255 5′-CTTTGGAGTTGCAGTTGATG-3′ (Wilson et al., 1995);
L15989-16008 5′-CCCAAAGCTAAGATTCTAAT-3′;
H16130-16152 5′-AGGTGGTCAAGTATTTATGGTAC-3′ (Eichmann and Parson, 2008);
L16094-16122 5′-TCGTACATTACTGCCAGYC-3′;
H16228-16248 5′-GTTGCAGTTGATGTGTGATAG-3′ (Eichmann and Parson, 2008);
L16190-16209 5′-CCCCATGCTTACAAGCAAGT-3′;
H16380-16398 5′-CAAGGGACCCCTATCTGAG-3′ (Poinar et al., 2001);

HVR II:
L8-29 5′-GGTCTATCACCCTATTAACCAC-3′;
H408-429 5′-CTGTTAAAAGTGCATACCGCC-3′ (Vigilant et al., 1989)

Haplogroup A:
L610-633 5′-TGAAAATGTTTAGACGGCCTCACA-3′;

H712-730 5'-CCAGTGAGTTCACCCTCTA-3' (Parr et al., 1996).
Haplogroup B:
L8196-8215 5'-ACAGTTTCATGCCCATCGTC-3';
H8297-8316 5'-CTGTAAAGCTAACTTAGCAT-3' (Wrischnik et al., 1987);
Haplogroup C:
L13198-13213 5'- GCAGCAGTCTGCGCCC -3';
H13384-13403 5'- ATATCTTGTTCATTGTTAA -3' (Lorenz and Smith, 1996)
(1996)
Haplogroup D:
L5101-5120 5'-TAACTACTACCGCATTCCTA-3';
H5230-5249 5'-TGCCCCCGCTAACCGGCTTT-3' (Stone and Stoneking, 1993)

All amplifications were carried out in a GeneAmp® PCR System 9700 thermocycler with the following profile: 5 min at 94°C, followed by 40 cycles of 1 min at 94°C, 1 min at 59°C (haplogroups A, D and RHVs) or 55 °C (haplogroups B and C), and 1 min at 72°C, with a final extension of 10 min at 72°C. At least one PCR blank was amplified alongside each batch.

The PCR products were visualised on 2% agarose gels with ethidium bromide, and all positive products were purified using the QIAquick kit (Qiagen) and sequenced using the BigDye® Terminator v3.1 kit (Applied Biosystems) in an ABI PRISM 310 genetic analyser.

2.4 Data analysis

Phylogenetic analysis. The sequences of the pre-Hispanic PCR products from the HVI segment were aligned with representative Amerindian mtDNA control-region sequences (GenBank accession numbers: AY195760, EU719927, EU719811, EU719679, EU720004, EU720308, EU720078, EU719797, AY195749, EU720177, EU720123, EU719764, HQ012155, EU720242, HQ012184, HQ012164, HQ012134, AY195772, EU720073, EU720339, AY195759, EU720071, EU720336, EU720102, EU720202, HQ012188, HQ012198, EU720029, HQ012255, HQ012254, HQ012253, GQ449339, EU034320, AY195748, AF214088, DQ973581, AF478614) and two ancient sequences from a prehistoric Oneota population (Stone and Stoneking, 1998) using the Clustal W program (Thompson et al., 1994). Then, the phylogenetic tree was constructed with the Jukes-Cantor method, and the distances were obtained from a neighbour joining algorithm. Finally, the tree was optimised for maximum likelihood, using Hy-Phy software (Kosakovsky-Pond et al., 2005).

Haplotype network analysis. The median-joining (ε=0) networks (Bandelt et al., 1999) of haplotypes were constructed using the Network package, v4.5.1.0 (Fluxus Engineering). Sequences used were those described for phylogenetic analysis. This method is for constructing networks from recombination-free population data.

3. Examples of ancient DNA extraction procedures

An example of DNA extraction from a pre-Hispanic sample is depicted in Figures 1 and 2. Figure 1 displays the contaminants with different colours during the phenol-chloroform-isoamyl alcohol procedure. Figure 2 shows the DNA extracted by the phenol-chloroform-isoamyl alcohol technique from 0.25 g of two powdered bone samples from the same individual. In this figure, we observe the sample contaminants that are one of the major obstacles to these studies because they inhibit the Taq polymerase. The contaminants, such as Maillard products of reducing sugars (Pääbo, 1989) and humic acids with phenolic

groups, were observed by fluorescent stain in blue while the DNA degraded (which results in a smear pattern) is stained in pink by ethidium bromide (Figure 2, panel A). Figure 2, panel B shows the second sample in lane 3, which displays only contaminants. The DNA was not apparent. These compounds can be partially eliminated using kits such as the Amicon® Ultra-0.5 30 kDa columns. These results show variation in DNA yields between extracts taken from different samples of the same bone, even when using the same extraction method. We attribute such differences to heterogeneity within the bones.

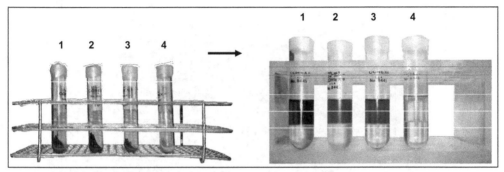

Fig. 1. Extraction of aDNA of bone samples from four different pre-Hispanic samples using the phenol-chloroform-isoamyl alcohol technique.

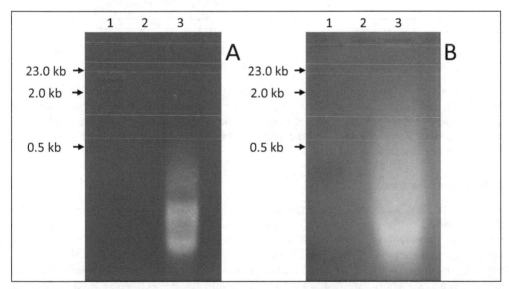

Fig. 2. Extraction of DNA of two independent samples (A, B) from the same pre-Hispanic individual by the technique of phenol-chloroform-isoamyl alcohol and ethanol precipitation. Lanes 1, molecular weight markers of *HindIII*; Lanes 2, no-sample; lane 3, DNA extracted from sample 1 of the Mexican pre-Hispanic population from Monte Alban.

Because DNA concentration is only possible with limited precision and concentrations of standard dilution series change over time in storage, we evaluated the relative performance of the DNA during PCR amplification using serial dilutions of the extracted DNA starting from 5 µl. Using this method, we were able to dilute the inhibitors of the Taq DNA polymerase.

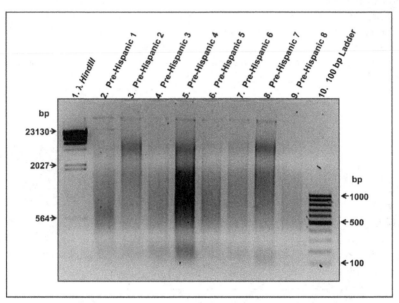

Fig. 3. Extraction of DNA of pre-Hispanic samples by the silica technique. Lanes 1 and 10, molecular weight markers of 23 kbp and 100 bp, respectively. Lanes 2 to 9, DNA extracted from different samples of Mexican pre-Hispanic populations.

Silica gel was also used to purify aDNA, results are shown in Figure 3. Each lane of this figure (2-9) displays aDNA extracted from 0.25 g of different pre-Hispanic bone samples. Lanes 1 and 10 show molecular weight markers. The quantity of Taq polymerase inhibitors is not evident, although we know that all ancient samples contain some of these inhibitors in different concentrations.

4. PCR performance

To study the effects of the Taq polymerase inhibitors, we added decrease quantities of the DNA extracted by the phenol-chloroform-isoamyl alcohol technique (shown in Figure 1) to the amplification reaction of the hypervariable segment I (15975-16420) using contemporary DNA. Figure 4, lane 1 displays the 100 bp molecular weight marker, lane 2 the negative control, lane 3 the PCR product of the contemporary DNA with the aDNA without dilution, lanes 4, 5, 6 and 7 show aDNA diluted 1:1, 1:2, 1:4 and 1:8, respectively, added to the PCR reaction mix and lane 8 contemporary DNA without any addition (positive control). Contaminated DNA allowed positive amplifications when aDNA was diluted at least 1:4 (Figure 4, lane 6). Therefore, an additional method to obtain the PCR product from the aDNA is by sample dilution.

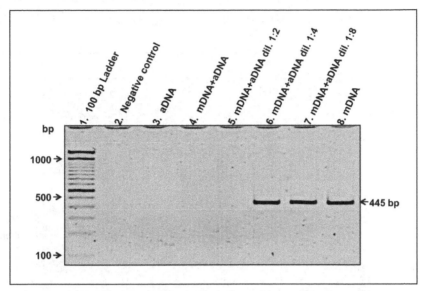

Fig. 4. Inhibition of the mitochondrial DNA hypervariable segment I amplification via inhibition of Taq polymerase by aDNA contaminants.

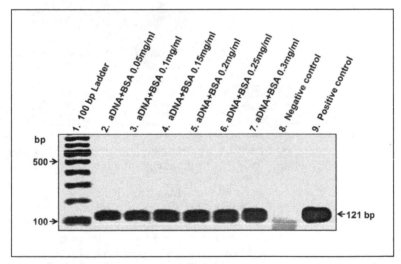

Fig. 5. Positive effect of bovine serum albumin (BSA) on the PCR performance of aDNA extracted by the phenol-chloroform-isoamyl alcohol procedure.

Sometimes aDNA dilution is not enough to obtain the PCR products, so we tested the effect of BSA addition by increasing the concentration of BSA in the reaction from 0.1 to 0.25 mg/ml. Figure 5 shows the positive effect of BSA on the PCR of aDNA extracted by the phenol-chloroform-isoamyl alcohol procedure. Increased amplification was observed in the PCR experiments when BSA was added in increasing concentrations (Figure 5, lanes 2-7, BSA at a concentration of 0.05, 0.1, 0.15, 0.2, 0.25 and 0.3 mg/ml, lane 8, negative control and lane 9, positive control). Based on these results, we added 0.25 mg/ml BSA to all PCR experiments.

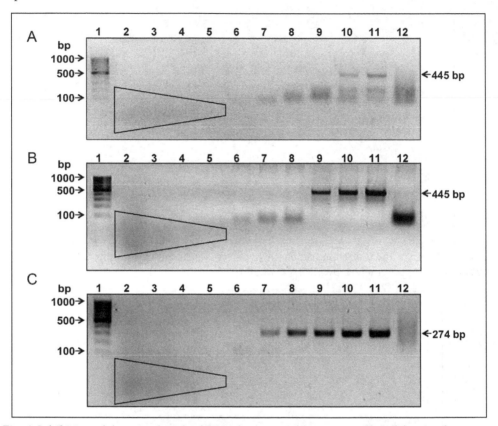

Fig. 6. Inhibition of the mitochondrial DNA hypervariable segment I amplification by inhibition of the Taq polymerase with aDNA contaminants and the effect of PVP.

Soils with high organic contents have humic acids with phenolic groups that denature biological molecules by bonding to N-substituted amides or oxidise to form a quinone that bonds to DNA or proteins (Young et al., 1994). Because aDNA contains these Taq polymerase inhibitors from soil, we tested the effect of Polyvinylpyrrolidone (PVP) during DNA extraction as has been suggested previously (Young et al., 1994; Rohland and Hofreiter, 2007). In addition, to make sure that PVP did not inhibit the PCR experiment, the reagent was added directly to the amplification mix. Figure 6 shows that 2% PVP added during the DNA extraction had a positive effect on DNA amplification. PCR amplification of contemporary DNA containing different dilutions of aDNA (1:1, 1:2, 1:3, 1:4, 1:8, 1:16, 1:32, 1:64, 1:128; Figure 6, lanes 3 to 10, respectively) and 0.25 mg/ml BSA is shown in Figure 6, panel A, and amplifications using the same conditions in the presence of 2% PVP during aDNA extraction is shown in Figure 6, panel B. The use of 2% PVP during aDNA extraction resulted in amplification at an aDNA dilution of 1:64, in contrast to the aDNA sample without PVP in which amplification is only observed at 1:128 dilution or beyond.

The HVI mtDNA segment of 445 bp was amplified in Figure 6, panels A and B with the primers L15975-15996 and H16401-16420. When the primers L15975-15996 and H16228-16248 were used, the PCR product is shorter (273 bp), and the presence of PVP makes evident the PCR fragment at a dilution of 1:16 (Figure 6, panel C). In addition, PVP at a concentration of 0.4% in the PCR experiment did not inhibit amplification, as was previously published (Young et al., 1993). The molecular weight marker is in lane 1; positive control with contemporary DNA alone is in lane 11; and the negative control with no DNA is in lane 12. Nevertheless, the positive effect was not evident in all aDNA bone samples, likely because the amount of Taq polymerase inhibitors is different in each sample.

Contemporary DNA is very easy to amplify. However, when working with aDNA, the PCR reaction efficiency is greatly reduced. For example, in Figure 7, we show a PCR-amplified fragment of mDNA using the human-specific primers L15975-15996 and H16236-16255 where 4 of the 7 bone samples from the pre-Hispanic populations displayed the PCR product (Figure 7, lanes 2 to 8). Positive and negative controls are shown in lanes 9 and 10, respectively, and the molecular weight marker is shown in lane 1.

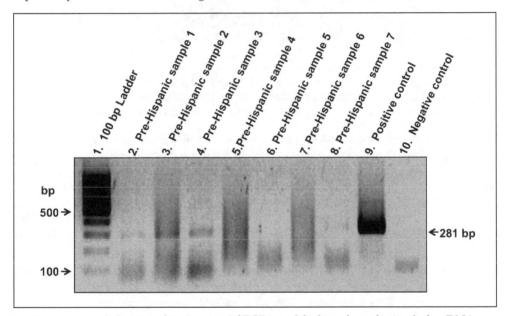

Fig. 7. Agarose gel showing the amounts of PCR-amplified product obtained after DNA extraction with the phenol-chloroform-isoamyl alcohol procedure.

Using the procedure indicated in the methods section, we purified and amplified the DNA from 14 bone samples of pre-Hispanic Native Americans to type them for haplogroup A described for Amerindians. The PCR products obtained were digested by the restriction enzyme *HaeIII*. Haplogroup A was detected in the 10 samples typed (Figure 8, lanes 2-11). Partial restriction digestion was observed in all of the ancient PCR products. This finding suggests the presence of the Amerindian polymorphism; however, we must sequence these amplification fragments or use real-time PCR to confirm the presence of the specific polymorphism because the partial restriction observed.

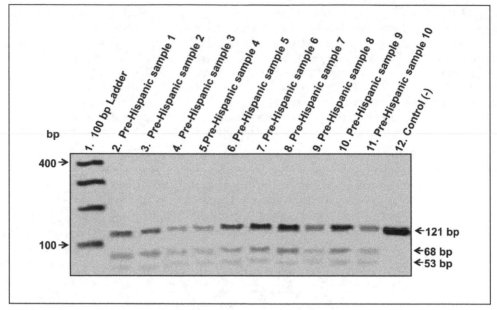

Fig. 8. Agarose gel showing the amounts of PCR-amplified product obtained after DNA extraction with the phenol-chloroform-isoamyl alcohol procedure and digested with the restriction enzyme *HaeIII*. Primers for amplification were specific to type haplogroup A.

Amplification of DNA extracted by Chelex was tested in the samples from pre-ceramic bones (Figure 9). Although experiments that compared the phenol-chloroform and Chelex method concluded that the Chelex method was simple and fast, inhibitory substances had not been eliminated in most of the cases (Kalmár et al., 2002). In our experience, DNA extracted by the phenol-chloroform method followed by Chelex treatment may improve DNA purification. Nevertheless, the silica method was better overall in our experience. The amplification products are observed at DNA dilutions of 1:30 in all samples, as shown in Figure 9.

Figure 10 displays the PCR amplification fragments using the specific primers L15975-15996 and H16236-16255 producing a fragment of 281 bp in panel A and L16140-16159 and H16380-16398 producing a fragment of 259 bp in panel B. We compared the amplification of aDNA extracted by the silica procedure and phenol-chloroform-isoamyl alcohol. Our results showed that aDNA extraction with the silica procedure was better than the phenol-chloroform-isoamyl alcohol method in this specific sample from pre-Hispanic populations because the amplification was observed exclusively in samples in which the DNA was extracted by the silica method. However, this may not be the case for all types of samples, and it is important to consider that when one method does not give good results, other methods may be useful.

Figure 10 shows aDNA from pre-Hispanic samples extracted by the silica gel method (lanes 2 and 3) compared with the phenol-chloroform-isoamyl alcohol procedure followed by concentration of aDNA with filter units (Centricon®) (lanes 4-5).

Ancient DNA was added to the PCR mix without any dilution (lane 2), diluted 1:10 (lane 3); aDNA phenol-chloroform-isoamyl alcohol extracted (lane 4); same procedure but diluted 1:10 (lane 5); washing buffer of the Centrifugal filter units (Centricon®) that were used to purify and concentrate aDNA in the phenol-chloroform-isoamyl alcohol purification procedure (lane 6) and diluted 1:10 (lane 7); negative control without DNA (lane 8); positive control with contemporary DNA (lane 9); and no sample (lane 10).

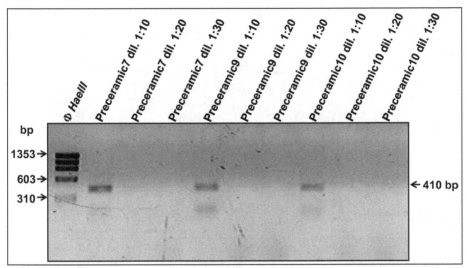

Fig. 9. Amplification of mtDNA HVI segment of 410 bp with specific primers (L15975-15996 and H16380-16398) using aDNA from pre-ceramic samples extracted with phenol-chloroform-isoamyl alcohol procedure followed by treatment with Chelex. Lanes 1 and 10, molecular weight markers φ. Lanes 2-4 pre-ceramic 7 diluted 1:10; 1:20 and 1:30; lanes 1-7, pre-ceramic 9 same dilutions as in lanes 2-4; lanes 8-10, pre-Ceramic 10, same dilutions as lanes 2-4.

When extracting DNA from small, degraded forensic samples or degraded ancient samples, the final concentration of DNA is usually too low for subsequent amplification. Consequently, we concentrated the aDNA samples extracted by the phenol-chloroform-isoamyl alcohol with filter units (lanes 2-3 and 4-5). Figure 10 shows clearly how aDNA was amplified using the DNA that was concentrated with the filter units. Furthermore, the washing buffer did not show any amplification, confirming in part that we do not have DNA contaminating our assays.

Next, we wanted to test all these methods with tissue from different mummies and determine the differences in using internal tissue and skin. From our results, we observe that when the mummy tissue is compact and from an internal organ the quantity of aDNA is very high compared with that obtained from bone samples. In addition the aDNA from the internal tissue was better as far as content is concerned. We had the opportunity to obtain DNA from the internal tissue of the mummy called Pepita that was intact and had no contamination by contemporary DNA. We were able to amplify the HVI segment using the specific primers for a mtDNA fragment of 445 bp (L15975-15996 and

H16401-16420, Figure 11, lane 8) and a fragment of 281 bp (L15975-15996 and H16236-16255, Figure 11, lane 1) or to amplify the specific second segment of HVI (L16140-16159 and H16401-16420, lane 5). The aDNA from this mummy was very well conserved. We have previously published DNA extraction from Mexican mummies with different origin and age (López-Armenta et al., 2008; Bustos-Ríos et al., 2008; Herrera-Salazar et al., 2008).

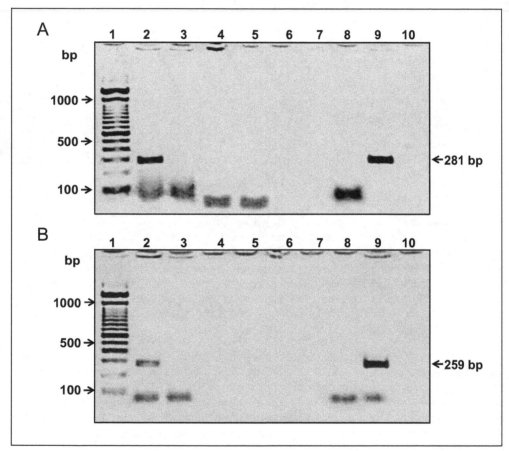

Fig. 10. Amplification of mtDNA HVI segments with the specific primers L15975-15996 and H16236-16255 producing a fragment of 281 bp (A); and L16140-16159 and H16380-16398 producing a fragment of 259 bp (B).

To examine the relationships between mtDNA lineages found in ancient and contemporary Native Americans, phylogenetic trees were constructed with the Jukes-Cantor method, and the distances were obtained from a neighbour joining algorithm and optimised for maximum likelihood, using Hy-Phy software (Kosakovsky-Pond et al., 2005). A total of 290 bp (nucleotides 16104–16394) of the HVI common to all sequences were used for these analyses. Sequences from Monte Alban and Teotihuacán from this study as well as those from the Oneota population were clustered in the haplogroup D linage.

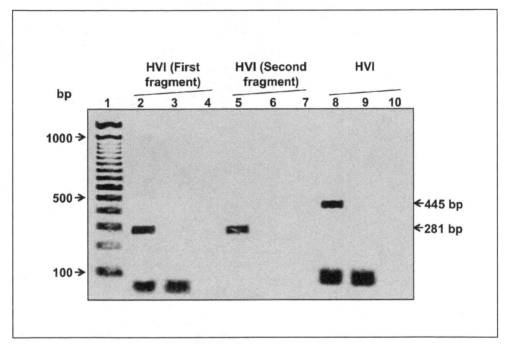

Fig. 11. Amplification of aDNA extracted from a Mexican mummy.

Haplotype network analyses were carried out on 290 bp of the mtDNA HVI from nucleotides 16104 to 16394. These networks were constructed using the Network package, v4.5.1.0 (Fluxus Engineering). These analyses included sequences from our own work and from other authors. The accession numbers of the sequences included in this network analyses were mentioned in the data analysis section. The pre-Hispanic DNA sequences included two ancient sequences from the prehistoric Oneota population (Stone and Stoneking, 1998), two sequences from Monte Alban, Oaxaca, Mexico and one from Teotihuacán, Mexico. Interestingly, the haplotype from the Oneota sequence may be derived from the Teotihuacán haplotype. The sequences from Monte Alban were grouped in the same haplotype as the more frequent haplotype from Native American populations. These results showed the potential to know the relationship among all Mexican pre-Hispanic populations or other populations as well as some haplotypes that were lost through the time.

It is important to mention that we never observed contaminant fragments with the specific HVR-1 mutations carried by the excavators or the geneticists. Therefore we are confident that following the procedures recommended by previous authors and our laboratory generates authentic sequences. Problems arise when the samples come from museums or collections where the researcher does not know how they were managed. In these conditions, additional controls are recommended for all of the procedures.

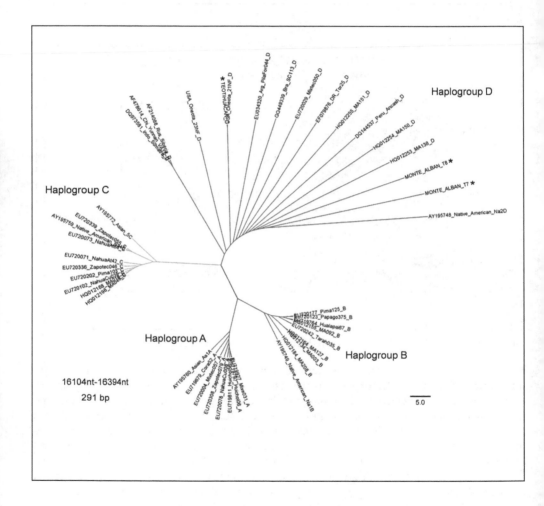

Fig. 12. Phylogenetic analyses of American native populations including five sequences from samples of pre-Hispanic populations. Tree of Native American and ancient pre-Hispanic Amerindian, constructed with the Jukes-Cantor method, and the distances were obtained from a neighbour joining algorithm and optimised for maximum likelihood using Hy-Phy software (Kosakovsky-Pond et al., 2005). The lanes in different colours indicate the haplogroup designation of lineages. Sequences of this study are marked with an asterisk.

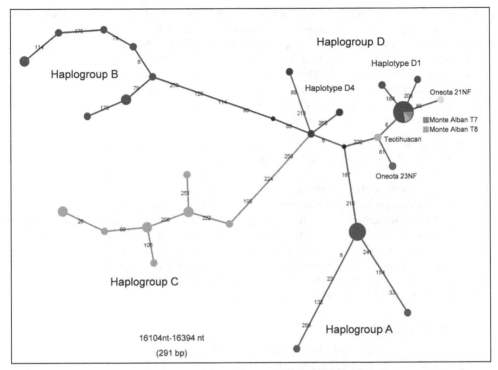

Fig. 13. Haplotype network of American native populations including five sequences from samples of pre-Hispanic populations. Each haplotype is represented by a circle in which the square radius (surface) is proportional to its population frequency. Circle colours show the site location as represented by the indicated colours. Dark circles without number in the network indicate mutational steps between haplotypes (theoretically extinct or unrepresented in the sample). Numbers between haplotypes represent mutational steps.

5. Conclusion

This review offers a direct overview of the different methods of aDNA extraction, including all special conditions needed in the laboratory to avoid contamination by contemporary DNA. It reveals the complexity involved in demonstrating the authenticity of human aDNA because the risk of contamination is very high. However, exogenous DNA contamination can be avoided if the necessary care is taken. In our experience and the experience of other laboratories, obtaining the ancient sample with coat, gloves and mask, and maintaining it in sterile conditions without human contact reduces the chances of sample contamination. It is also very important to test all reagents to verify that they are free of contemporary DNA. In addition, we also recommend performing negative control PCR experiments with at least 45 cycles to convincingly demonstrate the absence of contemporary DNA contamination. In our experience the best method to purified aDNA is phenol-chloroform-isoamyl alcohol with concentration using Amicon® Ultra-0.5 30 kDa columns (Millipore, Billerica, USA) or the Silica gel method using the QIAquick (Qiagen) columns. We also prefer to include the EDTA in the extraction buffer to optimise the aDNA extraction. This is supported by recent publications that have demonstrated that some DNA may be lost during decalcification (Campos et al., 2011). It is also important to keep DNA at -70°C in aliquots to maintain its integrity. Maintaining bone tissue samples at -70°C during aDNA extraction is useful to avoid additional DNA degradation. In our point of view, the best method will be that containing the least sample manipulation because this will avoid DNA contamination. Finally, there will be always risk of contamination by contemporary human DNA; however, next generation sequencing methods do provide a greatly improved means of measuring the degree of contamination in a sample.

Sequencing of the PCR products from aDNA as well as phylogenetic and network analyses of remains from America would allow testing of the hypotheses concerning single versus multiple waves of migration to the New World. This analysis will also reveal new haplotypes that were lost through time because not all migrations were successful in terms of leaving descendants among contemporary populations. Furthermore, the development of next generation sequencing is revolutionising aDNA research. The examples presented in Figure 13 and 14 display the relationship between the Oneota sample and that from Teotihuacán showing different haplotypes. There were also two ancient samples from Monte Albán that were grouped with the more frequent haplotype in the D1 haplogroup. Further analysis of more pre-Hispanic human samples will give us more detailed information about the history of these populations.

6. Acknowledgment

This work was supported by Instituto de Ciencia y Tecnología del Distrito Federal, México grant Clave: PICSA10-189 (01/11/2010-2011).

7. References

Anderung, C., Persson, P., Bouwman, A., Elburg, R., & Götherström, A. (2008). Fishing for ancient DNA. *Forensic Sci. Int. Genet. Vol.* 2, pp. 104-107.

Balter, M. (2010). Human evolution. Ancient DNA from Siberia fingers a possible new human lineage. *Science.* Vol. 327, No. 5973, pp. 1566-1567.

Bailliet G, Rothhammer F, Carnese FR, Bravi CM, & Bianchi NO. (1994). Founder mitochondrial haplotypes in Amerindian populations. *Am. J. Hum. Genet.* Vol. 54, No. 1, pp. 27-33.

Bandelt, H. J., Forster, P., & Röhl, A. (1999). Median-Joining Networks for Inferring Intraspecific Phylogenies. *Mol. Biol. Evol.* Vol. 16, No. 1, pp. 37-48.

Bustos-Ríos, D., López-Armenta, M., Moreno-Galeana, M. A., Herrera-Salazar, A., Pérez-Campos, E. M., Chávez-Balderas, X., & Muñoz, M. L. (2008). Purification of DNA from an ancient child mummy from Sierra Gorda, Queretaro. Mummies and Science. *World Mummies Research/Proceediñngs of the VI World Congress on Mummy Studies (Teguise, Lanzarote=; Pablo Atoche,* Conrado Rodríguez and Ma. Angeles Ramírez (eds.) Santa Cruz de Tenerife: Academia Canaria de la Historia (etc.), p 700. ISBN: 978-84-612-5647-1 Edition: Academia Canaria de la Historia, Ayuntamiento de Teguise, Cabildo Insular de Lanzarote, Caja Canarias, Fundción Canaria Mapfre Guanarteme, Universidad de las Palmas de Gran Canaria. First Edition, pp. 259-266.

Callaway, E. (2011). Ancient DNA reveals secrets of human history. *Nature*. Vol. 476, No. 7359, pp. 136-137. doi:10.1038/476136a

Campos, P. F., Craig, O. E., Turner-Walker, G., Peacock, E., Willerslev, E., & Gilbert, M. T. (2011). DNA in ancient bone – Where is it located and how should we extract it? *Ann. Anat.* [Epub ahead of print]

Cooper, A., Mourer-Chauviré, C., Chambers, G.K., van Haeseler, A. Wilson, A.C., & Pääbo, S. (1992). Independent origin of New Zealand Moas and kiwis. *Proc. Natl. Acad. Sci. USA*. Vol. 89, pp. 8741-8744.

Cooper, A., & Poinar, H. N. (2000). Ancient DNA: do it right or not at all. *Science*. Vol. 289, No. 5482, pp. 1139.

Deguilloux, M. F., Ricaud, S., Leahy, R., & Pemonge, M. H. (2011). Analysis of ancient human DNA and primer contamination: one step backward one step forward. *Forensic Sci. Int.* Vol. 210 No. 1-3, pp. 102-109.

Eichmann, C., & Parson, W. (2008). 'Mitominis': multiplex PCR analysis of reduced size amplicons for compound sequence analysis of the entire mtDNA control region in highly degraded samples. *Int. J. Legal. Med.* Vol. 122, No. 5, pp. 385-388.

Faerman, M., Filon, D., Kahila, G., Greenblatt, C. L., Smith, P., & Oppenheim, A. (1995). Sex identification of archaeological human remains based on amplification of the X and Y amelogenin alleles. *Gene*. Vol. 167, No. 1-2, pp. 327-332.

Green, R. E., Krause, J., Briggs, A.W., Maricic, T., Stenzel, U., Kircher, M., Patterson, N., Li, H., Zhai, W., Fritz, M. H., Hansen, N. F., Durand, E. Y., Malaspinas, A. S., Jensen, J. D., Marques-Bonet, T., Alkan, C., Prüfer, K., Meyer, M., Burbano, H. A., Good, J. M., Schultz, R., Aximu-Petri, A., Butthof, A., Höber, B., Höffner, B., Siegemund, M., Weihmann, A., Nusbaum, C., Lander, E. S., Russ, C., Novod, N., Affourtit, J., Egholm, M., Verna, C., Rudan, P., Brajkovic, D., Kucan, Z., Gusic, I., Doronichev, V. B., Golovanova, L. V., Lalueza-Fox, C., de la Rasilla, M., Fortea, J., Rosas, A., Schmitz, R. W., Johnson, P. L., Eichler, E. E., Falush, D., Birney, E., Mullikin, J. C., Slatkin, M., Nielsen, R., Kelso, J., Lachmann, M., Reich, D., & Pääbo S. (2010). A draft sequence of the Neandertal genome. *Science*. Vol. 328, No. 5979, pp. 710-722.

Greenwood, A. D., Capelli, C., Possnert, G., & Pääbo, S. (1999). Nuclear DNA sequences from late pleistocene megafauna. *Mol Biol Evol*. Vol. 16, No. 11, pp. 1466-1473.

Hagelberg, E., & Clegg, J. B. (1991). Isolation and characterization of DNA from archaeological bone. *Proc Biol Sci*. Vol. 244, No. 1309, pp. 45-50.

Hänni, C. T., Brousseau, V., Laudet, D. & Stehelin, D. (1995). Isopropanol precipitation removes PCR inhibitors from ancient bone extracts. *Nucleic Acids Res*. Vol. 23, No. 5, pp. 881-882.

Herrera-Salazar, A., Bustos, Ríos, D., López-Armenta, M., Moreno-Galeana, M. A., Martínez-Meza, A., & Muñoz, M. L. (2008). Mitochondrial DNA analysis of mummies from the North of México. *World Mummies Research/Proceediñngs of the VI World Congress on Mummy Studies (Teguise, Lanzarote; Pablo Atoche*, Conrado Rodríguez and Ma. Angeles Ramírez (eds.) Santa Cruz de Tenerife: Academia Canaria de la Historia (etc.), p 700. ISBN: 978-84-612-5647-1. Edition: Academia Canaria de la Historia, Ayuntamiento de Teguise, Cabildo Insular de Lanzarote, Caja Canarias, Fundción Canaria Mapfre Guanarteme, Universidad de las Palmas de Gran Canaria. First Edition, pp. 417-422.

Higuchi R, Bowman B, Freiberger M, Ryder OA, & Wilson AC. (1984). DNA sequences from the quagga, an extinct member of the horse family. *Nature*. Vol. 312, No. 5991, pp. 282-284.

Horai, S., Murayama, K., Hayasaka, K., Matsubayashi, S., Hattori, Y., Fucharoen, G., Harihara, S., Park, K. S., Omoto, K., & Pan, I.H. (1996). mtDNA polymorphism in East Asian populations, with special reference to the peopling of Japan. Am. J. Hum. Genet. Vol. 59, No. 3, 579-590.

Höss, M. & Pääbo, S. (1993). DNA extraction from Pleistocene bones by a silica-based purification method. Nucleic Acids Res. Vol. 21, No. 16, pp. 3913-3914.

Hughes, S., Hayden, T. J., Douady, C. J., Tougard, C., Germonpré, M., Stuart, A., Lbova, L., Carden, R. F., Hänni, C., & Say, L. (2006). Molecular phylogeny of the extinct giant deer, Megaloceros giganteus. Mol. Phylogenet. Evol. Vol. 40, pp. 285-291.

Jasny, B. R., & Zahn, L. M. (2011). Genome-sequencing anniversary. A celebration of the genome, part I. Science. Vol. 331, No. 6017, pp. 546.

Kalmár, T., Bachrati, C. Z., Marcsik, A., & Raskó, I. (2000). A simple and efficient method for PCR amplifiable DNA extraction from ancient bones. Nucleic Acids Res. Vol. 28, No. 12, E67.

Knapp, M., Clarke, A. C., Horsburgh, K. A., & Matisoo-Smith, E. A. (2011). Setting the stage –building and working in an ancient DNA laboratory. Ann. Anat. [Epub ahead of print].

Kosakovsky-Pond, S. L., Frost, S. D. W., & Muse, S. V. (2005). HyPhy: hypothesis testing using phylogenies. Bioinformatics. Vol. 21, pp. 676-679.

Krause, J., Fu, Q., Good, J. M., Viola, B., Shunkov, M. V., Derevianko, A. P., & Pääbo, S. (2010). The complete mitochondrial DNA genome of an unknown hominid from southern Siberia. Nature. Vol. 464, No. 7290, pp. 894-7.

Krings, M., Stone, A., Schmitz R. W., Krainitzki, H., Soneking, M., & Pääbo, S. (1997). Neandertal DNA sequences and the origin of modern humans. Cell. Vol. 90, No. 1, pp. 19-30.

Lander, E. S. (2011). Initial impact of the sequencing of the human genome. Nature. Vol. 470, No. 7333, pp. 187-197.

Lorenz, J., & Smith, D. G. (1996). Distribution of four founding mtDNA haplogroups among native North Americans. Am. J. Phys. Anthropol. Vol. 101, No. 1, pp. 307-323.

López-Armenta, M., Bustos, Ríos, D., Moreno-Galeana, M. A., Herrera-Salazar, A., Pérez-Campos, E. M., Chávez-Balderas, X., & Muñoz, M. L. (2008) Genetic origin of a mommy from Queretaro (Pepita). Mummies and Science. World Mummies Research/Proceediñngs of the VI World Congress on Mummy Studies (Teguise, Lanzarote=; Pablo Atoche, Conrado Rodríguez and Ma. Angeles Ramírez (eds.) Santa Cruz de Tenerife: Academia Canaria de la Historia (etc.), p 700. ISBN: 978-84-612-5647-1. Edition: Academia Canaria de la Historia, Ayuntamiento de Teguise, Cabildo Insular de Lanzarote, Caja Canarias, Fundción Canaria Mapfre Guanarteme, Universidad de las Palmas de Gran Canaria. First edition, pp. 251-258.

Maniatis, T., Fritsch, E. F., & Sambrook, J. (1989). Molecular Cloning. Cold Spring Harbor Laboratory Press.

Miller, W., Drautz, D. I., Ratan, A., Pusey, B., Qi, J., Lesk, A. M., Tomsho, L. P., Packard, M. D., Zhao, F., Sher, A., Tikhonov, A., Raney, B., Patterson, N., Lindblad-Toh, K., Lander, E. S., Knight, J. R., Irzyk, G. P., Fredrikson, K. M., Harkins, T. T., Sheridan, S., Pringle, T., & Schuster, S. C. (2008). Sequencing the nuclear genome of the extinct woolly mammoth. Nature. Vol. 456, No. 7220, pp. 387-390.

Muñoz, M. L., Moreno-Galeana, M., Díaz-Badillo, A., Loza-Martínez, I., Macías-Juárez, V. M., Márquez-Morfin L., Jiménez-López, J. C. & Martínez-Meza, A. (2003).Análisis de DNA mitocondrial de una población prehispánica de Monte Albán, Oaxaca, México. In: *Antropología y Biodiversidad, Vol. 2, Ediciones Bellaterra S.L., Nava de Tolsa*, 289 bis. 08026 Barcelona, España. M. Pilar Aluja, Asunción Malgosa y Ramon M.a Nogués (eds.) ISBN: 84-7290-206-4. P.p. 170-182.

Parr, R. L., Carlyle, S. W., & O'Rourke, D. H. 1996. Ancient DNA analysis of Fremont Amerindians of the Great Salt Lake Wetlands. *Am. J. Phys. Anthropol.* Vol. 99, No. 4, pp. 507-518.

Pääbo, S. (1985). Molecular cloning of ancient Egyptian mummy DNA. *Nature.* Vol. 314, No. 6012, pp. 644-645,

Pääbo, S., Poinar, H., Serre, D., Jaenicke-Despres, V., Hebler, J., Rohland, N., Kuch, M., Krause, J., Vigilant, L., & Hofreiter, M. (2004). Genetic analyses from ancient DNA. *Annu. Rev. Genet.* Vol. 38, pp. 645-679.

Poinar, H. N., Kuch, M., Sobolik, K. D., Barnes, I., Stankiewicz, A. B., Kuder, T., Spaulding, W. G., Bryant, V. M., Cooper, A., & Pääbo, S. (2001). A molecular analysis of dietary diversity for three archaic Native Americans. *Proc Natl. Acad. Sci. U S A.* Vol. 98, No. 8, pp. 4317-22.

Rasmussen, M., Li, Y., Lindgreen, S., Pedersen, J. S., Albrechtsen, A., Moltke, I., Metspalu, M., Metspalu, E., Kivisild, T., Gupta, R., Bertalan, M., & Nielsen, K. (2010). Ancient human genome sequence of an extinct Palaeo-Eskimo. *Nature.* Vol. 463, No. 7282, pp. 757-62.

Reich, D., Green, R. E., Kircher, M., Krause, J., Patterson, N., Durand, E. Y., Viola, B., Briggs, A. W., Stenzel, U., Johnson, P. L., Maricic, T., Good, J. M., Marques-Bonet, T., Alkan, C., Fu, Q., Mallick, S., Li, H., Meyer, M., Eichler, E. E., Stoneking, M., Richards, M., Talamo, S., Shunkov, M. V., Derevianko, A. P., Hublin, J. J., Kelso, J., Slatkin, M., & Pääbo, S. (2010). Genetic history of an archaic hominin group from Denisova Cave in Siberia. *Nature.* Vol. 468, No. 7327, pp. 1053-1060.

Richards, M., Côrte-Real, H., Forster, P., Macaulay, V., Wilkinson-Herbots, H., Demaine, A., Papiha, S., Hedges, R., Bandelt, H. J., & Sykes, B. (1996). Paleolithic and Neolithic lineages in the European mitochondrial gene pool. *Am. J. Hum. Gent.* Vol. 59, No. 3, pp. 185-203.

Rohland, N., & Hofreiter, M. (2007). Comparison and optimization of ancient DNA extraction. *Bio Techniques.* Vol. 42, No. 3, pp. 343-352.

Sachidanandam, R., Weissman, D., Schmidt, S. C., Kakol, J. M., Stein, L. D., Marth, G., Sherry, S., Mullikin, J. C., Mortimore, B. J., & Willey, D. L. (2001). International SNP Map Working Group. A map of human genome sequence variation containing 1.42 million single nucleotide polymorphisms. *Nature.* Vol. 409, No. 6822, pp. 928-933.

Schurr, T. G., Ballinger, S. W., Gan, Y.-Y., Hodge, J. A., Merriwether, D. A., Lawrence, D. N., Dnower, W. C., Weiss, K. M., & Wallace, D. C. (1990). Amerindian mitochondrial DNAs have rare Asian mutations at high frequencies, suggesting they derived from four primary maternal lineages. *Am. J. Hum. Genet.* Vol. 46, No. 3, pp. 768-765.

Stone, A. C., & Stoneking, M. (1993). Ancient DNA from a pre-Columbian Amerindian population. *Am J Phys Anthropol.* Vol. 92, No. 4, pp. 463-471.

Stone, A. C., & Stoneking, M. (1998). mtDNA analysis of a prehistoric oneota population: Implications for the Peopling of the New World. *Am. J. Hum. Genet.* Vol. 62, No. 5, pp. 1153-1170.

Stoneking M. & Krause, J. (2011). Learning about human population history from ancient and modern genomes. *Nat. Rev. Genet.* Vol. 12, No. 9, pp. 603-614. doi: 10.1038/nrg3029.

Thomas, M. G., & Moore, L. J. (1997). Preparation of bone samples for DNA extraction: a nuts and bolts approach. *Biotechniques.* Vol. 22, No. 3, pp. 402

Thompson, J. D., Higgins, D. G. & Gibson, T. J. (1994). CLUSTAL W: improving the sensitivity of progressive multiple sequence alignment through sequence weighting, positions-specific gap penalties and weight matrix choice. *Nucleic. Acids Res.* Vol. 22, pp. 4673-4680.

Torroni, A., Chen, Y.-S., Semino, O., Santachiara-Beneceretti, A. S., Scott, C. R., Lott, M.T., Winter, M., & Wallace, D. C. 1994a. mtDNA and Y-chromosome polymorphisms in four Native American Populations from Southern Mexico. *Am. J. Hum. Gene.* Vol. 54, No. 2, pp. 303-318.

Torroni, A., Neel, J. V., Barrantes, R., Schurr, T. G., & Wallace, D. C. 1994b. Mitochondrial DNA "clock" for the Amerinds and its implications for timing their entry into North America. *Proc. Natl. Acad. Sci. U. S.A.* Vol. 91, No. 3, pp. 1158-1162.

Torroni, A., Schurr, T. G., Cabell, M. F., Brown, M. D., Neel, J. V., Larsen, M., Smith, D. G., Vullo, C. M., & Wallace, D. C. 1993. Asian affinities and continental radiation of the four founding Native American mtDNAs. *Am. J. Hum. Gene.* Vol. 53, No. 3, 563-590.

Torroni, A., Schurr, T. G., Yang, C.-C., Szathmary, E. J., Williams, R. C., Schanfield, M. S., Troup, G. A., Knowler, W. C., Lawrence, D. N., Weiss, K. M., & Wallace, D. C. 1992. Native American mitochondrial DNA analysis indicates that the Amerind and the NaDene populations were founded by two independent migrations. *Genetics.* Vol. 130, pp. 153-162.

Venter, J. C., Adams, M. D., Myers, E. W., Li, P. W., Mural, R. J., Sutton, G. G., & Smith, H. O. (2001).The sequence of the human genome. *Science.* Vol. 291, No. 5507, pp. 1298-1302.

Vigilant, L., Pennington, R., Harpending, H., Kocher, T. D., & Wilson, A. C. (1989). Mitochondrial DNA sequences in single hairs from a southern African population. *Proc Natl Acad Sci U S A.* Vol. 86, No. 23, pp. 9350-9354.

Wallace, D. C., Garrison, K., & Knowler, W. C. (1985). Dramatic founder effects in Amerindian mitochondrial DNAs. *Am. J. Phys. Anthropol.* Vol. 68, No. 2, pp. 149-155.

Wallace, D. C, & Torroni, A. (1992). American Indian prehistory as written in the mitochondrial DNA: A review. *Hum. Biol.* Vol. 64, No. 3, pp. 403-416.

Walsh, P. S., Metzger, D. A., & Higuchi, R. (1991). Chelex 100 as a medium for simple extraction of DNA for PCR-based typing from forensic material. *Biotechniques.* Vol. 10, No. 4, pp. 506–513.

Willerslev, E., & Cooper, A. (2005). Ancient DNA. *Proc. R. Soc. B.* Vol. 272, pp. 3-16.

Wilson, M. R., DiZinno J. A., Polanskey, D., Replogle, J., & Budowle, B. (1995). Validation of mitochondrial DNA sequencing for forensic casework analysis. *Int. J. Legal. Med.* Vol. 108, No. 2, pp. 68-74.

Wrischnik, L. A., Higuchi, R. G., Stoneking, M., Erlich, H. A., Arnheim, N., & Wilson, A. C. (1987). Length mutations in human mitochondrial DNA: direct sequencing of enzymatically amplified DNA. *Nucleic Acids Res.* Vol. 15, No. 2, pp. 529-42.

Young, C. C., Burghoff, R. L., Keim, L. G., Minak-Bernero, V., Lute, J. R., & Hinton, S. M. (1993). Polyvinylpyrrolidone-Agorose Gel Electrophoresis Purification of Polymerase Chain Reaction-Amplifiable DNA from Soils. *Appl Environ Microbiol. J.* Vol. 59, No. 6, pp.1972-1974.

Permissions

The contributors of this book come from diverse backgrounds, making this book a truly international effort. This book will bring forth new frontiers with its revolutionizing research information and detailed analysis of the nascent developments around the world.

We would like to thank Sameh Magdeldin, MVSc, PhD, for lending his expertise to make the book truly unique. He has played a crucial role in the development of this book. Without his invaluable contribution this book wouldn't have been possible. He has made vital efforts to compile up to date information on the varied aspects of this subject to make this book a valuable addition to the collection of many professionals and students.

This book was conceptualized with the vision of imparting up-to-date information and advanced data in this field. To ensure the same, a matchless editorial board was set up. Every individual on the board went through rigorous rounds of assessment to prove their worth. After which they invested a large part of their time researching and compiling the most relevant data for our readers. Conferences and sessions were held from time to time between the editorial board and the contributing authors to present the data in the most comprehensible form. The editorial team has worked tirelessly to provide valuable and valid information to help people across the globe.

Every chapter published in this book has been scrutinized by our experts. Their significance has been extensively debated. The topics covered herein carry significant findings which will fuel the growth of the discipline. They may even be implemented as practical applications or may be referred to as a beginning point for another development. Chapters in this book were first published by InTech; hereby published with permission under the Creative Commons Attribution License or equivalent.

The editorial board has been involved in producing this book since its inception. They have spent rigorous hours researching and exploring the diverse topics which have resulted in the successful publishing of this book. They have passed on their knowledge of decades through this book. To expedite this challenging task, the publisher supported the team at every step. A small team of assistant editors was also appointed to further simplify the editing procedure and attain best results for the readers.

Our editorial team has been hand-picked from every corner of the world. Their multi-ethnicity adds dynamic inputs to the discussions which result in innovative outcomes. These outcomes are then further discussed with the researchers and contributors who give their valuable feedback and opinion regarding the same. The feedback is then collaborated with the researches and they are edited in a comprehensive manner to aid the understanding of the subject.

Apart from the editorial board, the designing team has also invested a significant amount of their time in understanding the subject and creating the most relevant covers. They scrutinized every image to scout for the most suitable representation of the subject and create an appropriate cover for the book.

The publishing team has been involved in this book since its early stages. They were actively engaged in every process, be it collecting the data, connecting with the contributors or procuring relevant information. The team has been an ardent support to the editorial, designing and production team. Their endless efforts to recruit the best for this project, has resulted in the accomplishment of this book. They are a veteran in the field of academics and their pool of knowledge is as vast as their experience in printing. Their expertise and guidance has proved useful at every step. Their uncompromising quality standards have made this book an exceptional effort. Their encouragement from time to time has been an inspiration for everyone.

The publisher and the editorial board hope that this book will prove to be a valuable piece of knowledge for researchers, students, practitioners and scholars across the globe.

List of Contributors

Claudia M. d'Avila-Levy and Patrícia Cuervo
Instituto Oswaldo Cruz, Fundação Oswaldo Cruz, Rio de Janeiro, Brazil

Marta H. Branquinha and André L. S. Santos
Departamento de Microbiologia Geral, Instituto de Microbiologia Paulo de Góes, Universidade Federal do Rio de Janeiro, Rio de Janeiro, Brazil

José Batista de Jesus
Universidade Federal de São João Del Rei, São João Del Rei, Brazil

Reyna Lucero Camacho Morales, Vanesa Zazueta-Novoa, Carlos A. Leal-Morales, Alberto Flores Martínez, Patricia Ponce Noyola and Roberto Zazueta-Sandoval
University of Guanajuato, México

Romano-Bertrand Sara, Parer Sylvie, Lotthé Anne and Jumas-Bilak Estelle
University Montpellier 1, Equipe pathogènes et environnements, UMR 5119 ECOSYM, France
University Hospital of Montpellier, hospital hygiene and infection control team, France

Colson Pascal
University Hospital of Montpellier, Cardio-thoracic intensive care unit, France

Albat Bernard
University Hospital of Montpellier, Cardio-thoracic surgery unit, France

Darren J. Bauer, Gary B. Smejkal and W. Kelley Thomas
Hubbard Center for Genome Studies, University of New Hampshire, USA

Soundarapandian Kannan, Mohanan V. Sujitha and Shenbagamoorthy Sundarraj
Proteomics and Molecular Cell Physiology Lab, Department of Zoology, Bharathiar University, Coimbatore, India

Ramasamy Thirumurugan
Department of Animal Science, Bharathidasan University, Tiruchirappalli, India

Helen Karlsson, Stefan Ljunggren, Maria Ahrén, Bijar Ghafouri, Kajsa Uvdal, Mats Lindahl and Anders Ljungman
Linköping University; County Council of Östergötland, Sweden

Eiji Tanesaka, Naomi Saeki, Akinori Kochi and Motonobu Yoshida
Kinki University, Japan

Pınar Erkekoglu
Hacettepe University, Faculty of Pharmacy, Department of Toxicology, Ankara, Turkey

Lev Elkonin, Julia Italianskaya and Irina Fadeeva
Agricultural Research Institute for South-East Region, Russia

Veronika Kocurová
Nuclear Physics Institute, Academy of Sciences of the Czech Republic, 250 68 Rᵛ ež near Prague, Czech Republic

Maria de Lourdes Muñoz, Miguel Moreno-Galeana, Alvaro Díaz-Badillo, Gerardo Pérez-Ramirez and Alma Herrera-Salazar
Department of Genetics and Molecular Biology, Centro de Investigación y de Estudios Avanzados del Instituto, Politécnico Nacional; Mexico

Mauro Lopez-Armenta
Department of Genetics and Molecular Biology, Centro de Investigación y de Estudios Avanzados del Instituto, Politécnico Nacional; Mexico
Posgrado en Ciencias Genómicas, Universidad Autónoma de la Ciudad de México, Mexico

Elizabeth Mejia-Pérez-Campos
Instituto Nacional de Antropología e Historia, Querétaro, Mexico D.F., Mexico

Sergio Gómez-Chávez
Teotihuacan, Mexico

Adrián Martínez-Meza
Mexico City, Mexico